DRUG DELIVERY APPROACHES AND NANOSYSTEMS

Volume 1: Novel Drug Carriers

DRUG DELIVERY APPROACHES AND NANOSYSTEMS

Volume 1: Novel Drug Carriers

Edited by

Raj K. Keservani, MPharm

Anil K. Sharma, MPharm

Rajesh K. Kesharwani, PhD

AAP APPLE ACADEMIC PRESS

Apple Academic Press Inc.
3333 Mistwell Crescent
Oakville, ON L6L 0A2 Canada

Apple Academic Press Inc.
9 Spinnaker Way
Waretown, NJ 08758 USA

ISBN 13: 978-1-77188-583-6 (hbk)
ISBN 13: 978-1-315-22537-1 (ebk)

Library and Archives Canada Cataloguing in Publication

Drug delivery approaches and nanosystems / edited by Raj K. Keservani, MPharm, Anil K. Sharma, MPharm, Rajesh K. Kesharwani, PhD.
Includes bibliographical references and indexes.
Contents: Volume 1. Novel drug carriers -- Volume 2. Drug targeting aspects of nanotechnology.
Issued in print and electronic formats.
ISBN 978-1-77188-583-6 (v. 1 : hardcover).--ISBN 978-1-77188-585-0 (set : hardcover).--ISBN 978-1-77188-584-3 (v. 2 : hardcover).--
ISBN 978-1-31522537-1 (v. 1 : PDF).--ISBN 978-1-31522-536-4 (v. 2 : PDF)
1. Drug delivery systems. 2. Nanotechnology. 3. Nanostructures. I. Kesharwani, Rajesh Kumar, 1978-, editor II. Sharma, Anil K., 1980-, editor III. Keservani, Raj K., 1981-, editor
RS420.D78 2017 615.1'9 C2017-902014-5 C2017-902015-3

Library of Congress Cataloging-in-Publication Data

Names: Keservani, Raj K., 1981- editor. | Sharma, Anil K., 1980- editor. | Kesharwani, Rajesh Kumar, 1978- editor.
Title: Drug delivery approaches and nanosystems / editors, Raj K. Keservani, Anil K. Sharma, Rajesh K. Kesharwani.
Other titles: Novel drug carriers. | Drug targeting aspects of nanotechnology. Description: Toronto ; New Jersey : Apple Academic Press, 2017. | Includes bibliographical references and index.
Identifiers: LCCN 2017012335 (print) | LCCN 2017013149 (ebook) | ISBN 9781315225371 (ebook) | ISBN 9781771885836 (hardcover ; v. 1 : alk. paper) | ISBN 9781771885843 (hardcover ; v. 2 : alk. paper) | ISBN 9781771885850 (hardcover ; set : alk. paper) | ISBN 9781315225371 (eBook)
Subjects: | MESH: Drug Delivery Systems | Nanotechnology | Nanostructures
Classification: LCC RS420 (ebook) | LCC RS420 (print) | NLM QV 785 | DDC 615.1/9--dc23
LC record available at https://lccn.loc.gov/2017012335

Apple Academic Press also publishes its books in a variety of electronic formats. Some content that appears in print may not be available in electronic format. For information about Apple Academic Press products, visit our website at **www.appleacademicpress.com** and the CRC Press website at **www.crcpress.com**

The Present Book Is Dedicated To

Our Beloved

Aashna,

Atharva

Vihan

&

Vini

CONTENTS

LIST OF CONTRIBUTORS

Nour A. H. Alhalabi
Department of Pharmaceutics, RAK College of Pharmaceutical Sciences, RAK Medical & Health Sciences University, UAE

Onur Alpturk
Department of Chemistry, Istanbul Technical University, 34469, Maslak, Istanbul, Turkey

Ghufran A. R. Alsalloum
Department of Pharmaceutics, RAK College of Pharmaceutical Sciences, RAK Medical & Health Sciences University, UAE

Angel Barranco
CSIC, Institute of Material Science, Seville, Spain

Elena Campano-Cuevas
School of Dentistry, University of Seville, Seville, Spain

Gabriel Castillo-Dalí
School of Dentistry, University of Seville, Seville, Spain

Yaser Dahman
Department of Chemical Engineering, Ryerson University, Toronto, Ontario, M5B 2K3, Canada

Marcus Vinícius Dias-Souza
Biological Sciences Institute, Federal University of Minas Gerais, Belo Horizonte, Brazil; Integrated Pharmacology and Drug Interactions Research Group (GPqFAR), Brazil

Pierre Dramou
China Pharmaceutical University, Nanjing, China

Ahmed Shaker Eltahan
Laboratory of Controllable Nanopharmaceuticals, CAS Key Laboratory for Biomedical Effects of Nanomaterials and Nanosafety, National Center for Nanoscience and Technology of China, Beijing, China

Surya Prakash Gautam
CT Institute of Pharmaceutical Sciences, Shahpur Campus, Jalandhar, India

Tapsya Gautam
CT Institute of Pharmaceutical Sciences, Shahpur Campus, Jalandhar, India

Arun Kumar Gupta
RKDF Institute of Pharmaceutical Sciences, Indore, India

Revati Gupta
RKDF Institute of Pharmaceutical Sciences, Indore, India

Hua He
China Pharmaceutical University, Nanjing, China, Key Laboratory of Drug Quality Control and Pharmacovigilance, Ministry of Education, China Pharmaceutical University, Nanjing, China

Hamideh Hosseinabadi
Department of Mechanical and Industrial Engineering, Ryerson University, Toronto, Ontario, M5B 2K3, Canada

Dušica Ilić
Faculty of Technology, University of Niš, Bulevar oslobodjenja 124, 16000 Leskovac, Republic of Serbia

Snežana Ilić-Stojanović
Faculty of Technology, University of Niš, Bulevar oslobodjenja 124, 16000 Leskovac, Republic of Serbia

Raj K. Keservani
Faculty of Pharmaceutics, Sagar Institute of Research and Technology-Pharmacy, Bhopal–462041, India

Rajesh K. Kesharwani
Department of Biotechnology, NIET, NIMS University, Shobha Nagar, Jaipur, Rajasthan–303121, India

Xing-Jie Liang
Laboratory of Controllable Nanopharmaceuticals, CAS Key Laboratory for Biomedical Effects of Nanomaterials and Nanosafety, National Center for Nanoscience and Technology of China, Beijing, China

Leonardo Quintana Soares Lopes
Microbiology Laboratory Research, Centro Universitário Franciscano, Santa Maria-RS, Brazil

Ana Mora-Boza
CSIC, Institute of Material Science, Seville, Spain

Ljubiša Nikolić
Faculty of Technology, University of Niš, Bulevar oslobodjenja 124, 16000 Leskovac, Republic of Serbia

Vesna Nikolić
Faculty of Technology, University of Niš, Bulevar oslobodjenja 124, 16000 Leskovac, Republic of Serbia

Chuong Pham-Huy
Faculty of Pharmacy, University of Paris V, 4 avenue de l'Observatoire, 75006 Paris, France

Lien Ai Pham-Huy
Department of Pharmacy, Lucile Packard Children's Hospital Stanford, Palo Alto, CA, USA

Ivan S. Ristić
Faculty of Technology, University of Novi Sad, Bulevar Cara Lazara 1, 21000 Novi Sad, Republic of Serbia

Agustín Rodríguez-Gonzalez-Elipe
CSIC, Institute of Material Science, Seville, Spain

Satyajeet S. Salvi
Hampshire College, Amherst, MA 01002, USA

Renan Martins dos Santos

Faculty of Pharmacy, Federal University of Minas Gerais, Belo Horizonte, Brazil; Integrated Pharmacology and Drug Interactions Research Group (GPqFAR), Brazil

Roberto Christ Vianna Santos
Microbiology and Parasitology Department, Federal University of Santa Maria, Santa Maria-RS, Brazil

Ceyda Tuba Sengel-Turk
Department of Pharmaceutical Technology, Faculty of Pharmacy, Ankara University, 06100, Ankara, Turkey

María-Angeles Serrera-Figallo
School of Dentistry, University of Seville, Seville, Spain

Anil K. Sharma
Delhi Institute of Pharmaceutical Sciences and Research, Formerly College of Pharmacy, University of Delhi, Pushp Vihar, Sector III, New Delhi–110017, India

Ranjita Shegokar
Free University of Berlin, Kelchstr. 31, Berlin, Germany

Maninder Pal Singh
CT Institute of Pharmaceutical Sciences, Shahpur Campus, Jalandhar, India

Márcia Ebling de Souza
Microbiology Laboratory Research, Centro Universitário Franciscano, Santa Maria-RS, Brazil

Ana Tačić
Faculty of Technology, University of Niš, Bulevar oslobodjenja 124, 16000 Leskovac, Republic of Serbia

Daniel Torres-Lagares
School of Dentistry, University of Seville, Seville, Spain

Ruslan G. Tuguntaev
Laboratory of Controllable Nanopharmaceuticals, CAS Key Laboratory for Biomedical Effects of Nanomaterials and Nanosafety, National Center for Nanoscience and Technology of China, Beijing, China

Rodrigo de Almeida Vaucher
Microbiology Laboratory Research, Centro Universitário Franciscano, Santa Maria-RS, Brazil

Deli Xiao
China Pharmaceutical University, Nanjing, China

Hemant K. S. Yadav
Department of Pharmaceutics, RAK College of Pharmaceutical Sciences, RAK Medical & Health Sciences University, UAE

LIST OF ABBREVIATIONS

a-CDs	a-cyclodextrins
AAO	anodized aluminum oxide
AAS	*Aconitum sinomontanum*
ACV	acyclovir
AD	Alzheimer's disease
AFM	atomic force microscopy
AGNPS	silver nanoparticles
AIST	National Institute of Advanced Industrial Science and Technology
ALG	alginate
APP	amyloid precursor protein
ASGP-R	asialoglycoprotein receptors
BALF	bronchoalveolar lavage fluid
BBB	blood–brain barrier
bFGF	basic fibroblast growth factor
BG	bioactive glass
BLPs	biotinylated liposomes
BRCAA1	breast cancer-associated antigen 1
BSA	bovine serum albumin
CAFs	cancer-associated fibroblasts
CF	cystic fibrosis
CMC	critical micelle concentration
CMs	cardiomyocytes
CNS	central nervous system
CNTs	carbon nanotubes
Col	collagen
CP	capillary electrophoresis
CPT	camptothecin
CS	chitosan
CS	chondroitin sulfate

CVD	chemical vapor deposition
DA	deoxycholic acid
DC	dendritic cells
DDAB	dimethyl-dioctadecylammonium bromide
DDS	drug delivery system
DEX	dexamethasone
DLS	dynamic light scattering
DM	diabetes mellitus
DMAP	4-(dimethylamino) pyridine
DMSA	dimercaptosuccinic acid
DOEPC	1,2-dioleoyl-sn-glycerol-3-ethylphosp hocholine
DOX	doxorubicin hydrochloride
DPP-4	dipeptidyl peptidase-4
DPPC	dipalmitoylphosphatidylcholine
DPSCs	dental pulp stem cells
DSC	differential scanning colorimeter
DSPE	1,2-distearoyl-sn-glycero-3-phosphatidyl ethanolamine
DW	distilled water
ECM	extracellular matrix
ECN	econazole
EE	entrapment efficiency
EGF	epidermal growth factor
EGFR	epidermal growth factor receptor
EMT	epithelial–mesenchymal transition
EndMT	endothelial-mesenchymal transition
EPC	egg yolk phosphatidylcholine
EPO	erythropoietin
EPR	enhanced permeability and retention
ESR	electron spin resonance
FA	folic acid
FAP	fibroblast-activated-protein
FD&C Act	Federal Food, Drug, and Cosmetic Act
FDA	Food and Drug Administration
FGF	fibroblast growth factor
FUS	focused ultrasound
GAG	glycosaminoglycan

GF	growth factor
GFP	green fluorescence protein
GI	gastrointestinal
GLP	glucagon-like peptide
GSH	glutathione
GTP	green tea polyphenol
HA	hydroxyapatite
HAART	highly active antiretroviral therapy
$HAuCl_4$	hydrogen tetrachloroaurate
HB-EGF	heparin-binding EGF
HCV	hepatitis C virus
HER	human epidermal receptor
HER-2	human epidermal receptor-2
HGF	hepatocyte growth factor
HI	hydrophobic interaction
HIF	hypoxia-inducible transcription factors
HLB	hydrophilic-lipophilic balance
HPH	high-pressure homogenization
HPLC	high-performance liquid chromatography
HPMA	N-(2-hydroxypropyl)-methacrylamide copolymer
IBM	International Business Machines
IGF1	insulin-like growth factor 1
IGF2	insulin-like growth factor 2
IND	Investigational New Drug
INS	insulin
IR	infrared
IVDs	intervertebral discs
JAK	Janus activated kinase
LA	lappacontine
LCST	lower critical solution temperature
LOX	lysyl oxidase
LRT	loratadine
MBC	minimal bactericidal concentration
MCAO	middle cerebral artery occlusion
MCNT	magnetic carbon nanotube
MED	minimal erythema dose

MIC	minimum inhibitory concentration
MMA	methyl methacrylate
MMIPs	magnetic molecularly imprinted polymers
MMPs	matrix metalloproteases
MNPs	magnetic nanoparticles
MP	macrophages
MPLA	monophosphoryl lipid A
MPO	myeloperoxidase
MPS	mononuclear phagocyte system
MRI	magnetic resonance imaging
MSCs	mesenchymal stem cells
MV	multilamellar vesicles
MWCNTs	multiwalled carbon nanotubes
MWNTs	multi-walled carbon nanotubes
NCs	nanocomplexes
NDDS	novel drug delivery systems
NF-κB	nuclear factor-kappa B
NIR	near infrared
NLC	nanostructured lipid carriers
NMR	nuclear magnetic resonance
NMs	Nanomaterials
NP	nanoparticles
NSCs	neural stem cells
NSERC	Natural Sciences and Engineering Research Council of Canada
NSP	nanosilver particles
OV	oligolamellar vesicles
PAMAM	polyamidoamine
PAMAMOS	poly (amidoamine) organosilicon
PAs	peptide-amphiphiles
PCL	poly(ε-caprolactone)
PDGF	platelet-derived growth factor
PDLA	poly(D-lactide)
PEG	poly(ethylene glycol)
PEI	polyethylenimine
PEO	polyethylene oxide

PET	polyethylene terephthalate
PET	positron emission tomography
PEUU	poly(ester-urethane)urea
PGA	poly L-glutamic acid
PGA	poly(glycolic acid)
PHB	poly(hydroxybutyrate)
PK	pharmacokinetic
PLA	poly(lactic acid)
PLGA	poly(lactic-coglycolic) acid
PLLA	poly(L-lactic acid)
PMMA	polymethyl methacrylate
PP	pancreatic polypeptide
PPO	polypropylene oxide
PR	polyrotaxane
PVA	poly(vinyl alcohol)
PVP	poly N-vinyl-2-pyrrolidone
QDs	quantum dots
RAN	ranaconitine
RES	reticuloendothelial system
RSV	respiratory syncytial virus
SELEX	systemic evolution of ligands by exponential enrichment
SEM	scanning electron microscopy
SFRP1	secreted frizzled-related protein 1
SLN	solid lipid nanoparticle
SPE	solid-phase extraction
SPIO	super-paramagnetic iron oxide
SSMs	sterically stabilized micelles
STAT3	signal transducer activator of transcription3
SWCNT	single-walled carbon nanotubes
TAM	tumor-associated macrophages
TCV	tumor cell vaccine
TDDS	transdermal drug delivery system
TEG	triethylene glycol
TEM	transmission electron microscopy
TGA	thermal gravimetric analysis
TGF-α	transforming growth factor-α

TGF-β	transforming growth factor-β
UV	ultraviolet spectrophotometry
UV	unilamellar vesicles
VCAM	vascular cell adhesion molecules
VEGF	vascular endothelial growth factor
VEGF-2	vascular endothelial growth factor receptor 2
VSM	vibrating sample magnetometer
XRD	x-ray diffraction

PREFACE

This edited book, *Drug Delivery Approaches and Nanosystems,* comprised of two volumes—*Volume I: Novel Drug Carriers* and *Volume II: Drug Targeting Aspects of Nanotechnology*, presents a full picture of the state-of-the-art research and development of actionable knowledge discovery in the real-world discovery of drug delivery systems using nanotechnology and its applications.

The book is triggered by the ubiquitous applications of nanotechnology, or nano-sized materials, in the medical field, and the real-world challenges and complexities to the current drug delivery methodologies and techniques.

As we have seen, and as is often addressed at different conferences and seminars, many methods have been used but very few of them have been validated in medical use.

A major reason for the above situation, we believe, is the gap between academia and research and the gap between academic research and real-time clinical applications and needs.

This book, *Drug Delivery Approaches and Nanosystems: Novel Drug Carriers*, includes 12 chapters that contain information on the preparation and characterization of nanocomposite materials used in drug delivery systems; advanced research of carbon nanotubes; nanocomposite materials and polymer-clay, ceramic, silicate glass-based nanocomposite materials; and the functionality of graphene nanocomposites.

This edited book provides a detailed application of nanotechnology in drug delivery systems in health care system and medical applications.

Chapter 1, *Introduction to Nanotechnology in Drug Delivery*, is written by Raj K. Keservani and colleagues, discusses general characteristics of several nanosystems that have applications in drug delivery. The discussion has been supported by suitable examples wherever necessary.

Chapter 2, *Nanoparticles: General Aspects and Applications*, written by Onur Alpturk and Ceyda Tuba Sengel-Turk, provides a discussion of the general features of nanoparticles encompassing preparation methods, evaluations parameters, and different polymers used in formulation. The commercial applications have also been listed in the chapter.

The role of nanotechnology in therapeutics, nowadays termed as 'nanomedicine', is explained in Chapter 3, *Nanotechnology in Medicine: Drug Delivery Systems*, written by Elena Campano-Cuevas and colleagues. The chapter describes how the robotic devices are emerging as a tool to provide numerous biomedical applications.

The diverse nanotechnology-based drug delivery systems have been presented in Chapter 4, *Polymeric Matrix Systems for Drug Delivery*, written by Snežana Ilić-Stojanović and associates. This chapter provides an overview of the drug delivery systems, (DDS), based on natural and/or synthetic polymers as carriers for the active substances, proteins, or cells. The materials used to design nanosystems have also been discussed with suitable instances.

Chapter 5, *Applications of Nanobiomaterials in Drug Delivery*, written by Yaser Dahman and Hamideh Hosseinabadi, deals with biomaterials used for drug delivery. The biomaterials of natural as well as synthetic ones are discussed in this chapter. The biocompatibility of these polymers imparts them preference over their counterparts.

Chapter 6, *Carbon Nanotubes Used as Nanocarriers in Drug and Biomolecule Delivery*, written by Hua He and associates, deals with the applications of CNTs used as nanocarriers in drug and biomolecule delivery for chemotherapeutic use and also studies the pharmacokinetics, metabolism and toxicity of different forms of CNTs. Finally, it discusses the prospect of this promising bio-nanotechnology in the future clinical exploitation.

An overview of dendrimers has been presented in Chapter 7, *Dendrimers: A Glimpse of History, Current Progress and Applications*, written by Surya Prakash Gautam and associates. The key features of dendrimers with therapeutic relevance have been discussed. The journey of these nanometric carriers from emergence to until today have been explained in chronological order. The utility of dendrimers beyond therapeutics has also been described.

Chapter 8, *Nanofibers: Production Techniques and Applications*, written by Hemant K. S. Yadav and colleagues, explains the general aspects of nanofibers and enumerates various preparation techniques. The authors then go on to give manifold applications of these nanometric architects. Biomimetic nanofibers as drug delivery devices that are responsive to different stimuli, such as temperature, pH, light, and electric/magnetic field for controlled release of therapeutic substances, are the new thrust area of research.

A historical map of liposomes and nanoparticles is presented in Chapter 9, *Drug and Food Applications of Liposomes and Nanoparticles: From Benchmark to Bedside?*, written by Marcus Vinícius Dias-Souza and Renan Martins dos Santos. Controlled and vectorized release of different compounds is among the main features that stimulate research on the use of liposomes and nanoparticles in health sciences.

The chapter that talks about the everlasting desire of majority of human beings to look beautiful is Chapter 10, *Nanotechnology for Cosmetic Herbal Actives: Is It a New Beauty Regime?*, written by Ranjita Shegokar. To state precisely, hair care and dermal care products are the everyday need of modern women. Men are not behind; today many men have become conscious of their looks and appearance also. Natural products are among the favorite and are traditionally used for beauty care. It is well known that Cleopatra used to apply donkey's milk to her skin. The science of cosmetology is believed to have originated in Egypt and India, but the earliest records of cosmetic substances and their application dates back to circa 2500 and 1550 B.C. to the Indus valley civilization.

The applications of nanotechnology-based products listed as carriers of antibacterial actives are provided in Chapter 11, *Antimicrobial Activity of Nanotechnological Products*, written by Leonardo Quintana Soares Lopes and colleagues. The effectiveness of antimicrobial therapy is usually low because of limited access to infected site or prevalence of dose-related adverse effects. Nanocarriers have enabled the researchers to reduce the dose of drug with improved control over the growth of microorganisms.

Chapter 12, *Drug Targeting: Principles and Applications*, written by Ruslan G. Tuguntaev and associates, explains the targeting potential of nanosystems. Drug targeting is a promising strategy for the efficient treatment of various serious diseases such as cancer. The primary goal of this concept is to provide precise delivery of therapeutic agents into pathological areas, while avoiding negative impacts on healthy tissues. Through this approach, high therapeutic efficacy of the drug can be attained while experiencing minimum side effects. Recently, nano-sized carriers have received great attention as delivery systems that can load, bring, and release the drugs to a localized area. Accurate drug delivery can be achieved either via passive targeting, which is based on enhanced vascular permeability into the affected zone, or active targeting, which can be achieved by decoration of nanocarriers surface with ligands that have high affinity toward the targeted area. In this chapter, general methods and means of drug targeting-based on nanomedicine have been discussed.

The book also provides detailed information on the application of nanotechnology in drug delivery systems in health care systems and medicine. The book describes how nanostructures are synthesized and draw attention to wide variety of nanostructures available for biological research and treatment applications.

This valuable volume provides a wealth of information that will be valuable to scientists and researchers, faculty, and students.

ABOUT THE EDITORS

Raj K. Keservani, MPharm
Faculty of Pharmaceutics, Sagar Institute of Research and Technology-Pharmacy, Bhopal, India

Raj K. Keservani, MPharm, has more than seven years of academic (teaching) experience from various institutes of India in pharmaceutical education. He has published 25 peer-reviewed papers in the field of pharmaceutical sciences in national and international journals, 15 book chapters, two co-authored books, and two edited books. He is also active as a reviewer for several international scientific journals. Mr. Keservani graduated with a pharmacy degree from the Department of Pharmacy Kumaun University, Nainital (UA), India. He received his Master of Pharmacy (MPharm) (specialization in pharmaceutics) from the School of Pharmaceutical Sciences, Rajiv Gandhi Proudyogiki Vishwavidyalaya, Bhopal, India. His research interests include nutraceutical and functional foods, novel drug delivery systems (NDDS), transdermal drug delivery/ drug delivery, health science, cancer biology, and neurobiology.

Anil K. Sharma, MPharm
Delhi Institute of Pharmaceutical Sciences and Research, University of Delhi, India

Anil K. Sharma, MPharm, is working as a lecturer at the Delhi Institute of Pharmaceutical Sciences and Research, University of Delhi, India. He has more than seven years of academic experience in pharmaceutical sciences. He has published 25 peer-reviewed papers in the field of pharmaceutical sciences in national and international journals as well as 12 book chapters. He received a bachelor's degree in pharmacy from the University of Rajasthan, Jaipur, India, and a Master of Pharmacy degree from the School of Pharmaceutical Sciences, Rajiv Gandhi Proudyogiki Vishwavidyalaya,

Bhopal, India, with a specialization in pharmaceutics. His research interests encompass nutraceutical and functional foods, novel drug delivery systems (NDDS), drug delivery, nanotechnology, health science/life science, and biology/cancer biology/neurobiology.

Rajesh K. Kesharwani, PhD
Faculty, Department of Biotechnology, NIET, NIMS University, India

Rajesh K. Kesharwani, PhD, has more than seven years of research and two years of teaching experience in various institutes of India, imparting bioinformatics and biotechnology education. He has received several awards, including the NASI-Swarna Jayanti Puruskar-2013 by The National Academy of Sciences of India. He has authored over 32 peer-reviewed articles and 10 book chapters. He has been a member of many scientific communities as well as a reviewer for many international journals. Dr. Kesharwani received a BSc in biology from Ewing Christian College, Allahabad, India, an autonomous college of the University of Allahabad; his MSc (Biochemistry) from Awadesh Pratap Singh University, Rewa, Madhya Pradesh, India; and MTech-IT (specialization in Bioinformatics) from the Indian Institute of Information Technology, Allahabad, India. He earned his PhD from the Indian Institute of Information Technology, Allahabad, and received a Ministry of Human Resource Development (India) Fellowship and Senior Research Fellowship from the Indian Council of Medical Research, India. His research fields of interest are medical informatics, protein structure and function prediction, computer-aided drug designing, structural biology, drug delivery, cancer biology, and next-generation sequence analysis.

CHAPTER 1

INTRODUCTION TO NANOTECHNOLOGY IN DRUG DELIVERY

RAJ K. KESERVANI,[1,*] RAJESH K. KESHARWANI,[2] and ANIL K. SHARMA[3]

[1]*Faculty of Pharmaceutics, Sagar Institute of Research and Technology-Pharmacy, Bhopal–462041, India, *E-mail: rajksops@gmail.com*

[2]*Department of Biotechnology, NIET, NIMS University, Shobha Nagar, Jaipur, Rajasthan–303121, India*

[3]*Delhi Institute of Pharmaceutical Sciences and Research, University of Delhi, New Delhi–110017, India*

CONTENTS

ABSTRACT

The science has been ever evolving and credit goes to the numerous researchers working with a goal to enhance the quality of life.

The nanoscience has become a boon to mankind by offering a number of advantages over conventional drug formulations. This has subsequently led to rethink the game plan to battle against diseases/disorders. The apparent benefit are reduction in adverse effects by offering site specific drug delivery, reduced dosing, access to small apertures too. Overall the patient compliance has been observed to be improved upon use of nanotechnology-based products. The present chapter focuses on general aspects of nanotechnology and applications embracing manifold routes of application.

1.1 INTRODUCTION

There's plenty of room at the bottom is the title of a lecture in 1959 by Richard Feynman, that introduced the concept of nanotechnology as an important field for future scientific researches (Feynman, 1960). Nanotechnology research can be developed to advances in communications, engineering, chemistry, physics, robotics, biology, and medicine. Nanotechnology has been used in medicine for therapeutic drug delivery and the development of treatments for a variety of diseases and disorders. So, there are very significant advances in these disciplines.

Nanoparticles used as drug delivery vehicles are generally <100 nm in at least one dimension, and consist of different biodegradable materials such as natural or synthetic polymers, lipids, or metals. Nanoparticles are taken up by cells more efficiently than larger micromolecules and therefore, could be used as effective transport and delivery systems. For therapeutic applications, drugs can either be integrated in the matrix of the particle or attached to the particle surface. A drug targeting system should be able to control the fate of a drug entering the biological environment. Nanosystems with different compositions and biological properties have been extensively investigated for drug and gene delivery applications (Brannon-Peppas and Blanchette, 2004; Pison et al., 2006; Schatzlein, 2006; Stylios et al., 2005; Yokoyama 2005).

Recent years have witnessed unprecedented growth of research and applications in the area of nanoscience and nanotechnology. There is increasing optimism that nanotechnology, as applied to medicine, will bring significant advances in the diagnosis and treatment of disease.

Anticipated applications in medicine include drug delivery, both in vitro and in vivo diagnostics, nutraceuticals and production of improved biocompatible materials (Duncan, 2003; De Jong et al., 2005; ESF, 2005; Ferrari, 2005).

Drug delivery systems can improve the properties of free drugs by increase their in vivo stability and biodistribution, solubility and even by modulation of pharmacokinetics, promoting the transport and even more important the release of higher doses of the drug in the target site in order to be efficient (Cai, 2008; De Jong and Borm, 2008; Drbohlavova et al., 2013; Ghosh, 2008). As far as drug delivery is concerned, the most important nanoparticle platforms are liposomes, polymer conjugates, metallic nanoparticles (e.g., AuNPs), polymeric micelles, dendrimers, nanoshells, and protein and nucleic acid-based nanoparticles (Davis et al., 2008; Pathak et al., 2009).

The aims for nanoparticle entrapment of drugs are enhanced delivery to, or uptake by, target cells and/or a reduction in the toxicity of the free drug to nontarget organs. Both situations will result in an increase of therapeutic index, the margin between the doses resulting in a therapeutic efficacy (e.g., tumor cell death) and toxicity to other organ systems. For these aims, creation of long-lived and target-specific nanoparticles is needed. Most of the compounds are biodegradable polymers resulting in drug release after degradation. One of the problems in the use of particulate drug carriers including nanomaterials is the entrapment in the mononuclear phagocytic system as present in the liver and spleen (Demoy et al., 1997; Gibaud et al., 1996; Lenaerts et al., 1984; Moghimi et al., 2001).

Nature is the ultimate in nanotechnology, producing nanostructures that offer functional proteins and many other compounds at cellular level of great significance to life on earth. It is thought that one of the functions of proteins and compounds that exist at cellular level is that of nanotechnological separations. Biological systems are thought by some scientists to have come about through a process of dynamic self-assembly comprising separation and compartmentalization of many substances into the desired pattern or device (Eijkel and Berg, 2006).

Research presently seeks systematic approaches to fabricate man-made objects at nanoscale and to incorporate nanostructures into macrostructures as nature does (Roco, 1999, 2003; Smith, 2006). Such approaches and concept – which may differ from the living systems in aqueous medium – as self-assembly, templating of atomic and molecular structures on other nano-structures, interaction on surfaces of various shapes, self-repair and integration on multiple length scales may be used as models (Roco, 1999, 2003).

Nanotechnology, being an interdisciplinary field, has three main extensively overlapping areas: nanoelectronics, nanomaterials and nanobiotechnology which find applications in materials, electronics, environment, metrology, energy, security, robotics, healthcare, information technology, biomimetics, pharmaceuticals, manufacturing, agriculture, construction, transport, and food processing and storage (Miyazaki and Islam, 2007; Ochekpe et al., 2009; Shea, 2005; Tratnyek and Johnson, 2006).

1.2 DELIVERY METHOD/ROUTES OF ADMINISTRATION FOR NANOPARTICLES

Several methods of drug administration were employed in delivery nanotechnological product such as oral, injectables, transdermal, topical, pulmonary, and nasal and implantable drug delivery system. Each delivery system described in the following subsections in detail.

1.2.1 ORAL ADMINISTRATION

Oral drug delivery is the most widely used route of administration among all the routes that have been explored for systemic delivery of drugs via pharmaceutical products of different dosage form. Oral route is considered most natural, convenient and safe due to its ease of administration, patient acceptance, and cost effective manufacturing process. Pharmaceutical products designed for oral delivery are mainly immediate release type or conventional drug delivery systems, which are designed for immediate release of drug for rapid absorption (Ummadi et al., 2013).

Oral administration is most preferred because of the various advantages over other routes of drug delivery. The advantages include patient convenience and compliance, which increase the therapeutic efficacy of the drug. Oral formulations are also cheaper to produce because they do not need to be manufactured under sterile conditions (Yun et al., 2013).

An understanding of the GI tract and of drug target sites offers an opportunity for targeted oral delivery of proteins (Wang, 1996). The GI tract has various proteolytic enzymes such as trypsin, chymotrypsin, and elastase which are endopeptidases. Carboxypeptidase A and aminopeptidase are exopeptidases which are also involved as proteolytic enzymes (Woodley, 1994).

Oral drug delivery is the choicest route for drug administration because of its noninvasive nature. The oral route presents the advantage of avoiding pain and discomfort associated with injections as well as eliminating contaminations. However, administered bioactive drugs like peptides and proteins must resist the hostile gastric and intestinal environments. They must then persist in the intestinal lumen long enough to adhere to cell apical surface and then, be transcytosed by intestinal cells. Therefore, peptides and proteins remain poorly bioavailable when administrated orally, mainly due to their low mucosal permeability and lack of stability in the gastrointestinal environment, resulting in degradation of the compound prior to absorption. For many years, many studies have been focused on the improvement of oral delivery of therapeutic peptides and proteins; various strategies have been thus developed to enhance drug and vaccine oral delivery (Bai et al., 1995; Fasano, 1998; Fix, 1996; Galindo-Rodriguez et al., 2005; Hamman et al., 2005; Kompella et al., 2001; Lee, 2002; Lehr, 1994; Russell-Jones, 1998; Sanders, 1990; Steffansen et al., 2004; Wang, 1996). Their association with colloidal carriers, such as polymeric nanoparticles, is one of several approaches proposed to improve their oral bioavailability.

The nature of polymers constituting the formulation significantly influences nanoparticles size and their release profile. Although natural polymers generally provide a relatively quick drug release, synthetic polymers enable extended drug release over periods from days to several weeks. Profile and mechanism of drug release depend on the nature of the polymer, and on all the ensuing physicochemical properties. Some polymers are less sensitive to processing conditions than others, which could be due to their chemical composition, molecular weight and crystallinity (Lemoine et al., 1996).

Insulin is the most effective medicine in lowering the glucose level of blood for the treatment of diabetes mellitus (Daneman, 2006). Early introduction of insulin can also protect islets from apoptosis and increase β-cell regeneration in type 2 diabetes (Sabetsky and Ekblom, 2010). Subcutaneous injections of insulin remain to be the preferred approach for diabetic patients but often result in poor patient compliance (Gowthamarajan and Kulkarni, 2003; Zambanini et al., 1999). Oral administration of insulin seems to be the most convenient way and can mimic endogenous production of insulin (Owens et al., 2003). However, a reliable insulin formulation for the oral delivery is encountered with some barriers in the gastrointestinal (GI) tract that include (a) enzymatic degradation in the GI tract and (b) poor insulin permeability through the GI system (Khafagy et al., 2007). The bioavailability of insulin solution delivered orally is less than 1% (Lowman et al., 1999).

1.2.2 INJECTABLE ROUTE

Parenteral application is a very wide field for SLN. Subcutaneous injection of drug loaded SLN can be employed for commercial aspect, e.g., erythropoietin (EPO), interferon-β. Other routes are intraperitoneal and also intraarticular. Intraperitoneal application of drug-loaded SLN will prolong the release because of the application area. In addition, incorporation of the drug into SLN might reduce irritancy compared to injecting drug micro particles (Gaumet et al., 2008).

Intravenous injection is appealing because it has the highest bioavailability (almost 100%) among all administration routes with superior advantages in immediate effect, targeting effect and overcoming the first pass effect (Gao et al., 2008). For example, the tumor targeting effect induced by intravenous administration has become a long-term interest of oncology (Maeda et al., 2001). Theoretically, hydrophobic drugs could be administrated by intravenous injection if they were formulated into particles sufficiently small enough to circulate in human vascular systems without causing immune reactions and embolism. Thus, nanotechnology may open the possibility for intravenous administration of hydrophobic drugs. Recently, a few groups have tried to intravenously deliver hydrophobic drugs using nanoparticle suspensions (Baba et al., 2007; Gradishar, 2005; Peters et al., 2000).

After in vivo nanoparticles administration, the systemic circulation can distribute them to all body organs and tissues. Precise characterization of nanoparticles distribution and accumulation in the different body parts in preclinical settings is required before any nanoparticles use, whether for diagnosis, photothermal therapy or drug delivery in humans (Varna et al., 2012).

When nanoparticles are administered intravenously, they are easily recognized by the body immune systems, and are then cleared by phagocytes from the circulation (Muller and Wallis, 1993). Apart from the size of nanoparticles, their surface hydrophobicity determines the amount of adsorbed blood components, mainly proteins (opsonins). This in turn influences the in vivo fate of nanoparticles (Brigger et al., 2002; Muller and Wallis, 1993).

The zeta potential of a nanoparticle is commonly used to characterize the surface charge property of nanoparticles (Couvreur et al., 2002).

Solid lipid nanoparticle (SLN) has been administered intravenously to animals. Pharmacokinetic studies of doxorubicin incorporated into SLN showed higher blood levels in comparison to a commercial drug solution after i.v. injection in rats. Regarding distribution, SLN were found to have higher drug concentrations in lung, spleen and brain, while the solution led to more distribution into liver and kidneys (Yang et al., 1999).

SLN were introduced at the beginning of the 1990 s as a new colloidal drug delivery system with advantages such as nontoxicity, excellent biocompatibility, and large scale production facilities, which made SLN interesting alternatives to liposomes, microemulsions, and other polymeric nanoparticles (Aji et al., 2011; Mehnert and Mader, 2012; Muller et al., 2006). Due to their solid matrix, solid lipid nanoparticles can protect the incorporated drug from chemical degradation in the gastrointestinal environment and have been extensively investigated as a promising drug delivery system for controlling the release of therapeutic agents (Gonzalez-Mira et al., 2011; Mehnert and Mader, 2012; Muller et al., 2000; Zur Muhlen et al., 1998).

1.2.3 TRANSDERMAL/TOPICAL ROUTE

Intracellular macromolecular matrix within the stratum corneum abounds in keratin, which does not contribute directly to the skin diffusive barrier but supports mechanical stability and thus intactness of the stratum corneum.

Transcellular diffusion is practically unimportant for transdermal drug transport (Cevc and Vier, 2010). Particle size and shape affect drug release, physical stability and cellular uptake of the nanoparticulate materials. The attachment of nanoparticles to cell membrane is affected by the surface charge of the particles. Variation of the particle surface charge could potentially control binding to the tissue and direct nanoparticles to cellular compartments both in vitro and in vivo. Cellular surfaces are dominated by negatively charged sulfated proteoglycans molecules that play pivotal roles in cellular proliferation, migration and motility (Bernfild et al., 2004).

Transdermal delivery provides convenient and pain-free self-administration for patients. It eliminates frequent dosing administration and plasma level peaks and valleys associated with oral dosing and injections to maintain constant drug concentrations and a drug with a short half-life can be delivered easily. All this leads to enhanced patient compliance, especially when long-term treatment is required, as in chronic pain treatment and smoking cessation therapy (Chandak and Verma, 2008; Dnyanesh and Vavia, 2003; Valenta and Auner, 2004). Transdermal route permits the use of a relatively potent drug with minimal risk of system toxicity (Mundargi et al., 2007; Mutalik and Udupa, 2004).

Transdermal route of drug administration have unique advantages drug bypass the first pass metabolism and reaches in the systemic circulation. Painless, noninvasive, and patient-friendly application of patches offers good patient compliance and patches are also easy to remove in the event of hyperinsulinemia (Mugumu, 2006).

Transdermal delivery involves application of a pharmacologically active compound on to the skin to achieve therapeutic blood levels in order to treat diseases remote from the site of application. Ever since the approval of Transderm-Scop, the first transdermal drug delivery system (TDDS) in 1981, there has been explosive research in the field of transdermal therapeutics for treatment of a variety of clinical conditions (Gordon and Peterson, 2003).

For topical or transdermal administration, MNPs can be classified as a type of microreservoir-dissolution-controlled system that can be tailored to deliver drugs topically (skin being the site of action) or transdermally (systemic availability). The physicochemical properties of MNP formulations can be tailored for a given route of administration (Lee et al., 2010).

In order to optimize the locoregional release of therapeutics, the topical route has been one of the most promising noninvasive delivery options, ameliorating patient compliance, improving the pharmacokinetics of degradable compounds and reducing frequently occurring side effects (Bolzinger et al., 2012; DeLouise, 2012; Mathias and Hussain, 2010). Nevertheless, drug topical administration remains a challenge in pharmaceutics because of the difficulties to adjust the skin penetration and to determine and reproduce the exact amount of drug reaching the skin layers at the desired depth (Labouta et al., 2011; Sonavane et al., 2008).

The second generation of H1 receptor blocking agents including loratadine (LRT) can be used for inhibiting histamine release from the basophilic granulocytes and the mastocytes where histamine is stored in tissues (Hadzijusufovic et al., 2010). LRT displays various side effects including several allergic reactions including rash, hives, itching, difficulty in breathing, tightness in the chest, swelling of the mouth or face and dizziness during systemic administration by oral route. Thus, topical preparations of LRT for skin application would be advantageous for the management of skin reactions. However, LRT has a poor penetration through the skin. This limitation may be overcome by incorporation of LRT into colloidal drug carrier systems like solid lipid nanoparticles (SLN) and nanostructured lipid carriers (NLC) (Uner, 2006; Wissing et al., 2001).

1.2.4 PULMONARY/NASAL DELIVERY

Drug delivery through the pulmonary route offers several advantages, including increased local concentration of drug, improved pulmonary receptor occupancy, increased absorption due to vast surface area, reduced dose, local and systemic delivery of drug and decreased systemic adverse effect (Mansour et al., 2009; Pison et al., 2006). However, drug delivery through the pulmonary route continues to pose challenges like mucociliary clearance and phagocytosis by alveolar macrophage, which can cause drug degradation at the site of absorption (Mansour et al., 2009; Sahib et al., 2011). Large molecules dissolve in the bronchoalveolar lavage fluid (BALF) and diffuse across the alveolar epithelium; alveolar macrophage presence here can cause drug degradation that will result in reduced bioavailability. Alternately, small molecules are rapidly absorbed through lung epithelium that can be advantageous for immediate release but might not be useful for sustained release. Both cases will end up in increasing dosing frequency which might lead to noncompliance to treatment (Ryan et al., 2013).

Several drugs have been successfully formulated into inhalable non-PEGylated proliposomal dry powders (Awasthi et al., 2003), which include but are not limited to budesonide, beclomethasone and formoterol, salbutamol sulfate, ketotifen fumarate, dapsone, amiloride hydrochloride, tacrolimus, and cyclosporine A. Curcumin has been shown to have anti-inflammatory and antioxidant properties that may be promising in the treatment of cystic fibrosis, asthma, COPD, and other pulmonary diseases (Egan et al., 2004; Mukhopadhyay et al., 1982; Reddy and Lokesh, 1994).

Intranasal administration has been investigated due to its accessibility and noninvasive nature. The nasal route allows for both local delivery to the upper respiratory tract (i.e., nasal region, nasal tissue, and nasal fluids), noninvasive systemic delivery, and noninvasive CNS (central nervous system) delivery of drugs, due to the large surface area and highly vascularized nature of the nasal cavity and direct access to the olfactory region (Illum et al., 2007). Drug administration through the nasal route has shown its success throughout the years, allowing for the avoidance of first-pass effect, reducing systemic side effects, bypassing blood–brain barrier (BBB) (Zhang et al., 2014) and increasing bioavailability (Illum et al., 2007; Kim et al., 2012).

PEGylated drug delivery through the nasal route has considerable potential in regards to the transport of drugs directly to the brain through the olfactory region (Chekhonin et al., 2008). PEG–poly(lactic-coglycolic acid) PEG–PLGA () nanoparticles have demonstrated effectiveness in delivering basic fibroblast growth factor (bFGF) directly to the brain for treatment of Alzheimer's disease.

Das and co-workers studies and reveals that pulmonary administration of nanoparticles containing voriconazole could be a better therapeutic choice even as compared to the i.v. route of administration of the free drug and/or the drug loaded nanoparticles (Das et al., 2015). Pulmonary and nasal routes are the other mucosal pathways that are attracting considerable attention as alternative routes for peptide and protein delivery since they involve very large surface areas and less intracellular and extracellular enzymatic degradation. Amidi et al. showed the potential of N-trimethylchitosan nanoparticles as a new delivery system for the transport of proteins through the nasal mucosa (Amidi et al., 2006). Hybrid chitosan–cyclodextrin nanoparticles have also demonstrated their potential for enhancing the transport of complex molecules across the nasal barrier (Teijeiro-Osorio et al., 2009).

1.2.5 IMPLANTABLE DELIVERY

The term implant is used for devices that replace or act as a fraction of or the whole biological structure. Currently, implants are being used in many different parts of the body for various applications such as orthopedics, pacemakers, cardiovascular stents, defibrillators, neural prosthetics or drug delivery system (Regar et al., 2001).

Implantable drug-delivery devices are often relied upon to meet these specialized dosing challenges and provide a means to achieve the drug delivery required by these novel pharmaceutical agents and their atypical drug regimen. Implantable drug pumps are routinely used in pharmacokinetic and pharmacodynamics studies, which are now an integral process in drug discovery and preclinical and clinical evaluation of efficacy (Maas et al., 2007).

Biomedical implants obviously provide a wide range of medical cures for many of the disorders, such as cardiovascular diseases. Vascular grafts, defibrillators, heart valves, pacemakers and stents are the most common cardiovascular implantable devices used in the medical field. However, the present implant technology is facing a major difficulty of being perceived by the human body as foreign substances. Nanotechnology provides a medical solution to revolutionize the biomedical implant technology exactly by modifying and designing their structures thereby to overcome these problems (Arsiwala, 2013).

Implantable and portable biosensors for drug delivery offer self-monitoring and increased patients' compliance (Sershen and West, 2002). Integrated biosensors and drug delivery devices can offer a continuous diagnosis, prognosis and efficient therapeutic management.

Most of the microfabricated devices are in the form of biosensors. There is a time limitation to the use of microfabricated implantable biosensors due to their short time of functionality. Designing an implantable biosensor that has long-term functionality can be a critical component of the ideal closed-loop drug delivery or monitoring system, without considering issue of implant biocompatibility and biofouling which must be addressed in order to achieve long-term in vivo sensing (Grayson et al., 2004).

The implant at the target site mainly functions to deliver the drug dose from the external reservoir to the target site and may contain other functions such as sensing and flow control. The external reservoir contains a pump or a type of actuation mechanism which enables it to deliver the drug to the implant at the target site via a cannula/catheter connected between the two parts. This approach has been used to treat diabetes mellitus over the past few years as described by Hanaire et al. (2008), Lenhard and Reeves (2001) and Staples et al. (2006), where the reservoir of an insulin pump is located in the subcutaneous region of the patient rather than implanting it at the target site, allowing it to have more drug available for treatment.

A novel bioengineered corneal implant with Acyclovir (ACV) loaded silica nanoparticle carriers was fabricated for the controlled release of the drug during the corneal transplantation surgery. The drug release from the biosynthetic implants was sustained over 10 days, in comparison with free ACV incorporated directly into the hydrogel constructs. This enabled effective prevention of virally induced cell death, which could not be observed with the free drug (Bettina et al., 2010).

1.3 CONCLUSION

Nanoparticles hold tremendous potential as an effective drug delivery system. It appears that nanodrug delivery systems hold great potential to overcome some of the barriers to efficient targeting of cells and molecules in inflammation and cancer. These nanomaterials are capable to provide a high degree of biocompatibility before and after conjugation to biomolecules for specific function so as to translate into nanomedicines and clinical practice. To optimize this drug delivery system, greater understanding of the different mechanisms of biological interactions, and particle engineering, is still required. Nanomaterials provide for a favorable blood half-life and physiologic behavior with minimal off-target effects. Furthermore, because nanosystems increase efficiency of drug delivery, the doses may need recalibration. Finally, in this chapter we discuss about the nanoparticle drug delivery routes, different technology and its application.

KEYWORDS

- **bio-distribution**
- **biological environment**
- **drug targeting**
- **nanotechnology**
- **particulate carriers**
- **routes of drug administration**
- **therapeutics**

REFERENCES

Aji Alex, M. R.; Chacko, A. J.; Jose, S.; Souto, E. B. Lopinavir loaded solid lipid nanoparticles (SLN) for intestinal lymphatic targeting. *Eur. J. Pharm. Sci.* **2011**, *42*, 11–18.

Amidi, M.; Romeijn, S. G.; Borchard, G.; Junginger, H. E.; Hennink, W. E.; Jiskoot, W. Preparation and characterization of protein-loaded N-trimethyl chitosan nanoparticles as nasal delivery system. *J Control Rel.* **2006**, *111*, 107–116.

Arsiwala, A. M.; Raval, A. J.; Patravale, V. B. Nanocoatings on implantable medical devices. *Pharm Pat Anal.* **2013**, *2*, 499–512.

Awasthi, V. D.; Garcia, D.; Goins, B. A.; Phillips, W. T. Circulation and biodistribution profiles of long-circulating peg-liposomes of various sizes in rabbits. *Int. J. Pharm.* **2003**, *253*, 121–132.

Baba, K.; Pudavar, H. E.; Roy, I.; Ohulchanskyy, T. Y.; Chen, Y.; Pandey, R. K.; Prasad, P. N. New method for delivering a hydrophobic drug for photodynamic therapy using pure nanocrystal form of the drug. *Mol. Pharm.* **2007**, *4*(2), 289–297.

Bai, J. P.; Chang, L. L.; Guo, J. H. Targeting of peptide and protein drugs to specific sites in the oral route. *Crit. Rev. Ther. Drug Carr. Syst.* **1995**, *12*, 339–371.

Bareiss, B.; Ghorbani, M.; Fengfu, Li.; Blake, J. A.; Scaiano, J. C.; Jin Z.; Chao D.; Merrett, K.; Harden, J. L.; Diaz-Mitoma, F.; Griffith, M. Controlled release of acyclovir through bioengineered corneal implants with silica nanoparticle carriers. *Open Tissue Eng Regen Med J.* **2010**, *3*, 10–17.

Bernfild, M.; Gotte, M.; Park, P. W.; Reizes, O.; Fitzgerald, M. L.; Lincecum, J.; Zako, M. Functions of cell surface heparan sulfate proteoglycans. *Annu. Rev. Biochem.* **1999**, *68*, 729–777.

Bolzinger, M. A.; Briançon, S.; Pelletier, J.; Chevalier, Y. Penetration of drugs through skin, a complex rate-controlling membrane. *Curr Opin Colloid Interface Sci.* 2012, *17*, 156–165.

Brannon-Peppas, L.; Blanchette, J. Q. Nanoparticle and targeted systems for cancer therapy. *Adv. Drug Deliv. Rev.* **2004**, *56*, 1649–1659.

Brigger, I.; Dubernet, C.; Couvreur, P. Nanoparticles in cancer therapy and diagnosis. *Adv. Drug Deliv. Rev.* **2002**, *54*, 631–651.

Cai, W.; Gao, T.; Hong, H.; Sun, J. Applications of gold nanoparticles in cancer nanotechnology. *Nanotechnol Sci Appl.* **2008**, *1*, 17–32.

Cevc, G.; Vier, I. U. Nanotechnology and the transdermal route. A state-of-the-art review and critical appraisal. *J. Control. Rel.* **2010**, *141*, 277–299.

Chandak, A. R.; Verma, P. R. P. Development and evaluation of HPMC based matrices for transdermal patches of tramadol. *Clin. Res. Reg. Affairs.* **2008**, *25*, 13–30.

Chekhonin, V. P.; Gurina, O. I.; Ykhova, O. V.; Ryabinina, A. E.; Tsibulkina, E. A.; Zhirkov, Y. A. Polyethylene glycol-conjugated immunoliposomes specific for olfactory ensheathing glial cells. *Bull. Exp. Biol. Med.* **2008**, *145*, 449–451.

Couvreur, P.; Barratt, G.; Fattal, E.; Legrand, P.; Vauthier, C. Nanocapsule technology: a review. *Crit Rev Ther Drug Carrier Syst.* **2002**, *19*, 99–134.

Daneman, D. Type 1 diabetes. *Lancet.* **2006**, *367*(9513), 847–858.

Das, P. J.; Paul, P.; Mukherjee, B.; Mazumder, B.; Mondal, L.; Baishya, R.; Chatterjee, M. D.; Dey, S. K. Pulmonary Delivery of Voriconazole Loaded Nanoparticles Providing a Prolonged Drug Level in Lungs: A Promise for Treating Fungal Infection. *Mol. Pharm.* **2015**, *12*(8), 2651–2664.

Davis, M. E.; Chen, Z. G.; Shin, D. M. Nanoparticle therapeutics: an emerging treatment modality for cancer. *Nat Rev Drug Discov.* **2008**, *7*(9), 771–782.

De Jong, W. H.; Geertsma, R. E.; Roszek, B. Possible risks for human health. Report 265001002/2005. National Institute for Public Health and the Environment (RIVM). Nanotechnology in medical applications. Bilthoven, The Netherlands. 2005.

DeJong, W. H.; Borm, P. J. A. Drug delivery and nanoparticles: Applications and hazards. *Int. J. Nano.* **2008**, *3*(2), 133–149.

DeLouise, L. A. Applications of nanotechnology in dermatology. *J Invest Dermatol.* **2012**, *132*, 964–975.

Demoy, M.; Gibaud, S.; Andreux, J. P. et al. Splenic trapping of nanoparticles: complementary approaches for in situ studies. *Pharm Res.* **1997**, *14*, 463–468.

Dnyanesh, N. T.; Vavia, P. R. Acrylate-based transdermal therapeutic system of nitrendipine. *Drug Dev. Ind. Pharm.* **2003**, *29*, 71–78.

Drbohlavova, J.; Chomoucka, J.; Adam, V.; Ryvolova, M.; Eckschlager, T.; Hubalek, J.; Kizek, R. Nanocarriers for Anticancer Drugs-New Trends in Nanomedicine. *Curr Drug Metab.* **2013**, *14*(5), 547–564.

Duncan, R. The dawning era of polymer therapeutics. *Nat Rev Drug Disc.* **2003**, *2*, 347–360.

Egan, M. E.; Pearson, M.; Weiner, S. A.; Rajendran, V.; Rubin, D.; Glockner-Pagel, J.; Canny, S.; Du, K.; Lukacs, G. L.; Caplan, M. J. Curcumin, a major constituent of turmeric, corrects cystic fibrosis defects. *Science*. **2004**, *304*, 600–602.

Eijkel, J. C. T.; Berg, A. V. The promise of nanotechnology for separation devices – from a top-down approach to nature-inspired separation devices. *Electrophoresis*. **2006**, *27*, 677–685.

European Science Foundation. 2005. Policy Briefing (ESF), ESF Scientific Forward Look on Nanomedicine. IREG Strasbourg, France, 2005.

Fasano, A. Modulation of intestinal permeability: an innovative method of oral drug delivery for the treatment of inherited and acquired human diseases. *Mol. Genet. Metab.* **1998**, *64*, 12–18.

Ferrari, M. Cancer nanotechnology: opportunities and challenges. *Nat Rev Cancer.* **2005**, 5,161–171.

Feynman, R. P. There is plenty of room at the bottom. *Engineering & Science Magazine*. Volume XXIII No.5, California Institute of Technology, Pasadena, USA, 1960.

Fix, J. A. Oral controlled release technology for peptides: status and future prospects. *Pharm. Res.* **1996**, *13*, 1760–1764.

Galindo-Rodriguez, S. A.; Allemann, E.; Fessi, H.; Doelker, E. Polymeric nanoparticles for oral delivery of drugs and vaccines: a critical evaluation of in vivo studies. *Crit. Rev. Ther. Drug Carr. Syst.* **2005**, *22*, 419–464.

Gao, L.; Zhang, D.; Chen, M. Drug nanocrystals for the formulation of poorly soluble drugs and its application as a potential drug delivery system. *J. Nanopart. Res.* **2008**, *10*(5), 845–862.

Gaumet, M.; Vargas, A.; Gurny, R.; Delie, F. Nanoparticles for drug delivery: The need for precision in reporting particle size parameters. *Euro. J. Pharm. Biopharm.* **2008**, *69*, 1–9.

Ghosh, P.; Han, G.; De, M.; Kim, C. K.; Rotello, V. M. Gold nanoparticles in delivery applications. *Adv. Drug Deliv. Rev.* **2008**, *60*(11), 1307–1315.

Gibaud, S.; Demoy, M.; Andreux, J. P.; Weingarten, C.; Gouritin, B.; Couvreur, P. Cells involved in the capture of nanoparticles in hematopoietic organs. *J. Pharm. Sci.* **1996**, *85*, 944–950.

Gonzalez-Mira, E.; Egea, M. A.; Souto, E. B.; Calpena, A. C.; Garcia, M. L. Optimizing flurbiprofen-loaded NLC by central composite factorial design for ocular delivery. *Nanotechnology*. **2011**, *22*, 045101.

Gowthamarajan, K.; Kulkarni, G. Oral insulin: fact or fiction? *Resonance*. **2003**, *8*(5), 38–46.

Gradishar, W. J. Phase III trial of nanoparticle albumin-bound paclitaxel compared with polyethylated castor oil-based paclitaxel in women with breast cancer. *J. Clin. Oncol.* **2005**, *23*(31), 7794–7803.

Grayson, A. C.; Shawgo, R. S.; Johnson, A. M.; Flynn, N. T.; Li, Y.; Cima, M. J.; Langer, R. A. BioMEMS review: MEMS technology for physiologically integrated devices. *Proc. IEEE.* **2004**, *92*, 6–21.

Hadzijusufovic, E.; Peter, B.; Gleixner, K. V.; Schuch, K.; Pickl, W. F.; Thaiwong, T.; Yuzbasiyan-Gurkan, V.; Mirkina, I.; Willmann, M.; Valent, P. H1-receptor antagonists terfenadine and loratadine inhibit spontaneous growth of neoplastic mast cells. *Exp Hematol.* **2010**, *38*(10), 896–907.

Hamman, J. H.; Enslin, G. M.; Kotze, A. F. Oral delivery of peptide drugs: barriers and developments. *BioDrugs*. **2005**, *19*, 165–177.

Hanaire, H.; Lassmann-Vague, V.; Jeandidier, N.; Renard, E.; Tubiana-Rufi, N.; Vambergue, A.; Raccah, D.; Pinget, M.; Guerci, B. Treatment of diabetes mellitus using an external insulin pump: the state-of-the-art. *Diabetes Metab.* **2008**, *34*, 401–423.

Illum, L. Nanoparticulate systems for nasal delivery of drugs: A real improvement over simple systems. *J. Pharm. Sci.* **2007**, *96*, 473–483.

Khafagy, E. S.; Morishita, M.; Onuki, Y.; Takayama, K. Current challenges in noninvasive insulin delivery systems: a comparative review. *Adv. Drug Deliv. Rev*. **2007**, *59*(15), 1521–1546.

Kim, T. H.; Park, C. W.; Kim, H. Y.; Chi, M. H.; Lee, S. K.; Song, Y. M.; Jiang, H. H.; Lim, S. M.; Youn, Y. S.; Lee, K. C. Low molecular weight (1 kDa) polyethylene glycol conjugation markedly enhances the hypoglycemic effects of intranasally administered exendin-4 in type 2 diabetic db/db mice. *Biol. Pharm. Bull.* **2012**, *35*, 1076–1083.

Kompella, U. B.; Lee, V. H. Delivery systems for penetration enhancement of peptide and protein drugs: design considerations. *Adv. Drug Deliv. Rev.* **2001**, *46*, 211–245.

Labouta, H. I.; El-Khordagui, L. K.; Krausc, T.; Schneider, M. Mechanism and determinants of nanoparticle penetration through human skin. *Nanoscale*. **2011**, *3*, 4989–4999.

Lee, H. J. Protein drug oral delivery: the recent progress. *Arch. Pharm. Res.* **2002**, *25*, 572–584.

Lee, R. W.; Shenoy, D. B.; Rajiv Sheel, R. Micellar Nanoparticles: Applications for Topical and Passive Transdermal Drug Delivery. In: *Handbook of Non-Invasive Drug Delivery Systems*; Kulkarni, V. S., Ed.; William Andrew, Elsevier, Norwich, NY, 2010; pp. 37–58.

Lehr, C. M. Bioadhesion technologies for the delivery of peptide and protein drugs to the gastrointestinal tract. *Crit Rev. Ther. Drug Carrier Syst.* **1994**, *11*, 119–160.

Lemoine, D.; Francois, C.; Kedzierewicz, F.; Preat, V.; Hoffman, M.; Maincent, P. Stability study of nanoparticles of poly (epsilon-caprolactone), poly(DL-lactide) and poly(D, L-lactide-coglycolide). *Biomaterials*. **1996**, *17*, 2191–2197.

Lenaerts, V.; Nagelkerke, J. F.; Van Berkel, T. J.; Couvreur, P.; Grislain, L.; Roland, M.; Speiser, P. In vivo uptake of polyisobutyl cyanoacrylate nanoparticles by rat liver Kupffer, endothelial, and parenchymal cells. *J Pharm Sci.* **1984**, *73*, 980–982.

Lenhard, M. J.; Reeves, G. D. Continuous subcutaneous insulin infusion: A comprehensive review of insulin pump therapy. *Arch. Intern. Med.* **2001**, *161*, 2293–2300.

Lowman, A. M.; Morishita, M.; Kajita, M.; Nagai, T.; Peppas, N. A. Oral delivery of insulin using pH-responsive complexation gels. *J Pharm Sci.* **1999**, *88*(9), 933–937.

Maas, J.; Kamm, W.; Hauck, G. An integrated early formulation strategy-from hit evaluation to preclinical candidate profiling. *Eur. J. Pharm. Biopharm.* **2007**, *66*(1), 1–10.

Maeda, H. The enhanced permeability and retention (EPR) effect in tumor vasculature: the key role of tumor-selective macromolecular drug targeting. *Adv. Enzyme Regul.* **2001**, *41*, 189–207.

Mansour, H. M.; Rhee, Y. S.; Wu, X. Nanomedicine in pulmonary delivery. *Int. J. Nanomed.* **2009**, *4*, 299–319.

Mathias, N. R.; Hussain, M. A. Non-invasive systemic drug delivery: develop ability considerations for alternate routes of administration. *J. Pharm. Sci.* **2010**, *99*, 1–20.

Mehnert, W.; Mader, K. Solid lipid nanoparticles: Production, characterization and applications. *Adv. Drug Deliv. Rev.* **2012**, *64*, 83–101.

Miyazaki, K.; Islam N. Nanotechnology systems of innovation – An analysis of industry and academia research activities. *Technovation.* **2007**, *27*, 661–671.

Moghimi, S. M.; Hunter, A. C.; Murray, J. C. Long-circulating and target specific nanoparticles: theory and practice. *Pharmacol Rev.* **2001**, *53*, 283–318.

Mugumu, H. Transdermal delivery of $CaCo_3$-Nanoparticles Containing Insulin. *Diabetes Technol Ther.* **2006**, *8*(3), 369–374.

Mukhopadhyay, A.; Basu, N.; Ghatak, N.; Gujral, P. K. Anti-inflammatory and irritant activities of curcumin analogs in rats. *Agents Act.* **1982**, *12*, 508–515.

Muller, R. H.; Mader, K.; Gohla, S. Solid lipid nanoparticles (SLN) for controlled drug delivery: a review of the state-of-the-art. *Eur. J. Pharm. Biopharm.* **2000**, *50*, 161–177.

Muller, R. H.; Runge, S.; Ravelli, V.; Mehnert, W.; Thunemann, A. F.; Souto, E. B. Oral bioavailability of cyclosporine: Solid lipid nanoparticles (SLN) versus drug nanocrystals. *Int. J. Pharm.* **2006**, *317*, 82–89.

Muller, R. H.; Wallis, K. H. Surface modification of i.v. injectable biodegradable nanoparticles with poloxamer polymers and poloxamine 908. *Int. J. Pharm.* **1993**, *89*, 25–31.

Mundargi, R. C.; Patil, S. A.; Agnihotri, S. A.; Aminabhavi. T. M. Evaluation and controlled release characteristics of modified xanthan films for transdermal delivery of atenolol. *Drug Dev. Ind. Pharm.* **2007**, *33*, 79–90.

Mutalik, S.; Udupa, N. Glibenclamide transdermal patches: Physicochemical, pharmacodynamic and pharmacokinetic evaluations. *J. Pharm. Sci.* **2004**, *93*, 1577–1594.

Ochekpe, N. A.; Olorunfemi, P. O.; Ngwuluka, N. C. Nanotechnology and Drug Delivery Part 1: Background and Applications. *Trop J Pharm Res.* **2009**, *8*(3), 265–274.

Owens, D. R.; Zinman, B.; Bolli, G. Alternative routes of insulin delivery. *Diabet Med.* **2003**, *20*(11), 886–898.

Pathak, Y.; Thassu, D. Drug Delivery Nanoparticles Formulation and Characterization. *Rijeka: PharmaceuTech Inc.* **2009**, 1–393.

Peters, K.; Leitzke, S.; Diederichs, J. E.; Borner, K.; Hahn, H.; Müller, R. H.; Ehlers, S. Preparation of a clofazimine nanosuspension for intravenous use and evaluation of its therapeutic efficacy in murine Mycobacterium avium infection. *J. Antimicrob. Chemother.* **2000**, *45*(1), 77–83.

Pison, U.; Welte, T.; Giersing, M.; Groneberg, D. A. Nanomedicine for respiratory diseases. *Eur J Pharmacol.* **2006**, *533*, 341–350.

Reddy, A. C.; Lokesh, B. R. Effect of dietary turmeric (Curcuma longa) on iron-induced lipid peroxidation in the rat liver. *Food Chem. Toxicol.* **1994**, *32*, 279–283.

Regar, E.; Sianos, G.; Serruys, P. W. Stent development and local drug delivery. *Br Med Bull.* **2001**, *59*, 227–248.

Roco, M. C. Nanotechnology: Convergence with modern biology and medicine. *Curr Opinion Biotech.* **2003**, *14*, 337–346.

Roco, M. C. Towards a US National Nanotechnology Initiative. *J Nanoparticle Res.* **1999**, *1*, 435–438.

Russell-Jones, G. J. Use of vitamin B12 conjugates to deliver protein drugs by the oral route. *Crit. Rev. Ther. Drug Carr. Syst.* **1998**, *15*, 557–586.

Ryan, G. M.; Kaminskas, L. M.; Kelly, B. D.; Owen, D. J.; McIntosh, M. P.; Porter, C. J. Pulmonary administration of pegylated polylysine dendrimers: Absorption from the lung versus retention within the lung is highly size-dependent. *Mol. Pharm.* **2013**, *10*, 2986–2995.

Sabetsky, V.; Ekblom, J. Insulin: a new era for an old hormone. *Pharmacol Res.* **2010**, *61*(1), 1–4.

Sahib, M. N.; Darwis, Y.; Peh, K. K.; Abdulameer, S. A.; Tan, Y. T. Rehydrated sterically stabilized phospholipid nanomicelles of budesonide for nebulization: Physicochemical characterization and in vitro, in vivo evaluations. *Int. J. Nanomed.* **2011**, *6*, 2351–2366.

Sanders, L. M. Drug delivery systems and routes of administration of peptide and protein drugs. *Eur. J. Drug Metab. Pharmacokinet.* **1990**, *15*, 95–102.

Schatzlein, A. G. Delivering cancer stem cell therapies – a role for nanomedicines. *Eur J Cancer.* **2006**, *42*, 1309–1315.

Sershen, S.; West, J. Implantable, polymeric systems for modulated drug delivery. *Adv. Drug Deliv. Rev.* **2002**, *54*, 1225–1235.

Shea, C. M. Future management research directions in nanotechnology: A case study. *J. Eng. Technol. Manage.* **2005**, *22*, 185–200.

Smith, A. Nanotechnology: lessons from Mother Nature. *Chemistry International.* **2006**, *28*(6), 10–11.

Sonavane, G.; Tomoda, K.; Sano, A.; Ohshima, H.; Terada, H.; Makino, K. In vitro permeation of gold nanoparticles through rat skin and rat intestine: Effect of particle size. *Colloids Surf B.* **2008**, *65*, 1–10.

Staples, M.; Daniel, K.; Cima, M.; Langer, R. Application of micro and nano-electromechanical devices to drug delivery. *Pharm. Res.* **2006**, *23*, 847–863.

Steffansen, B.; Nielsen, C. U.; Brodin, B.; Eriksson, A. H.; Andersen, R.; Frokjaer, S. Intestinal solute carriers: an overview of trends and strategies for improving oral drug absorption. *Eur. J. Pharm. Sci.* **2004**, *21*, 3–16.

Stylios, G. K.; Giannoudis, P. V.; Wan, T. Applications of nanotechnologies in medical practice. *Injury.* **2005**, *36*, S6–S13.

Teijeiro-Osorio, D.; Remunan-Lopez, C.; Alonso, M. J. New Generation of Hybrid Poly/Oligosaccharide Nanoparticles as Carriers for the Nasal Delivery of Macromolecules. *Biomacromolecules.* **2009**, *10*, 243–249.

Tratnyek, P. G.; Johnson, R. L. Nanotechnologies for environment cleanup. *Nano Today.* **2006**, *1*(2), 44–48.

Ummadi, S.; Shravani, B.; Rao, N. G. R.; Reddy, M. S.; Nayak, B. S. Overview on Controlled Release Dosage Form. *International Journal of Pharma Sciences.* **2013**, *3*(4), 258–269.

Uner, M. Preparation, characterization and physicochemical properties of solid lipid nanoparticles (SLN) and nanostructured lipid carriers (NLC): their benefits as colloidal drug carrier systems. *Pharmazie.* **2006**, *61*(5), 375–386.

Valenta, C.; Auner, B. G. The use of polymers for dermal and transdermal delivery. *Eur. J. Pharm. Biopharm.* **2004**, *58*, 279–289.

Varna, M.; Ratajczak, P.; Ferreira, I.; Leboeuf, C.; Bousquet, G.; Janin. A. In Vivo Distribution of Inorganic Nanoparticles in Preclinical Models. *J. Biomater Nanobiotechnol.* **2012**, *3*, 269–279.

Wang, W. Oral protein drug delivery. *J. Drug Target.* **1996**, *4*, 195–232.

Wissing, S. A.; Lippacher, A.; Muller, R. H. Investigations on the occlusive properties of solid lipid nanoparticles (SLN). *J. Cosmet Sci.* **2001**, *52*(5), 313–324.

Woodley, J. F. Enzymatic barriers for GI peptide and protein delivery. *Crit. Rev. Ther. Drug Carrier Syst.* **1994**, *11*, 61–95.

Yang, S. C.; Lu, L. F.; Cai, Y.; Zhu, J. B.; Liang, B. W.; Yang, C. Z. Body distribution in mice of intravenously injected camptothecin solid lipid nanoparticles and targeting effect on brain. *J Control Release*. **1999**, *59*, 299–307.

Yokoyama, M. Drug targeting with nano-sized carrier systems. *J. Artif. Organs.* **2005**, *8*, 77–84.

Yun, Y.; Cho, Y. W.; Park, K. Nanoparticles for oral delivery: Targeted nanoparticles with peptidic ligands for oral protein delivery, *Adv. Drug Deliv. Rev.* **2013**, *65*(6), 822–832.

Zambanini, A.; Newson, R. B.; Maisey, M.; Feher, M. D. Injection related anxiety in insulin-treated diabetes. *Diabetes Res. Clin. Pract.* **1999**, *46*(3), 239–246.

Zhang, C.; Chen, J.; Feng, C.; Shao, X.; Liu, Q.; Zhang, Q.; Pang, Z.; Jiang, X. Intranasal nanoparticles of basic fibroblast growth factor for brain delivery to treat Alzheimer's disease. *Int. J. Pharm.* **2014**, *461*, 192–202.

Zur Muhlen, A.; Schwarz, C.; Mehnert, W. Solid lipid nanoparticles (SLN) for controlled drug delivery-Drug release and release mechanism. *Eur. J. Pharm. Biopharm.* **1998**, *45*, 14–155.

CHAPTER 2

NANOPARTICLES: GENERAL ASPECTS AND APPLICATIONS

ONUR ALPTURK[1] and CEYDA TUBA SENGEL-TURK[2,*]

[1]*Istanbul Technical University, Department of Chemistry, 34469, Maslak, Istanbul, Turkey*

[2]*Ankara University, Faculty of Pharmacy, Department of Pharmaceutical Technology, 06100, Ankara, Turkey,*
**E-mail: ctsengel@pharmacy.ankara.edu.tr*

CONTENTS

2.1 NANOTECHNOLOGY AND NANOMATERIALS

Since 80's, we have spent enormous amount of money and effort to make everything smaller, and even in nanoscales, if possible. Smaller cell phones, smaller computer, smaller robots, smaller transistors. It is not because small things are more convenient, but it is because we are convinced that small is better. But why? In 1959, Richard Feynman described a process in which the manipulation of individual atoms was feasible. When this process eventually became a reality, it allowed us to control atoms and to expand or alter their characteristics. In other words, we acquired a chance to generate new materials with superior properties, by making everything much smaller (Peterson, 2004).

Coined by Professor Norio Taniguchi, nanotechnology is based on the observation that the smaller is the better. This is so because miniaturization yields faster, lighter, and cheaper systems using less energy and fabricating less amount of waste, on top of a better control of molecular architecture. As far as medicine and biological sciences are concerned, miniaturization seemed as an opportunity to govern how molecules and atoms interact with surrounding cells and tissues. Surely, this notion translates into the discovery of drugs with improved therapeutic action with minimal side-effects. In conclusion, nanotechnology became platform to design novel therapeutic agents and nanomaterials, whose overall properties are distinctly different from bulk materials (Silva, 2004).

2.2 LIPOSOMES AND POLYMERIC NANOPARTICLES

Diseases and health concerns lead to tremendous amount of work on the discovery of novel therapeutic agents each year. Whilst most of these agents show promise under in vitro conditions, only a few of them make it to clinical studies and/or trials. This stems from the fact that commercialization of any chemical entity is a highly challenging and demanding path compared to simple in vitro studies, which is often sufficient for academic research and publications. For the sake of commercial use, therapeutic agents should exhibit maximum therapeutic efficiency in a particular site, as simultaneously minimizing side-effects, prior to its degradation or elimination from the body. Noteworthy, this description seems to match the notion of "magic bullet" by Paul Ehrlich who first proposed the concept of targeted drug delivery (Strebhardt and Ullrich, 2008).

In that regard, this chapter is dedicated to nanoparticles, which are nanomaterials frequently used for this purpose. Accordingly, the first section of this chapter is devoted to liposomes and the second part covers polymeric nanoparticles. It is needless to say that inorganics like quantum dots, iron oxide nanoparticles, silver nanoparticles and so forth, are not to be ruled out as nanomaterials; however, we choose to focus solely on liposomes and polymeric nanoparticles due to constraint in space.

2.3 WHAT ARE LIPOSOMES?

Liposomes were discovered by Dr. Alec D. Bangham by serendipity, in 1961 (Bangham and Horne, 1964). All began when Bangham was observing phospholipids under electron microscope. He noticed that phospholipids form vesicular structures, once they were hydrated or in touch with water. As Deamer wrote in "his memoirs on Bangham," these vesicles were causally called "Banghasomes" at first, because of an obvious reason. However, the name "liposomes" was suggested by Gerald Weissmann, thereafter (Deamer, 2010).

In water, hydrophobic effect forces phospholipids to assemble in a non-covalent manner; thereby affording lipid leaflets (Figure 2.1). Nonetheless, these lipid leaflets are not thermodynamically favored intermediates, since their hydrocarbon chains are considerably exposed to water. To overcome this thermodynamic handicap, they tend to bend into a curved bilayer structure, which forms a vesicle with closed edges (Balazs and Godbey, 2011). The ultimate vesicular structure wherein an aqueous

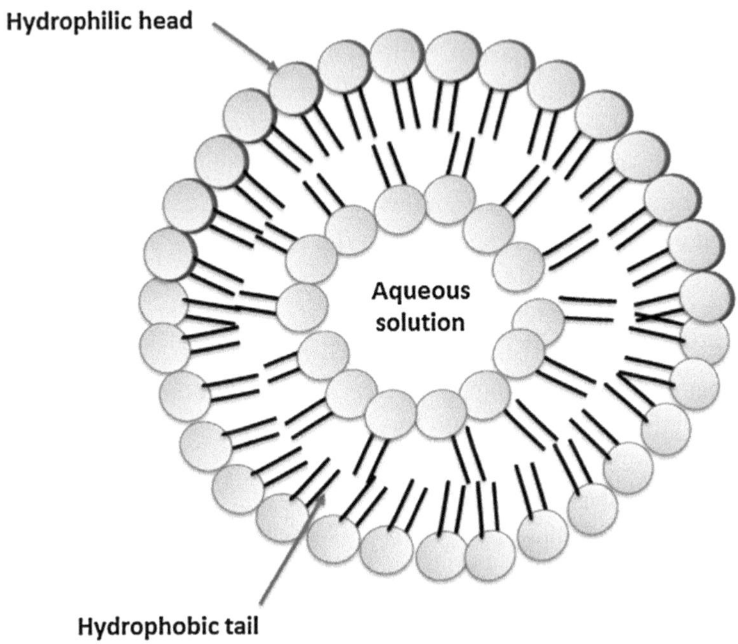

FIGURE 2.1 The structure of liposomes (with permission from Badri et al., 2016).

media gets trapped is what we call "liposomes," and it is particularly stable due to hydrophilic interactions between polar groups on phosphate residues and water, as well as van der Waals interactions between hydrocarbons moieties. The diameter of these nanoparticles varies from nm level (as low as 20 nm) to µm level or even greater, having a dimension comparable to that of living cells (Frezard, 1999). In overall, liposomes are best described as "vesicular architectures structures comprising one or multiple lipid layers (i.e., lamella) (Deamer, 2010), surrounding an equal numbers of aqueous compartments" (Weiner et al., 1989).

Conventionally, liposomes are classified regarding their number of lamellas, and size, as summarized in Table 2.1 (Pattni et al., 2015). However, it is quite conspicuous that these two criteria interweave to some extent. Regarding the number of lamella, liposomes are described as "unilamellar vesicles" (UV), "oligolamellar vesicles" (OV), and "multilamellar vesicles" (MV) (Table 2.1). The term "UVs" refers to liposomes with one single lamella and has subclasses as small (SUVs),

TABLE 2.1 Types of Vesicles-Based on Their Size, and Their Number of Lipid Layers

Vesicle type	Abbreviation	Diameter size	No. of lipid bilayer
Small Unilamellar	SUV	20–40 nm	One
Medium Unilamellar	MUV	40–80 nm	One
Large Unilamellar	LUV	More than 100 nm	One
Giant Unilamellar	GUV	More than 1 micrometer	One
Oligolamellar	OLV	0.1–1 micrometer	5
Multilamellar	MLV	More than 0.5 micrometer	5–25
Multi vesicular	MV	More than 1 micrometer	Multi-compartmental structure

*Adapted with permission from Pattni, B. S.; Chupin, V. V.; Torchilin, V. P. New Developments in Liposomal Drug Delivery. Chem. Rev., **2015**, *115*, 10938–10966. Copyright 2015, American Chemical Society.

medium (MUVs), large (LUVs), and gigantic unilamellar vesicles (LUVs). In opposition to ULVs, OLVs and MLVs possess multiple lamellas each of which is separated by an additional aqueous milieu, generating an onion-like structure. As for multivesicular liposomes, this is an extreme case where the liposome has a multiple concentric aqueous chambers surrounded by a network of colloidal membranes (Lasic and Papadjopoulos, 1995) (Figure 2.2).

2.4 ENCAPSULATION WITHIN LIPOSOMES

Published in 1964, Bangham's discovery of liposomes was accompanied by the finding that negative stain was entrapped within these nanostructures. This realization quickly conduced towards the possibility that "the need for a carrier within which drugs could be entrapped and directed to target tissues may be satisfied by liposome" (Gregoriadis, 1973). In subsequent years, this possibility was proven right by Gregoriadis (1971) who demonstrated that the targeted delivery of drugs and enzymes were achievable with liposomes in a way to eliminate or diminish unfortunate results of direct administration. All these efforts pioneered by Bangham and Gregoriadis made it markedly apparent that liposomes (and thus nanoparticles) offer an asset to manufacture the magic bullet of Ehrlich.

FIGURE 2.2 The structures of unilamellar (left) and multilamellar liposomes (right) (with permission from Badri et al., 2016).

Notwithstanding that early reports on encapsulation were limited to dyes and small molecules, subsequent researches rapidly expanded this repertoire to micro-molecules (like anticancer drugs, antifungal agents, nutrients, and antibiotics) and macro-molecules (like peptides, hormones, genetic material, and proteins) (Uhumwangho and Okor, 2005). Hence, liposomes have taken part in numerous industrial applications, including adjuvant in vaccination, carriers of medical diagnostics, solubilizer for ingredients, signal enhancers, penetration enhancer in cosmetics, imaging, and pharmaceutical research, as a "bio-friendly cargo system" (Allison and Gregoriadis, 1974; Gregoriadis, 1976; Papahadjopoulos, 1978; Petersen et al., 2012; Rahimpour and Hamishehkar, 2012).

One issue that needs to be addressed is how liposomes encapsulate compounds with different polarities (i.e., hydrophobic and hydrophilic). It is conceivable because bizonal structure of liposomes allows two modes of encapsulation: aqueous entrapment and lipid entrapment. The factor that dictates the site of encapsulation is the polarity of the compounds, in the sense that hydrophilic compounds are entrapped in the core section of liposomes (i.e., aqueous entrapment) and the hydrophobic ones are located in the lamella (i.e., lipid entrapment). As for the amphipathic compounds, they are encapsulated in the interphase region between the core and lipid bilayer, where hydrophilicity and hydrophobicity coincide (Sharma and Sharma, 1997).

2.5 LIPOSOMAL FORMULATIONS

It is of paramount significance to elucidate that liposomes have been evolved or modified, in compliance with how efficiently they encapsulate and how they interact with surrounding environment. Hence, the net physiochemical properties of lipids (membrane fluidity, charge density, and steric hindrance, to name few) have been reshaped on numerous occasions within this scope, and liposomal formulations underwent substantial changes since 1961 (Gabizon et al., 2001). We begin by the remark that the building blocks of liposomes are pure lipids or a mixture of lipids. These lipids may be natural like phospholipids or synthetic like dipalmitoylphosphatidylcholine (DPPC) or polymer bearing lipids. Phospholipids (such as phosphatidyl choline, phosphatidylethanolamine, phosphatidyl serine and so forth) are the most regularly employed lipids in liposomal formulations. To elaborate on the structure of phospholipids, a phosphate group is found on C1 of glycerol backbone and it may be further esterified to a large variety of alcohols such as choline, serine, and inositol. Due to pK_a of oxygens being 3 and 7, phosphate moiety harbors a negative charge slightly higher than one, but not quite two. As for the C2 and C3 of glycerol, hydroxyl groups are esterified to long fatty acids of 10–24 carbons with varying degree of saturation (Weiner et al., 1989).

Another common component of lamella is cholesterol or cholesterol-like compounds. Cholesterol is an amphipathic steroid and can be incorporated into lamella in very high concentrations. In nature, cholesterol is found in cellular membranes for three reasons: (i) to enhance the fluidity or microviscosity of the bilayer, (ii) to reduce the permeability of the membrane towards water-soluble materials, and (iii) to stabilize the membrane in biological fluids (Senior and Gregoriadis, 1982; Weiner et al., 1989). In essence, the one and only role of cholesterol and cholesterol-like compounds is to better the packing of the hydrocarbon chains by filling the space among phospholipids. At first glance, a better packing and reduced permeability may be interpreted as a protection against the uncontrolled flaw of materials into liposomes, but it also means to preserve what is inside. In fact, early reports have corroborated that encapsulated compounds might actually leak from liposomes into aqueous milieu by virtue of the fluidity in lamella or exposure to serum proteins (e.g., albumin, transferrin, macroglobulin) and that the degree of this leakage could be lowered by incorporating cholesterol into lamella (Allen and Cleland, 1980; Juliano and Stamp, 1978; Storm et al., 1987).

2.6 CONVENTIONAL LIPOSOMES VERSUS 2ND GENERATION OF LIPOSOMES

To date, liposomes made up of neutral and/or negatively charged lipids and cholesterol are recognized as 1st generation or "conventional" liposomes. Although the conception of cholesterol was a game-changer to conventional liposomal, two major obstacles such as binding to serum components and uptake by the cells of mononuclear phagocyte system remained unresolved (Gregoriadis and Neerunjun, 1974). In 1975, Juliano and Stamp reported two critical results: (i) neutral and positively charged liposomes have more favorable clearance rate than negatively charged liposomes, and (ii) the smaller the vesicles are, the less rapidly they are cleared (Juliano and Stamp, 1975). As much as these results manifested a link from physiochemical properties to efficiency in drug target, they prompted researchers to reconsider and reexamine the structure or structural components of liposomes, leading to 2nd generation of liposomes. Consequently, several approaches like imparting a weak surface negative charge or using a small fraction of some glycolipids in conjunction with phospholipids and cholesterol have reported in the literature (Gabizon and Papahadjopoulos, 1992).

Whilst these approaches overcame the problems related to liposomes with some success, so-called "stealth effect" has been an answer to our problems (Drummond et al., 1999). When hydrophilic polymers like polyethylene glycol (PEG, for short) or biomolecules like GM1 are coated around the liposomes, they tend to expel macromolecules simply by occupying the space in vicinity of liposomal surface. Consequently, liposomes are rendered invisible to blood plasma opsonins, as a result of which they acquire long circulation half-lives. Soon enough, stealth effect was culminated with variations like grafting PEG on to surface or altering hydrophilic molecule (i.e., using egg phsophaditylcholine or sphingomyelin) (Allen and Chonn, 1987). All the results assured the superiority of 2nd generation of liposomes over conventional liposomes, and sparked the inception of the clinical trials (James et al., 1994).

2.7 BENEFITS AND LIMITATIONS OF NANOPARTICULATE FORMULATIONS

Because liposomes and polymeric nanoparticles have similar role as "nanocarriers," they have comparable benefits and limitations. This is why the topic of "benefits and limitations" will not be covered for liposomes *per se*, but for nanoparticles in general. Amongst the fields of applications, pharmaceutical sciences, which vastly profit from nanoparticles formulations, deserve a special merit. As depicted earlier, therapeutic agents have to meet certain criteria prior to their commercial use: they are expected to exert a bioactivity in minimum concentration and in a particular site with minimum amount of side-effects. To do so, drug of subject should reach its destination in an intact form and should be taken up by the tissue with a high efficiency. All these criteria merge under the term "therapeutic efficiency," which fundamentally define how well drugs work (Sharma and Sharma, 1997).

Suitably, one compelling example to validate the profits of liposomes to pharmaceutical sciences has been well documented in the context of AIDS (Ramana et al., 2010). Nowadays, the standard treatment for HIV relies on Highly Active Antiretroviral Therapy (HAART), whose working principle is the use of a combination of at least three drugs to manage the chronic infection. Because the nature of these drugs and their mode of action remains beyond the scope of this chapter, we solely note that numerous complications related to this therapy (such as short half-life, toxic side effects of drugs and so forth) severely compromise both the treatment and the therapeutic efficiency of the drugs. On the other hand, some promising results concerning nevirapine (a non-nucleoside reverse transcriptase inhibitor) has proven that liposomal formulations seemed to drastically improve the delivery of this antiretroviral drug to selected compartments and cells with minimum side-effects. In light of this result and some others, the advantages of nanoparticles are as follows (Akbarzadeh et al., 2013; Date et al., 2007; Gabizon et al., 1983; Grottkau et al., 2013; Rahman et al., 2007):

1. Nanoparticles are reported to be fully friendly to biosystems and in vivo conditions as they are nontoxic, highly biocompatible, and non-immunogenic.
2. With their ability of encapsulation, they serve as an essential tool of targeted delivery. The compounds they deliver range in size (i.e., drugs versus proteins) or in polarity (i.e., hydrophilic versus lipophilic compounds).
3. Regardless of the nature of chemical agents, encapsulation protects compounds against elimination and degradation, while preserving their chemical stability.
4. Nanoparticles improve the cellular penetration and cellular uptake of molecules.
5. Delivery in nanoparticles provides solubility to molecules or pharmaceutical agents (like porphyrins, anthracycline) which otherwise display limited solubility in aqueous media. As for hydrophilic compounds, encapsulation may serve as an approach to enhance their solubility. For instance, anticancer agents like Doxorubicin or Acyclovir can attain concentrations well above their aqueous solubility, when harbored within liposomes (Nassander et al., 1993).
6. Also, targeted delivery of nanoparticles evidently magnifies the effective concentration of the pharmaceutical agents at a specific organ or tissue, upon diminished interaction to some organs (such as like kidney, heart, brain, nervous system, in case of liposomes). Hence, liposome-based delivery greatly enhances the therapeutic efficiency of pharmaceutical agents, keeping drug-related side-effects in minimum levels.
7. The applications of nanoparticulate formulations are feasible in various forms, including (colloidal) solution, aerosol, or in (semi) solid forms.
8. Nanoparticles can be given to the body via numerous routes, such as parenteral, topical and pulmonary routes of administration.

9. From standpoint of enzymes, liposomal formulation is overly significant because it both preserves the activity the enzymes and protects the enzymes against proteolytic cleavage. Furthermore, the permeability of lipid layer may help to regulate enzyme activity by controlling the transport of materials in and out of liposomes (Nasseau et al., 2001).

Of course, there is always the flip side of coin, as liposomal formulations are not free of some limitations (Bangham et al., 1962):

1. The phospholipids of liposomes are prone to oxidation.
2. High production cost disfavors of nanoparticles their manufacture.
3. Liposomes suffer from short half-life, less stability, and low solubility.
4. Leakage and fusion of molecules from liposomes have been well documented.
5. Allergic reactions are likely with liposomal constituents.
6. Nanoparticles experience quick uptake by R.E.S. cells.

2.8 CELLULAR UPTAKE MECHANISMS OF LIPOSOMES

So far, we had our attention on the encapsulation potential of liposomes, with some emphasis on targeted delivery. The missing link from encapsulation to targeted delivery is the cellular uptake, which is central to the fate of the drug and its therapeutic efficiency. Related mechanistic studies have indicated four pathways of cellular uptake (Pagano and Weinstein, 1978). Simple adsorption, whose name is self-explanatory, refers the adsorption of the liposomes to cell wall without any internalization. The event of adsorption may be mediated by either by noncovalent interactions or by specific components such as receptors and ligands. Endocytosis is a process where the entire vesicle is taken up into endocytotic vesicles and is canalized to lysosomes afterwards. However, this mechanism is unsought regarding the integrity of the therapeutic agents because the content of liposomes can escape from the lysosomes, only if it manages to survive the acidity and the digestive enzymes of endosomes and lysosomes. Fusion with cell membrane, which is much rarer, is the merging of lipid bilayer with cellular membrane. Despite that cytoplasma is the prime location for the concomitant release of lysosomal content upon fusion, some of the material may subsequently leak to the other cellular compartments through some secondary processes. And the last one is lipid transfer which concerns the transfer of lipid molecules from liposome to cellular membrane.

One should not overlook the fact that cellular uptake supplements the targeted delivery and selectivity of liposomes to a considerable extent. In that regard, receptor-mediated endocytosis (also known as active targeting) has been advantageous over conventional targeting (i.e., passive targeting) since the former assures a point-shot delivery of drugs. By definition, this form of endocytosis is mediated by the interaction between a ligand/group on liposomes and a receptor on cells. This ligand or group may be a variety of biomolecules such as antibody, glycopeptides, oligosaccharides, viral proteins and fusogenic residues and it may be attached to liposomes either covalently or non-covalently (Sihorkar and Vyas, 2001).

An early result that antibody-targeted liposomes enhance the selective toxicity of liposomal anticancer drugs of cultured cells has been quite promising and paved the way to development of novel methodologies to attach biomolecules to liposomes (Martin et al., 1981). Still, the manufacture of ligand-targeted liposomes remains as a major challenge to date, as they are infamous for being labor-demanding, hard to control and affording systems which are readily cleared from circulation by macrophages (Allen and Cullis, 2013).

Despite its manageable flaws with lots of room for improvement, the combination of ligand-liposome gives the feeling of a golden ticket and yet, some literature survey openly suggests that ligand-mediated liposomal delivery is far from being perfect. The whole problem scientific community encountered has surfaced because of controversial results between non-targeted liposomes and ligand-targeted liposomal delivery. As Allen and Cullis beautifully summarized, (Allen and Cullis, 2013) some results lean towards the superiority of ligand-targeted delivery over non-targeted delivery (Lopes de Menezes et al., 1998; Park et al., 2002) while some others document no difference in between (Goren et al., 1996; Vingerhoeds et al., 1996). Regardless, it looks like antibodies conjugated liposomes may not in reality alter or better the distribution of liposomes to the target under the conditions when both have similar circulation half-lives. Oddly enough, the amount of particles reaching the target tissue seems to be independent of the presence of the ligand and the ligands or groups on liposomes explicitly affects the amount of liposomes taken up by the tissues (Kirpotin et al., 2006). Surely, there is more to this story and these overall findings are decisive in the fate of ligand-mediated delivery and the direction this field is headed.

2.9 TRIGGERED DELIVERY OF NANOPARTICULATE CONTENT

In reality, cellular uptake by active or passive targeting is not the end of the path. This means that as long as the drug remains stuck within liposomes or it is not released in cellular media effectively, both the efficiency of cellular uptake and the therapeutic efficiency of the drug become completely meaningless. This is because the drug exhibits its biological activity only when it is released from the liposomes. Hence, the term "therapeutic efficiency" is, in a way, a combined consequence of both cellular uptake and release of the drug into the cellular media, and hence the significance of unloading liposomal content by any means should not be overlooked (Johnston et al., 2006).

To date, triggered release from nanoparticles has been achieved by external (such as light, magnetic field, temperature, and heat) or environmental stimuli (such as pH). In some instances, dual or triple combinations of stimuli have been reported in the literature, as well. The way these stimuli functions is to alter the environment surrounding nanoparticles, and hence the physicochemical properties of nanocarriers. The pH-mediated drug release is a well-relevant example wherein nanocarriers undergo destabilization upon pH-change and subsequently release their content. Given that different parts of body have different pH values (for instance, pH for tumor cells, endosomes and lysosomes are 6.5–7.2, 5.0–6.5, 4.5–5.0, respectively), it is perceptible that the targeted delivery to cancer site, and endo/lysosomal sites is feasible through pH-triggered delivery systems (Bennet and Kim, 2014).

2.10 ARE POLYMERIC NANOPARTICLES THE NEXT LIPOSOMES?

In spite of all the progress we made since 1961 and even commercialization of liposomal formulations to some extent, certain flaws concerning liposomes has remained unsolved to date *(vide infra)*. For instance, liposomes are not stable in harsh gastrointestinal environment, which prohibits their uptake through oral administration (Thanki et al., 2013). Consequently, we feel obligated to seek novel drug carrier systems, as we keep investigating on liposomes. As an outcome of this rationale, polymeric nanoparticles were introduced to scientific community almost 20 years after liposomes were. The textbook definition of the term "polymeric nanoparticles" is submicron-sized particles made up of wide variety of polymers. Despite its generic use, this term actually refers either of nanocapsules or nanospheres. These nanoparticles are commonly acknowledged as "drug carriers in nano-dimensions," as liposomes are (Thanki et al., 2013). Hence, we reasoned that it would be more beneficial if certain aspects of polymeric nanoparticles were covered by comparison to liposomes.

Despite their similarity in size and functions, these materials diverge lightly from each other, regarding their structure and mode of encapsulation. Nanocapsules are defined as heterogeneous reservoir systems in which the drug is caged in the inner cavity, which is composed of oil or aqueous core surrounded by a thin polymer membrane. On the other hand, nanospheres are homogenous matrix system wherein dispersed or dissolved active compound is adsorbed on the surface or entrapped within the polymeric matrix structure through the solid sphere (Harush et al., 2007) (Figure 2.3).

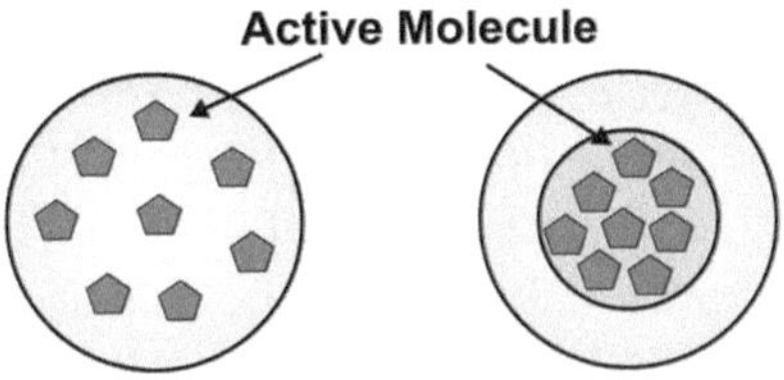

FIGURE 2.3 Schematic representations of nanospheres (left) and nanocapsules (right) into which drugs are encapsulated.

2.11 BUILDING BLOCK OF POLYMERIC NANOPARTICLES

Nanospheres and nanocapsules could be prepared from a large variety of polymers, ranging from natural polymers (such as alginates, chitosan, gelatin, and so on), to synthetic ones (such as poly(ε-caprolactone) (PCL), poly(lactic acid) (PLA), polymethyl methacrylate (PMMA), poly(lactic-co-glycolic) acid (PLGA), poly(glycolic acid) (PGA), to name few) (Kumari et al., 2010; Park et al., 2010). A large selection of ingredients certainly enables one to shape and to design the parameters of nanoparticles like the composition of the polymer, the architecture, the backbone stability, the water solubility, and hence the activity of nanocarriers.

The overall polymeric architecture is critical as it affects numerous factors such as the physicochemical characteristics, the incorporation capability, the drug release rate, and the bio distribution of the delivery system (Qui et al., 2006). For instance, a large degree of hydrophilicity on surface considerably enhances the circulation time and the water-solubility of the nanoparticles, while it concurrently diminishes its degradation through enzymes. In keeping with this precedent, the impact of hydrophilicity illustrates how tuning one single parameter reshapes overall properties and biocompatibility of nanoparticles. Therefore, we sought to describe general properties of some of these materials and discuss the outcomes of their utilization in polymeric nanoparticles.

2.11.1 CHITOSAN

Chitosan is a cationic polysaccharide obtained through the deacetylation of chitin, which is the primary component in the cell walls of fungi and the exoskeleton of shrimps and crabs. It consists of linear β-(1–4)-linked D-glucosamine (Agnihotri et al., 2004). The primary determinants of its physicochemical characteristics are the extent of deacetylation and its molecular weight, as the latter regulates parameters like solubility, viscosity and elasticity (Quasim et al., 2014). In acidic medium, it is fully water-soluble through the protonation of free amino groups, which grants a positive charge. Under this condition, chitosan can form highly viscoelastic polyelectrolyte complexes with anionic polymers such as hyaluronic acid. Owing to its versatile character, chitosan has what it takes to be an ingredient of nanoparticles and thus, it is generously used in drug formulations, including tablets, gels, and films (Bansal et al., 2011; Sinko, 2011).

2.11.2 GELATIN

Similarly to chitosan and alginates, gelatin is a natural and biodegradable protein. The main way to manufacture gelatin is the hydrolysis of collagen under basic or acidic conditions. Actually, it is a nonhomogeneous single- or multistranded polypeptides mixture and could be further degraded to amino acids under in vivo conditions. Gelatin has an amphoteric structure, consisting of both anionic and cationic functional groups (Kaul and Amiji, 2004). Although its characteristics such biocompatibility, biodegradability, and nature of chemical modifications makes it an appealing material for drug delivery applications, we must be cautious of the contamination of gelatin with transmissible spongiform encephalopathies. Hence, recombinant gelatin is more recommended for nanoparticulate formulations.

2.11.3 ALGINATES

Alginate is another natural polymer, exhibiting water-solubility and polyanionic character. It is a linear polysaccharide that mainly consists of two building blocks of mannuronic and gluronic acid moieties. Alginate exhibits pH-dependent anionic character, which enables it to interact with cationic proteoglycans and polyelectrolytes (Sinko, 2011). One important feature of alginates is that the molecular weight of alginates affects its mechanical properties, as well as the degradation rate of alginate-based nanoparticles. For instance, higher molecular weight minimizes the reactive position numbers existing for hydrolytic degradation, whereby to lower degradation rate (George and Abraham, 2006; Sun and Tan, 2013).

2.11.4 POLY(LACTIC ACID) (PLA)

PLA is biodegradable polyester which is of synthetic, thermoplastic and nontoxic aliphatic nature. It is retrieved from renewable facilities, such as corn starch, sugarcane or chips. Cyclic di-ester, lactide, and lactic acid are the main monomers of the structure. Ring opening polymerization of lactide is a conventional route to PLA in the presence of various metals. However, metal catalyzed reaction notoriously causes the racemization of the PLA, reducing its stereoregularity, in comparison to the starting material (Martin and Avérous, 2001). Its chirality yields to forms of PLA polymers – poly(L-lactide) (PLLA) and poly(D-lactide) (PDLA) (Lassalle and Ferreira, 2007). PLA is spontaneously degraded by simple hydrolysis through ester bonds without any need for a catalyst. However, the hydrolysis rate (or degradation rate) depends on numerous factors such as the size, the shape of the drug carriers, the isomer ratio of PLA, and the temperature of hydrolysis. The performance of PLA nanoparticles in the controlled delivery of various bioactive molecules such as conventional and antitumor drugs, peptides and genes for the treatment of a variety of illnesses have been extensively studied (Lassalle and Ferreira, 2007; Garlotta, 2001).

2.11.5 POLY(LACTIC-CO-GLYCOLIC) ACID (PLGA)

PLGA is a copolymer of PGA and PLA and is widely preferred in drug delivery systems thanks to its biocompatibility, biodegradability and ease of production. Regarding its design and performance, it is commonly known as one of the best biomaterials available for drug delivery (Jain et al., 1998). In fact, PLGA is currently used in some therapeutic devices, approved by Food and Drug Administration (FDA). Ring-opening copolymerization of lactic acid and glycolic acid constitutes the main synthesis of the polymer. Diverse forms of PLGA could be acquired in course of polymerization process depending on the ratio of lactide to glycolide, and by that, polymer is generally defined in terms of the molar ratio of monomers.

PLGA degrades through the hydrolysis reaction of the ester moieties. The time required for full degradation depends on the ratio of monomers. It has been reported that with the higher content of glycolide units cause, degradation process tends to proceeds the faster (Gentile et al., 2014; Samadi et al., 2013). PLGA is indeed quite advantageous as a biodegradable polymeric material, because the hydrolysis products are starting materials, lactic acid and glycolic acid. Since these two components are by-products of several pathways of metabolic process in the body, there is limited systemic toxicity related to the usage of PLGA for biomaterial applications or drug delivery systems (Pavot et al., 2014).

2.11.6 POLY(ε-CAPROLACTONE) (PCL)

PCL is another biodegradable, biocompatible and nontoxic synthetic polyester with excellent mechanical strength. It has been currently used for the production of implantable drug delivery systems (Lemarchand et al., 2003). Ring opening polymerization of ε-caprolactone with a catalyst such as stannous octoate is the main production method for PCL synthesis (Labet and Thielemans, 2009). The main reason of the PCL's great interest as an implantable biomaterial is the degradation process which occurs by the hydrolysis of its ester linkages under physiological conditions. Due to its degradation, which is even slower than that of PLA, PGA and PLGA, PCL is a very suitable polymer for the design of long-term implantable devices. The lipophilic nature of PCL is also supported the passive uptake processes (Bhavsar and Amiji, 2008; Haas et al., 2005).

2.11.7 POLYMETHYL METHACRYLATE (PMMA)

PMMA is a non-biodegradable synthetic copolymer of methyl methacrylate. It has rigid, hard but brittle structure. Interestingly, contact with water renders hydrophobic PMMA, rather partially hydrophilic. This polymer is routinely synthesized by emulsion, solution, and bulk polymerization techniques. Biomedical community considers PMMA as a nontoxic polymer with confidential toxicological data for various applications. Potential biomedical applications of PMMA-based nanoparticles include the carrier of several active molecules as antioxidants and antibiotics, and also adjuvant for vaccines through different administration routes (Bettencourt and Almeida, 2012).

2.12 CELLULAR UPTAKE MECHANISM OF POLYMERIC NANOPARTICLES

It appears that the immune system of the body consistently causes some trouble to nanomaterials (whether it's liposomes or polymeric nanoparticles).

Like liposomes, polymeric nanoparticles (especially hydrophobic ones) are regarded as aliens by immune system. Hence, they are eliminated from the blood stream by reticulo-endothelial system (RES) and are taken up through the spleen or the liver. This process constitutes the main biological barrier against the delivery role/function of polymeric nanoparticles (Kumari et al., 2010). Another notable obstacle is the opsonin proteins, as always. The opsonin proteins which locate in the blood stream immediately bind to the surface of the nanoparticles when the nano-sized drug carriers get into to the blood upon administration. Opsonized particles have attached to macrophages at once and their internalization has been occurred by phagocytosis (Hans and Lowman, 2002; Qwens III and Peppas, 2006).

To surpass the immune system, several approaches of surface modification have been formulated to guarantee that nanoparticles are not recognized by the carriers of RES macrophages. With that in mind, the trick is simply to coat a hydrophilic polymer (like PEG; chitosan, poloxamers, and poloxamines) over the surface of the particle to conceal its hydrophobicity. PEGylation of nanoparticles has some other benefits, as well: (i) it blocks the hydrophobic and electrostatic interactions that promote the binding of opsonins to the surface of the particles, and (ii) it elevates the blood circulation half-life of the nanoparticles through the size of the carrier (Danhier et al., 2012; Kumari et al., 2010). Hence, it is conclusive that "stealth effect" of PEGylation, which we have examined in the context of liposomes *(vide supra)*, is vital to polymeric nanoparticles, as well.

Surface charge of polymeric nanoparticles also is the key factor that shapes interactions between nano-sized particles and cell membrane, along with cellular uptake. Positive surface charge gives nanoparticles higher extent of internalization, in consequence of favorable ionic interactions between negatively charged cell membranes and positively charged nanoparticles (Foget et al., 2005; Vasir and Labhasetwar, 2008). Besides, positive surface charge gives nanoparticles the ability to easily escape form lysosomes upon internalization. Nanoparticles can localize perinuclear section whereas negatively and neutrally charged nanoparticles are retained in lysosomes. Clearly, this issue is reflected as a substantial increase in the uptake capacity of the nanoparticles, as it favors positively charged particles over the others (Yue et al., 2011). However, this conclusion does not necessarily infer that we should give up on the polymeric nanoparticles with inherently negative surface charges, for that matter; the cellular uptake properties of these particles could be restored or improved by yielding a positive or neutral charge over the surface through modifications by chemical agents including chitosans, PEGs, Vitamin E TPGS or didodecydimethyl ammonium bromide (Danhier et al., 2010; Sengel-Turk et al., 2014; Tahara et al., 2009).

Polymer-based nano-sized carriers are commonly internalized in cells by the way of endocytosis and partly via fluid phase pinocytosis. Once inside the cell, these nanoparticles can immediately run away from the endo-lysosomes and enter the cytoplasma after *ca.* 15 min of incubation. In comparison to liposomes, active-targeting of polymeric nanoparticles through surface attached ligands is possible as well, which allows an interaction between vesicular membranes and nanoparticles. Distinct from liposomes, targeting ligands are generally linked to the surface of the nanoparticles through the uptake on PEG chains (Betancourt et al., 2009). Once these ligands bind to the corresponding cellular receptors, the membrane is locally destabilized, resulting in the escape of nanoparticles into cytosol (Vasir and Labhasetwar, 2007). Herein, we can't stress enough that the process of PEG coating of particles should be handled with caution, as it may otherwise turn to a double-sided knife. While PEG coating is mandatory to avoid RES, it is very likely that this coating may shield the ligands, whereby to firmly damage the active transport (Wang and Thanou, 2010).

2.13 MANUFACTURING OF POLYMERIC NANOPARTICLES

The techniques to manufacture nanoparticles are segmented into two categories: (i) those taking advantage of preformed polymers, and (ii) those-based on polymerization. The choice of the method depends on various factors, such as (1) the aqueous solubility and the stability of the nanoparticles, (2) desired diameter of the particles, (3) the degree of biocompatibility, toxicity and biodegradability, (4) surface properties such as permeability and charge, and lastly (5) targeted drug release profiles and rates (Quintanar et al., 1998).

Today, there exist many approaches, including nanoprecipitation, double emulsion, emulsion/diffusion, emulsion coacervation, and layer-by-layer method to manufacture the nanocapsule type polymeric nanoparticles

(Danhier et al., 2012; Naahidi et al., 2013). Herein, our purpose is to briefly describe some of these approaches to give the readers a feel for how these materials are synthesized.

2.13.1 NANOPRECIPITATION

Nanocapsule production needs both a solvent phase (such as dioxane, acetone, ethanol, methylene chloride, or hexane) and non-solvent (usually aqueous phase) phase. As the name implies, nanoparticles are obtained through precipitation from a mixture of aqueous and organic phase. In this procedure, a water-miscible solvent is used to solve the polymer substance with intermediate polarity. Once the organic medium is added gradually onto an aqueous surfactant solution under gentle stirring, the polarity of the final medium forces the product to precipitate. The particle formation process occurs in three stages with this preparation technique: nucleation, growth, and aggregation. The particle size is dependent to the rate of each step and supersaturation is the impetus of this manufacturing process. During nanoprecipitation, the growth and the nucleation phase of the polymerization reaction are two stages, playing role in the uniformity of the particles. The critical parameters of this production process are the agitation rate of the aqueous phase, the injection rate of the organic phase, the ratio of organic to aqueous phase, and the method of organic phase addition (Quasim et al., 2014; Thanki et al., 2013).

2.13.2 DOUBLE EMULSION

Double emulsions are heterodisperse complex formations, referred to as "emulsions of emulsions." Double emulsions are categorized into two basic types: oil-water-oil emulsion (o/w/o), and water-oil-water emulsion (w/o/w). These two procedures are similar in the sense that both emulsions are formed by ultrasonication process to produce nanocapsules by this preparation process. However, they differ in the step of stabilization; in w/o/w procedure, w/o surfactant serves to stabilize the interface of the w/o internal emulsion, whereas the stabilization of nanoparticle dispersion is achieved by the supplementation of the surfactant molecules in the latter procedure. When the solvents are eventually removed through extraction or evaporation process by vacuum, hardened nanocapsules are obtained from the aqueous medium (Naahidi et al., 2013; Quintanar et al., 1998).

2.13.3 EMULSION–DIFFUSION

This production technology allows the encapsulation of active molecules of both lipophilic and hydrophilic nature into nanocapsules. A typical of emulsion–diffusion procedure involves three steps: organic, aqueous and dilution. Polymer, drug, and oil are initially dissolved into a partially water-miscible organic solvent. The subsequent step involves the emulsification of the organic phase in the aqueous phase under vigorous agitation. Afterwards, the addition of water causes the diffusion of the dissolver into the aqueous phase, resulting in the formation of nanocapsules. The diameter of the nano-capsules depends on the polymer concentration, shear rate used in the emulsification step, the chemical contents of the organic phase, and the primary emulsion's droplet size (Quasim et al., 2014; Quintanar et al., 1998).

2.13.4 LAYER-BY-LAYER

This production method relies on the alternate adsorption of oppositely charged materials, such as polyelectrolytes through electrostatic interactions. In accordance with this mechanism, nanocapsules are synthesized through attractiveness of irreversible electrostatic interaction. The adsorption of polyelectrolyte is thought to take place through this interaction at supersaturating bulk concentrations of polyelectrolytes. However, the need for a colloidal template, which is compatible with the active substance, renders this technology rather difficult (Quasim et al., 2014).

The production technologies of the nanosphere-type polymeric nanoparticles are generally divided to seven groups, as bellows: solvent displacement, solvent evaporation, emulsion-diffusion-evaporation, salting out, solvent diffusion, ionic gelation, and spray drying. Among them, solvent evaporation and solvent displacement are the most frequently preferred ones (Birnbaum et al., 2000; Mudgil et al., 2012; Li et al., 2011).

2.13.5 SOLVENT DISPLACEMENT

Polymer precipitation from an organic solution and the diffusion of this organic solution into the aqueous phase in the absence or presence of a suitable stabilizing agent constitute the mechanism of this methodology (Danhier et al., 2012). To precipitate the nanoparticles, a water-miscible solvent is used to dissolve the polymer. Dichloromethane, acetone are used to dissolve the polymers, and the active molecules. This organic medium is added into an aqueous surfactant solution under gentle stirring. When two phases are brought together, the organic solvent diffuses from the organic phase into the aqueous one. During diffusion process, organic solvent carries polymer chains concomitantly to the water phase, whereby they tend to aggregate, forming nanoparticles. The rapid diffusion of the solvent facilitates the instantaneous formation of the colloidal dispersion due to the deposition of the polymer on the interface between the organic solvent and water (Birnbaum et al., 2000; Li et al., 2011).

2.13.6 SOLVENT EVAPORATION

This method enables the incorporation of lipophilic drugs with high loading efficiencies through controlling the mixing speed and the mixing conditions. Volatile solvents such as ethyl acetate, chloroform or dichloromethane are typically used to dissolve the polymeric materials and an emulsion (either water in oil in water or oil in water) is generated high-speed mixing of the organic phase and aqueous phase. To reduce the particle size during evaporation process of the organic solvent, ultrasonication or high speed homogenization techniques are preferred. The organic solvent is removed from the system through gentle stirring at room temperature or by rotary evaporator. The desired nanoparticle dispersion is formed eventually, when the organic solvent is exactly moved away from the emulsion. Afterwards, solidified nanoparticles are collected with ultracentrifugation, and a final washing step is required with distilled water to remove additives such as residual polymers and the stabilizing agents. The final nanoparticle dispersion is lyophilized to yield dried powder (Birnbaum et al., 2000; Quintanar et al., 1998).

2.14 CHARACTERIZATIONS OF NANOPARTICLES

It has been well established that manufacturing methods have influence on the physicochemical properties of nanoparticles, such as size, charge, lamellarity (for liposomes) and so forth. Because these properties govern in vitro (shelf-life and sterilization) and in vivo performances of nanoparticles, these physicochemical properties need to be fully characterized in order to predict and in vitro and in vivo performances of the liposomes and fully understand their chemistry (Bennet and Kim, 2014). Hence, quality control tests to rapidly and accurately evaluate these properties are largely needed. In this section, we will briefly mention the methods, frequently employed in the literature to assess some of these parameters (Thanki et al., 2013).

The drug loading potency is defined as the ratio of drug to the total weight of the carrier. Because a lower loading capacity directly points out the inability of the system to serve as a carrier, the drug loading potency appears to be the central parameter that needs to be evaluated the first. The most commonly preferred approaches to evaluate the loading amount of encapsulated drug are high-performance liquid chromatography (HPLC), ultraviolet spectrophotometry (UV), and capillary electrophoresis (CP). Of them, HPLC and CP are documented to be more sensitive and precise to identify the active molecules and the excipients in the nanoparticle structure (Gumustas et al., 2013).

The size of the nanoparticles allows a control on the bioavailability of the particles in the body, especially in tumor tissues and penetration through biological barriers. In regard to the impact of the size, the common consensus is that small particles can pass through previously specified biological barriers, even when they are intact. Therefore, the size of nanoparticles is one of the most critical properties of the nano-sized drug delivery systems and is the key to design and fabricate efficient nanoparticulate drug delivery systems. The main technology to estimate both the average particle diameter and the PDI of the nano-sized particles is dynamic light scattering technique (also called "photon correlation spectroscopy"). This technique relies on the dispersion of the light caused by the particles' Brownian motion (Gaumet et al., 2008). Atomic force microscopy (AFM), scanning electron microscopy (SEM) and transmission electron microscopy (TEM) are some other approaches used for the same purpose. However, these visualization-based technologies are more informative on the account that the morphology and shape of the nanoparticles could be imparted too (Sengel-Turk et al., 2012).

Another parameter we will discuss is the surface charge that affects cellular uptake capacity, possible interaction with other biological environments and biodistribution profiles of the nanoparticles. On that account, we have elaborately examined how each charge influence the uptake behavior of nanoparticles. All of this means that attention should be paid to the charge of nanocarriers, in order to maximize the therapeutic efficiency of the active molecule. The technique to determine the charge involves the quantification of its zeta potential (ζ), where the charge of nanoparticles is correlated through their mobility monitored by an electrical potential. The values of surface charge of the systems can be varied in positive, neutral or negative, depending on the surface modification agents and the polymer (Soppimath et al., 2001). Chung et al. investigated the effects of heparin and chitosan over the surface functions of the PEG-PLGA nanoparticles through tumor targeting efficacy (Chung et al., 2010). The surface charge of unmodified PEG-PLGA nanoparticles was determined to be -20 ± 1 mV, whereas the surface charges were found to be -50 ± 2 mV and $+38\pm0$ mV for heparin-conjugated and chitosan-conjugated PEG-PLGA nanoparticles, respectively. A tumor model research has revealed that chitosan-functionalized PEG-PLGA nanoparticles accumulate into the tumor region on SCC7 tumor-bearing athymic mice 2.4 folds more than PEG-PLGA nanoparticles do. The accumulation profile of chitosan-functionalized PEG-PLGA nanoparticles in liver was found less than the control group. When heparin was functionalized to the nanoparticle surface, the accumulation was found to be 2.2 folds more than that of the PEG-PLGA nanoparticles, but the accumulation in liver was similar to that of the control. For this reason, the surface modification through the heparin or chitosan-functionalized PEG-PLGA will be a promising approach for the delivery of hydrophobic agents to better cancer treatment.

In order to examine the interior structure of the nanoparticles and analyze the elements; thermal analysis [thermal gravimetric analysis (TGA), differential scanning calorimetry (DSC)], and XRD can be used. XRD analysis gives to researchers some information for characterization of the composition and the physical structures (amorphous and/or crystalline) of the nano-sized particles. XRD is the main analysis technique for the analysis of solid-state active molecules, which comes handy on all drug development stages, production process and testing procedures (Chen et al., 2007). However, this technique can be time consuming and requires a large volume of sample. DSC and TGA, on the other hand, characterize the thermal properties of nanostructured particles (Chen et al., 2007; Darvishi et al., 2015). Thermal properties of nanoparticles essentially focuses on melting degrees, thermodynamics and kinetics for transition from nanocrystal to liquid and nanocrystal to glass. Nevertheless, the specific heat of nanoparticles and surface enthalpies are investigated via these methods (Gill et al., 2010; Wang et al., 2002). As for the thermoanalytical techniques, DSC is a popular one. It usually used in the characterization of drug delivery systems: (i) to quantify amorphous or crystalline phases in pharmaceutical nanoparticulate solids, (ii) to detect the presence of possible active molecule-excipients interactions, (iii) to identify thermal properties of nano-sized particles, (iv) to measure stability the of colloidal nanoparticles, and (iv) to determine the mechanism of the sorption for ion chelating nano carriers like chitosan (Gill et al., 2010).

2.15 COMMERCIALIZATION OF NANOPARTICULATE PRODUCTS

At the beginning of twentieth century, together with the importance of the pioneering research of synthetic polymer chemistry, the concept of nanomedicine gained importance. Indubitably, the superior performance of nanoparticulate formulation over free drugs initiated the era of commercialization and to date, there are a massive amount of nanoparticle-based commercial products employed in imaging and personalized medicine and some of these nanomedicines were even entered to routine clinical use/trials, too. At this point, we should remark that more commercial products clearly indicate a better chance to cure more diseases and to resolve circumstances wherein the nature of nanoparticles limits us. A list of some approved nanoparticle-based marketed products is tabulated below (Table 2.2).

TABLE 2.2 Nanoparticulate-Based Products Approved For Commercial Use

	Drug	Product	Indication
Polymer	Abraxane®	Paclitaxel	Metastatic Breast Cancer
	SandostatinLAR®	Octreotide acetate	Carcinoid Tumors, Vasoactive Intestinal Peptide Tumors
	Lupron Depot®	Leuprolide acetate	Prostate Cancer, Endometriosis
	Bydureon®	Exenatide	Diabetes Mellitus Type 2
	Zevalin®	Ibritumomabtiuxetan	Non-Hodgkin Lymphoma
	Bexxar®	Tositumomab	Follicular Cancer
	Mylotarg®	Gemtuzumab	Acute myeloid Leukemia
Liposome	Ambisome™	Amphoteracin B	Fungal infection
	DaounoXome™	Daunorubicin	Kaposi's sarcoma
	Doxil™	Doxorubicin	Refractory Kaposi's sarcoma, recurrent breast cancer and ovarian cancer
	Visudyne®	Verteporfin	Age-related macular degeneration, pathologic myopia and ocular Histoplasmosis
	Myocet®	Doxorubicin	Recurrent breast cancer
	DepoCyt®	Cytarabine	Neoplastic meningitis and lymphomatous meningitis
	DepoDur	Morphine sulfate	Pain

*Adapted with permission from Pattni, B. S.; Chupin, V. V.; Torchilin, V. P. New Developments in Liposomal Drug Delivery. Chem. Rev., **2015**, *115*, 10938–10966. Copyright 2015, American Chemical Society.

2.16 CONCLUDING REMARKS

Ever since Bangham saw liposomes under the electron microscope for the first time, we came a long way, without a doubt. Yet, we can't presume that this is it. Some of the complications related to the manufacture of liposomes are not fully resolved. These include stability issues, batch-to-batch reproducibility, particle size control, the production of large batch size and so forth (Sharma and Sharma, 1997). From the clinical perspective, the essential problem is that the success under in vitro conditions does not always stand for a similar success under in vivo conditions

(Bennet and Kim, 2014). With the discovery of scanning tunneling microscope, Feynman's vision became a reality. We sincerely hope that by inventing much more effective nanosystems free of these aforementioned problems, we can make Ehrlich's dream of "magic bullet" come true soon.

KEYWORDS

- **cellular uptake**
- **characterization technique**
- **liposomes**
- **nanoparticles**
- **polymeric biodegradable nanoparticles**
- **production technologies**

REFERENCES

Agnihotri, S. A.; Mallikarjuna, N. N.; Ambinabhavi, T. M. Recent advances on chitosan-based micro and nanoparticles in drug delivery. *J. Control. Release*, **2004**, *100*, 5–28.

Akbarzadeh, A.; Rezaei-Sadabady, R.; Davaran, S.; Joo, S. W.; Zarghamil, N.; Hanifehpour, Y.; Samiei, M.; Kouhi, M.; Nejati-Koshk, K. Liposome: classification, preparation, and Applications. *Nanoscale Res. Letts.*, **2013**, *8*, 102–110.

Allen, T. M.; Chonn, A. Large unilamellar liposomes with low uptake into thereticuloendothelial system. *FEBS Lett.*, **1987**, *223*, 42–46.

Allen, T. M.; Cleland, L. G. Serum-induced leakage of liposome contents. *Biochim. Biophys. Acta*, **1980**, *597*, 418–426.

Allen, T. M.; Cullis, T. M. Liposomal drug delivery systems: From concept to clinical Applications. *Adv. Drug Deliv. Rev.*, **2013**, *65*, 36–48.

Allison, A. G.; Gregoriadis, G. Liposomes as immunological adjuvants. *Nature*, **1974**, *252*, 252.

Badri, W.; Miladi, K.; Nazari, Q. A.; Greige-Gerges, H.; Fess, H.; Elaissari, A. Encapsulation of NSAIDs for inflammation management: overview, progress, challenges and prospects. *Int. J. Pharm.*, **2016**, *515*, 757–773.

Balazs, D. A.; Godbey, W. T. Liposomes for Use in Gene Delivery. *J. Drug Deliv.*, **2011**, Article ID 326497.

Bangham, A. D.; Horne, R. W. Negative staining of phospholipids and their structured modification by surface active agents as observed in the electron microscope. *J. Mol. Biol.*, **1964**, *8*, 660–668.

Bangham, A. D.; Horne, R. W.; Glauert, A. M.; Dingle, J. T.; Lucy, J. A. Action of saponin on biological cell membranes. *Nature*, **1962**, *196*, 952–955.

Bansal, V.; Kumar Sharma, P.; Sharma, N.; Pal, O. P.; Malviya, R. Applications of chitosan and chitosan derivatives in drug delivery. *Adv. Bio. Res.*, **2011**, *5*, 28–37.

Bennet D.; Kim, S. *Application of Nanotechnology in Drug Delivery*. Ed: Sezer, A. D. **2014**, *InTech*, p. 257310.

Betancourt, T.; Bryne, J. D.; Sunaryo, N.; Crowder, S. W.; Kadapakkam, M.; Patel, S.; Casciato, S.; Brannon-Peppas, L. PEGylation strategies for active targeting of PLA/PLGA nanoparticles. *J. Biomed. Mater. Res.*, **2009**, *91A*, 263–276.

Bettencourt, A.; Almeida, A. J. Poly(methyl methacrylate) particulate carriers in drug delivery. *J. Microencapsul.*, **2012**, *29*, 353–367.

Bhavsar, M. D.; Amiji, M. M. Development of novel biodegradable polymeric nanoparticles-in-microsphere formulation for local plasmid DNA delivery in the gastrointestinal tract. *AAPS PharmSciTech*, **2008**, *9*, 288–294.

Birnbaum, D. T.; Kosmala, J. D.; Brannon-Peppas, L. Optimization of preparation techniques for poly(lactic acid-co-glycolic acid) nanoparticles. *J. Nanopart. Res.*, **2000**, *2*, 173–181.

Chen, L.; Xie, Z.; Hu, J.; Chen, X.; Jing, X. Enantiomeric PLA-PEG block copolymers and their stereocomplex micelles used as rifampin delivery. *J. Nanopart. Res.*, **2007**, *9*, 777–785.

Chung, Y.; Kim, J. C.; Kim, Y. H.; Tae, G.; Lee, S. Y.; Kim, K.; Kwon, I. C. The effect of surface functionalization of PLGA nanoparticles by heparin- or chitosan-conjugated Pluronic on tumor targeting. *J. Control. Release*, **2010**, *143*, 374–382.

Danhier, F.; Ansorena, E.; Silva, J. M.; Coco, R.; Le Breton, A.; Préat, V. PLGA-based nanoparticles: an overview of biomedical applications. *J. Control. Release*, **2012**, *161*, 505–522.

Danhier, F.; Feron, O.; Preat, V. To exploit the tumor microenvironment: passive and active tumor targeting of nanocarriers for anticancer drug delivery. *J. Control. Release*, **2010**, *148*, 135–146.

Darvishi, B.; Manoochehri, S.; Kamalinia, G.; Samadi, N.; Amini, M.; Mostafavi, S. H.; Maghazei, S.; Atyabi, F.; Dinarvand, R. Preparation and antibacterial activity evaluation of 18-β-glycyrrhetinic acid loaded PLGA nanoparticles. *Iran J Pharm Res.*, **2015**, *14*, 373–383.

Date, A. A.; Joshi, M. D.; Patravale, V. B. Parasitic diseases: Liposomes and polymeric nanoparticles versus lipid nanoparticles. *Adv. Drug Deliv. Rev.*, **2007**, *59*, 505–521.

Deamer, D. W. From "banghasomes" to liposomes: a memoir of Alec Bangham, 1921–2010. *FASEB J.*; **2010**, *24*, 1308–1310.

Drummond, D. C.; Meyer, O.; Hong K.; Kirpotin, D. B.; Papahadjopoulos, D. Optimizing liposomes for delivery of chemotherapeutic agents to solid tumors. *Pharmacol. Rev.*, **1999**, *51*, 691–743.

Foget, C.; Brodin, B.; Frokjaer, S.; Sundblad, A. Particle size and surface charge affect particle uptake by human dendritic cells in an in vitro model. *Int. J. Pharm.*, **2005**, *298*, 315–322.

Frezard, F. Liposomes: from biophysics to the design of peptide vaccines. *Braz. J. Biol. Res.*, **1999**, *32*, 181–189.
Gabizon, A. A. Stealth Liposomes and Tumor Targeting: One Step Further in the Quest for the Magic Bullet. *Clin. Cancer Res.*, **2001**, *7*, 223–225.
Gabizon, A.; Goren, D.; Fuks, Z.; Barenholz, Y.; Dagan, A.; Meshorer, A. Enhancement of Adriamycin delivery to liver metastatic cells with increased tumoricidal effect using liposomes as drug carriers. *Cancer Res*, **1983**, *43*, 4730–4735.
Gabizon, A.; Papahadjopoulos, D. The role of surface charge and hydrophilic groups on liposome clearance in vivo. *Biochim. Biophys. Acta*, **1992**, *1103*, 94–100.
Garlotta, D. A literature review of poly(lactic acid). *J. Polym. Environ.*, **2001**, *9*, 63–84.
Gaumet, M.; Vargas, A.; Gurny, R.; Delie, F. Nanoparticles for drug delivery: the need for precision in reporting particle size parameters. *Eur. J. Pharm. Biopharm.*, **2008**, *69*, 1–9.
Gentile, P.; Chiono, V.; Carmagnola, I.; Hatton, P. V. An overview of poly(lactic-co-glycolic) acid (PLGA)-based biomaterials for bone tissue engineering. *Int. J. Mol. Sci.*, **2014**, *15*, 3640–3659.
George, M.; Abraham, T. E. Polyionic hydrocolloids for the intestinal delivery of protein drugs: alginate and chitosan – a review. *J. Control. Release*, **2006**, *114*, 1–14.
Gill, P.; Moghadam, T. T.; Ranjbar, B. Differential Scanning Calorimetry Techniques: Applications in Biology and Nanoscience. *Journal of Biomolecular Techniques: JBT*, **2010**. *21*, 167–193.
Goren, D.; Horowitz, A. T.; Zalipsky, S.; Woodle, M. C.; Yarden, Y.; Gabizon, A. Targeting of stealth liposomes to erB-2 (Her/2) receptor: in vitro and in vivo studies. *Br. J. Cancer*, **1996**, *74*, 1749–1756.
Gregoriadis, G. Drug Entrapment in Liposomes. *FEBS Lett.*, **1973**, *36*, 292–296.
Gregoriadis, G. The carrier potential of liposomes in biology and medicine. Part 1, *N. Engl. J. Med.*, **1976**, *295*, 704–710.
Gregoriadis, G.; Neerunjun, D. Control of the rate of hepatic uptake and catabolism of liposome-entrapped proteins injected into rats. Possible therapeutic applications. *Eur. J. Biochem.*, **1974**, *47*, 179–185.
Gregoriadis, G.; Ryman, B. E. Liposomes as Carriers of Enzymes or Drugs: a New Approach to the Treatment of Storage Diseases. *Biochem. J.*; **1971**, *124*, 58P.
Grottkau, B. E.; Cai, X.; Wang, J.; Yang, X.; Lin, Y. Polymeric Nanoparticles for a Drug Delivery System. *Curr. Drug Metab.*, **2013**, *14*, 840–846.
Gumustas, M.; Kurbanoglu, S.; Uslu, B.; Ozkan, S. A. UPLC versus HPLC on Drug Analysis: Advantageous, Applications and Their Validation Parameters. *Chromatographia*, **2013**, *76*, 1365–1427.
Haas, J.; Ravi Kumar, M. N. V.; Borchard, G.; Bakowsky, U.; Lehr, C. M. Preparation and characterization of chitosan and trimethyl-chitosan modified poly(ε-caprolactone) nanoparticles as DNA carriers. *AAPS PharmSciTech*, **2005**, *6*, Article 6.
Hans, M. L.; Lowman, A. M. Biodegradable nanoparticles for drug delivery and targeting. *Curr. Opin. Solid State Mater, Sci.*, **2002**, *6*, 319–327.
Harush, O.; Altschuler, Y.; Benita, S. The impact of surface charge on nanoparticle performance. in: Nanoparticles for Pharmaceutical Application. Eds.: Domb, A. J.; Tabata, Y.; Ravi Kumar, M. N. V.; Farber, S. **2007**, California, USA, *ASP American Scientific Publishers*, pp. 85–102.

Jain, R.; Shah, N. H.; Malick, A. W.; Rhodes, C. T. Controlled drug delivery by biodagradablepoly(ester) devices: different preparative approaches. *Drug Dev. Ind. Pharm.*, **1998**, *24*, 703–727.

James, N. D.; Coker, R. J.; Tomlinson, D.; Harris, J. R.; Gompels, M.; Pinching, A. J.; Stewart, J. S. Liposomal doxorubicin (Doxil): an effective new treatment for Kaposi's sarcoma in AIDS. *Clin. Oncol.*, **1994**, *6*, 294–296.

Johnston, M. J.; Semple, S. C.; Klimuk, S. K.; Edwards, K.; Eisenhardt, M. L.; Leng, E. C.; Karlsson, G.; Yanko, D.; Cullis, P. R. Therapeutically optimized rates of drug release can be achieved by varying the drug-to-lipid ratio in liposomal vincristine formulations. *Biochim. Biophys. Acta*, **2006**, *1758*, 55–64.

Juliano, R. L.; Stamp, D. Pharmacokinetics of liposome-entrapped antitumor drugs. *Biochem. Pharmacol.*, **1978**, *27*, 21–27.

Juliano, R. L.; Stamp, D. The effect of particle size and charge on the clearance rates of liposomes and liposome encapsulated drugs. *Biochem. Biophys. Res. Commun.*, **1975**, *63*, 651–658.

Kaul, G.; Amiji, M. Biodistribution and targeting potential of poly(ethylene glycol)-modified glatin nanoparticles in subcutaneous murine tumor model. *J. Drug Target.*, **2004**, *12*, 585–591.

Kirpotin, D. B.; Drummond, D. C.; Shao, Y.; Shalaby, M. R.; Hong, K.; Nielsen, U. B.; Marks, J. D.; Benz, C. C.; Park, J. W. Antibody targeting of long-circulating lipidic nanoparticles does not increase tumor localization but does increase internalization in animal models. *Cancer Res.*, **2006**, *66*, 6732–6740.

Kumari, A.; Yadav, S. K.; Yadav, S. C. Biodegradable polymeric nanoparticles-based drug delivery systems. *Colloids Surf. B Biointerfaces*, **2010**, *75*, 1–18.

Labet, M.; Thielemans, W. Synthesis of polycaprolactone: a review. *Chem. Soc. Rev.*, **2009**, *38*, 3484–3504.

Lasic, D. D.; Papadjopoulos, D. Liposomes revisited. *Science*, **1995**, *267*, 1275–1276.

Lassalle, V.; Ferreira, M. L. PLA nano- and microparticles for drug delivery: an overview of the methods of preparation. *Macromol. Biosci.*, **2007**, *7*, 767–783.

Lemarchand, C.; Couvreur, P.; Besnard, M.; Constantini, D.; Gref, R. Novel polyester-polysaccaride nanoparticles. *Pharm. Res.*, **2003**, *20*, 1284–1292.

Li, F. Q.; Yan, C.; Bi, J.; A novel spray-dried nanoparticles-in-microparticles system for formulating scopolamine hydrobromide into orally disintegrating tablets. *Int. J. Nanomed.*, **2011**, *6*, 897–904.

Lopes de Menezes, D. E.; Pilarski, L. M.; Allen, T. M. In vitro and in vivo targeting of immunoliposomal doxorubicin to human B-cell lymphoma. *Cancer Res.*, **1998**, *58*, 3320–3330.

Martin, F. J.; Hubbell, W. L.; Papahadjopoulos, D. Immunospecific targeting of Liposomes to cells: a novel and efficient method for covalent attachment of Fab' fragments via disulfide bonds. *Biochemistry*, **1981**, *20*, 4229–4238.

Martin, O.; Avérous, L. Poly(lactic acid): plasticization and properties of biodegradable multiphase systems. *Polymer*, **2001**, *42*, 6209–6219.

Mudgil, M.; Gupta, N.; Nagpal, M.; Pawar, P. Nanotechnology: a new approach for ocular drug delivery system. *Int. J. Pharm. Pharm. Sci.*, **2012**, *4*, 105–112.

Naahidi, S.; Jafari, M.; Edalat, F.; Raymond, K.; Khademhosseini, A.; Chen, P. Biocompatibility of engineered nanoparticles for drug delivery. *J. Control. Release*, **2013**, *166*, 182–194.

Nassander, U. K.; Steerenberg, P. A.; Poppe, P. A.; Storm, G.; Poeis, L. G.; de Jong, W. H.; Crommelin, D. J. A. In vivo targeting of OV-TL3 immunoliposomes to ascitic ovarian carcinoma cells (OVCAR- 3) in athymic nude mice. *Cancer Res.*, **1993**, *52*, 646–653.

Nasseau, M.; Boublik, Y.; Meier, W.; Winterhalter, M.; Fournier, D. Substrate-permeable encapsulation of enzymes maintains effective activity, stabilizes against denaturation, and protects against proteolytic degradation. *Biotech. Bioeng.*, **2001**, *75*, 615–618.

Pagano, R. E.; Weinstein, J. N. Interactions of Liposomes with Mammalian Cells. *Annu. Rev. Biophys. Bioeng.*, **1978**, *7*, 435–468.

Papahadjopoulos, D. (ed.): Liposomes and their use in biology and medicine. *Ann. NY Acad. Sci.*, **1978**, *308*, 1–412.

Park J. W.; Hong, K.; Kirpotin, D. B.; Colbern, G.; Shalaby, R.; Baselga, J.; Shao, Y.; Nielsen, U. B.; Marks, J. D.; Moore, D.; Papahadjopoulos, D.; Benz, C. C. Anti-HER2 immunoliposomes: enhanced efficacy attributable to targeted delivery. *Clin. Cancer Res.*, **2002**, *8*, 1172–1181.

Park, E. J.; Bae, E.; Yi, J.; Kim, Y.; Choi, K.; Lee, S. H.; Yoon, J.; Lee, B. C.; Park, K. Repeated-dose toxicity and inflammatory responses in mice by oral administration of silver nanoparticles. *Environ. Toxicol. Pharmacol.*, **2010**, *30*, 162–168.

Pattni, B. S.; Chupin, V. V.; Torchilin, V. P. New Developments in Liposomal Drug Delivery. *Chem. Rev.*, **2015**, *115*, 10938–10966.

Pavot, V.; Berthet, M.; Rességuier, J.; Legaz, S.; Handké, N.; Gilbert, S. C.; Paul, S.; Verrier, B. Poly(lactic acid) and poly(lactic-co-glycolic acid) particles as versatile carrier platforms for vaccine delivery. *Nanomedicine*, **2014**, *9*, 2703–2718.

Petersen, A. L.; Hansen, A. E.; Gabizon, A.; Andresen, T. L. Liposome imaging agents in personalized medicine. *Adv. Drug Deliver. Rev.*, **2012**, *64*, 1417–1435.

Peterson, C. L. Nanotechnology: From Feynman to the Grand Challenge of Molecular Manufacturing. *IEEE Technol. Soc. Mag.*, **2004**, 10–15.

Quasim, S. Z. U.; Naveed, A.; Athar, M. M.; Irfan, S.; Ali, M. I.; Ahmed, M. M.; Reddy, R. B. Materials for drug & gene delivery. In: Nanobiotechnology. Eds.: Phoenix, D. A.; Ahmed, W. **2014**, Manchester, UK, One Central Press Ltd, p. 32–67.

Qui, L.; Bae, Y.; Bae, Y.; Bae, H. Polymer architecture and drug delivery. *Pharm. Res.*, **2006**, *23*, 1–30.

Quintanar, G. D.; Allémann, E.; Fessi, H. Preparation techniques and mechanisms of formation of biodegradable nanoparticles from preformed polymers. *Drug Dev. Ind. Pharm.*, **1998**, *24*, 1113–1128.

Qwens III, D. E.; Peppas, N. A. Opsonization, biodistribution, and pharmacokinetics of polymeric nanoparticles. *Int. J. Pharm.*, **2006**, *307*, 93–102.

Rahimpour, Y.; Hamishehkar, H. Liposomes in cosmeceutics. *Expert Opin. Drug Deliv.*, **2012**, *9*, 443–455.

Rahman, A. M.; Yusuf, S. W.; Ewer, M. S. Anthracycline-induced cardiotoxicity and the cardiac-sparing effect of liposomal formulation. *Int. J. Nanomedicine*, **2007**, *2*, 567–583.

Ramana, L. N.; Sethuraman, S.; Ranga, U.; Krishnan, U. M. Development of a liposomal nanodelivery system for nevirapine. *J. Biomed Sci.*, **2010**, *17*, 57–65.

Samadi, N.; Abbadessa, A.; Di Stefano, A.; Van Nostrum, C. F.; Vermonden, T.; Rahimian, S.; Teunissen, E. A.; Van Steebergen, M. J.; Amidi, M.; Hennink, W. E. The effect of lauryl capping group on protein release and degradation of poly(D, L-lactic-co-glycolic acid) particles. *J. Control. Release*, **2013**, *172*, 436–443.

Sengel-Turk, C. T.; Hascicek, C.; Dogan. A. L.; Esendagli, G.; Guc, D.; Gönül, N. Preparation and in vitro evaluation of meloxicam-loaded PLGA nanoparticles on HT-29 human colon adenocarcinoma cells. *Drug Dev. Ind. Pharm.*, **2012**, *38*, 1107–1116.
Sengel-Turk, C. T.; Hascicek, C.; Dogan. A. L.; Esendagli, G.; Guc, D.; Gönül, N. Surface Modification and Evaluation of PLGA Nanoparticles: The Effects on Cellular Uptake and Cell Proliferation on the HT-29 Cell Line. *J. Drug Del. Sci. Technol.*, **2014**, *24*, 166–172.
Senior, J.; Gregoriadis, G. Stability of small unilamellar liposomes in serum and clearance from the circulation: the effect of the phospholipid and cholesterol components. *Life Sci.*, **1982**, *30*, 2123–2136.
Sharma, A.; Sharma, U. S. Liposomes in drug delivery: progress and limitations. *Int. J. Pharma.*, **1997**, *154*, 123–140.
Sharma, A.; Sharma, U. S. Liposomes in drug delivery: progress and limitations. *Int. J. Pharma.*, **1997**, *154*, 123–140.
Sihorkar, V.; Vyas, S. P. Potential of polysaccharide anchored liposomes in drug delivery, targeting and immunization. *J. Pharm. Pharmaceut.*, **2001**, *4*, 138–158.
Silva, G. A. Introduction to Nanotechnology and Its Applications to Medicine. *Surg. Neurol.*, **2004**, *61*, 216–220.
Sinko, P. J. Martin's physical pharmacy and pharmaceutical sciences. 6th ed., **2011**, Philadelphia, USA, *Lippincott Williams & Wilkins*, p. 512.
Soppimath, K. S.; Aminabhavi, T. M.; Kulkarni, A. R.; Rudzinski, W. E. Biodegradable polymeric nanoparticles as drug delivery devices. *J. Control. Release*, **2001**, *70*, 1–20.
Storm, G.; Roerdink, F. H.; Steerenberg, P. A.; de Jong, W. H.; Crommelin, D. J. Influence of lipid composition on the antitumor activity exerted by doxorubicin-containing liposomes in a rat solid tumor model. *Cancer Res.*, **1987**, *47*, 3366–3372.
Strebhardt, K.; Ullrich, A. Paul Ehrlich's magic bullet concept: 100 years of progress. *Nat. Rev. Cancer*, **2008**, *8*, 473–480.
Sun, J.; Tan, H. Alginate-based biomaterials for regenerative medicine applications. *Materials*, **2013**, *6*, 1285–1309.
Tahara, K.; Sakai, T.; Yamamoto, H.; Takeuchi, H.; Hirashima, N.; Kawashima, Y. Improved cellular uptake of chitosan-modified PLGA nanospheres by A549 cells. *Int. J. Pharm.*, **2009**, *382*, 198–204.
Teekamp, N.; Duque, L. F.; Frijlink, H. W.; Hinrichs, W. L. J.; Olinga, P. Production methods and stabilization strategies for polymer-based nanoparticles and microparticles for parenteral delivery of peptides and proteins. *Expert Opin. Drug Deliv.*, **2015**, *12*, 1311–1331.
Thanki, K.; Gangwal, R. P.; Sangamwar, A. T.; Jain, S. Oral delivery of anticancer drugs: challenges and opportunities. *J. Control. Release*, **2013**, *170*, 15–40.
Uhumwangho, M. U.; Okor, R. S. Current trends in the production and biomedical applications of liposomes: a review. *J. Med. Biomed. Res.*, **2005**, *4*, 9–21.
Vasir, J. K.; Labhasetwar, V. Biodegradable nanoparticles for cytosolic delivery of therapeutics. *Adv. Drug Deliv. Rev.*, **2007**, *59*, 718–728.
Vasir, J. K.; Labhasetwar, V. Quantification of the force of nanoparticle-cell membrane interactions and its influence on intracellular trafficking of nanoparticles. *Biomaterials*, **2008**, *29*, 4244–4252.

Vingerhoeds, M. H.; Steerenberg, P. A.; Hendriks, J. J.; Dekker, L. C.; Van Hoesel, Q. G.; Crommelin, D. J.; Storm, G. Immunoliposome-mediated targeting of doxorubicin to human ovarian carcinoma in vitro and in vivo. *Br. J. Cancer*, **1996**, *74*, 1023–1029.

Wang, M.; Thanou, M. Targeting nanoparticles to cancer. *Pharmacol. Res.*, **2010**, *62*, 90–99.

Wang, Z. L. Handbook of nanophase and nanostructured materials. Vol. 2: Characterization. **2002**, Springer-Verlag US.

Weiner, N.; Martin, F.; Riaz, M. Liposomes as drug delivery system. *Drug Dev. Ind. Pharm.*, **1989**, *15*, 1523–1554.

Yue, Z. G.; Wei, W.; Lv, P. P.; Yue, H.; Wang, L. Y.; Su, Z. G.; Ma, G. H. Surface charge affects cellular uptake and intracellular trafficking of chitosan-based nanoparticles. *Biomacromolecules*, **2011**, *12*, 2440–2446.

CHAPTER 3

NANOTECHNOLOGY IN MEDICINE: DRUG DELIVERY SYSTEMS

ELENA CAMPANO-CUEVAS,[1] ANA MORA-BOZA,[2]
GABRIEL CASTILLO-DALÍ,[1]
AGUSTÍN RODRÍGUEZ-GONZALEZ-ELIPE,[2]
MARÍA-ANGELES SERRERA-FIGALLO,[1]
ANGEL BARRANCO,[2] and DANIEL TORRES-LAGARES[1]

[1]*School of Dentistry, University of Seville, Seville, Spain*

[2]*CSIC, Institute of Material Science, Seville, Spain*

CONTENTS

3.1 INTRODUCTION TO NANOTECHNOLOGY

Nanotechnology is considered as a "new technological revolution" and has been largely explored over the last two decades. This field has potential applications in many research areas, such as physics, chemistry, engineering, biotechnology, and especially in health sciences and biomedicine (Navalakhe et al., 2007; Sahoo et al., 2007; Sanchez, 2010; Shrivastava et al., 2009). In this sense, it is expected that this extremely powerful technology has a substantial impact on tissue engineering and drug delivery.

Essentially, nanotechnology has been described as the development of engineered devices or materials at the molecular level in the nanometer range. The Greek prefix 'Nano' indicates the one-billionth part of a measure (10^{-9} = 0.000,000,001). To get an idea, typical carbon-carbon bond, or the space between these atoms in a molecule, is around 0.12 to 0.15 nanometers (nm) and a DNA double helix has a diameter of about 2 nm. Moreover, the smallest cellular life, a bacteria belonging to the Mycoplasma genus, is about 200 nm long (Allhoff et al., 2010).

More generally, nanotechnology includes many structures such as carbon nanotubes, nanocrystals, quantum dots, nanofibers, nanoporous filters, nanowires, metal oxide nanoparticles, dispersion of nanoparticles, polymer nanocomposites, etc. (Bawarski et al., 2008).

Scientists are currently debating the future of the implications of nanotechnology. Although this technology can lead to the production of new materials and systems with application in many fields, such as medicine, physics, chemistry and/or food industry, among others; nanotechnology also raises some concerns about the toxicity and environmental impact of nanomaterials (Buzea et al., 2007).

3.1.1 FROM THE BEGINNING TO THE FUTURE

Richard Feynman, Nobel Prize (Physics) in 1965, was the first to refer to the possibilities of nanoscience and nanotechnology in a speech at the Congress of the American Physical Society in 1959. He titled it as: "There's Plenty of Room at the Bottom," and he describes the idea of direct manipulation of individual atoms and molecules by using precision tools in order to build and operate them with tools of smaller proportions, and therefore at the nanoscale (Feynman, 1960).

Afterwards, the invention of the scanning tunneling microscope by Gerd Binnig and Heinrich Rohrer in 1981 provided an unprecedented display of atoms and individual bonds and, some years after, it was successfully used to manipulate individual atoms.

Despite the term "nanotechnology" was first used in 1974 refer to the ability to generate materials precisely at the nanometer level, it was not until 1986 when the term started to acquire relevancy because of the publication of the book Engines of Creation: The Coming Era of Nanotechnology, in which K. Eric Drexler proposed the idea of an "assembler" in the nanoscale level that would be able to build a copy of itself and other elements of arbitrary complexity with atomic level control (Drexler, 1986).

Also in 1986, Drexler cofounded The Foresight Institute to help increase public awareness and understanding of the concepts of nanotechnology and its implications.

In the early 2000's, the marketing of products-based on advances in nanoscale technologies began to emerge. In this sense, silver nanoparticles were studied as an antibacterial agent (Sondi et al., 2004) and solid lipid nanoparticles were used in sunscreens formulation (Wissing and Müller, 2003), to cite some examples. Therefore, it can be considered a very recent and still developing technology.

When applying nanotechnology to the different disciplines, some features triggered by the manipulation of materials at this very small scale must be taken into account. These materials can show different properties when they are reduced to the nanoscale, leading, in many cases, to important applications. For example, the copper is an opaque substance that becomes transparent when it is reduced to nanoscale. A material such as gold, which is chemically inert to normal level, can serve as a powerful chemical catalyst when it is reduced to nanoscale. Much of the fascination with nanotechnology result from these phenomena that matter exhibits at the nanoscale level (Lubick, 2008).

As it has happened in other sciences, this new technology has also been applied in the field of medicine. Current problems for nanomedicine involve understanding the consequences of toxicity and environmental impact of nanoscale materials (Freitas, 2005). Despite this, nanotechnology would have important revelations in this field. For example, it has been proposed to build small nano-robots (nanobots) that could be programd in order to find and destroy the responsible cells of cancer formation (Nie et al., 2007). It is envisaged that nanomedicine will provide new methods of diagnosis and screening of diseases, better systems for administering drugs and tools for monitoring certain biological parameters. In fact, according to a group of researchers from the University of Toronto, one of the most promising applications of nanotechnology is the creation of drug delivery systems.

3.1.2 NANOMEDICINE

The term Nanomedicine refers to the application of the nanotechnology to the diagnosis and treatment of diseases. In this field, the use of nanostructured materials, biostructures and nanodevices is studied to improve biomedical procedures and systems. In this context, the main areas that have attracted the attention of the scientific world are drug delivery, nanodiagnosis and regenerative medicine.

The human body is unable to absorb the total dose of a drug administered to a patient, so the one of the aims of the modern medicine is the use of nanotechnology to ensure that drugs are released in specific areas of the body with greater accuracy and improve their bioavailability (Foldvari et al., 2008; Moddaresi et al., 2010). One method we can use to improve the action of drugs is the controlled release of them. The idea is to supply the drug to produce its effect only on a specific destination, which will be the area we want to treat. Thus, we are able to provide fewer drugs without influencing the effectiveness of them, to decrease toxicity and adverse effects on tissues and cells that do not require drug treatment, as the drug would only act where it is necessary, and to generate a more stable therapeutic concentration of the drug compared with intermittent concentration changes during normal dosage forms.

In this sense, nanotechnology has been applied to the production of new medicaments (nanodrugs) and other components that improve the solubility and bioavailability of poorly soluble drugs; act as drug delivery vehicles for developing circulatory persistence and targeted delivery to specific sites and/or cells; carriers to promote controlled release; adjuvants for vaccines; diagnostics tools and drug delivery devices (Bai et al., 2011; Foldvari et al., 2008; Moddaresi et al., 2010).

3.1.3 NANOTECHNOLOGY AND DRUG DELIVERY

Nanotechnology applied to medicine and pharmacokinetic has allowed the creation of drug delivery nanosystems by using nanoscale devices capable of penetrating cell membranes and pores and releasing the drug in a minimally invasive manner. These nanodevices act as carriers of drugs through the body, contributing to increase the drug stability against degradation, facilitating its diffusion through biological barriers and therefore improving the access of the drug to the target cells. Thus, the drug's effectiveness is increased and the toxicity associated with the drug is diminished since the drug release is made in situ and does not affect organs, tissues and cells. These nanosystems are formed by the active substance of the medicinal product and a vehicle or carrier. The nanoparticles used in drug delivery are usually colloidal solids with a size between 2 nm and 1000 nm. Typically, the drug of interest is dissolved, encapsulated, adsorbed or bonded in or on nanocapsules (vesicular system in which the drug is located into a cavity surrounded by a polymer membrane) or nanospheres (matrix system where the drug is physically and uniformly dispersed) (Bawarski et al., 2008).

The nanoscale level size increases the rate of dissolution of these drug particulars, so, if the clearance rate is low, the saturation limit of the dissolution can be reached in situ. In other words, among the advantages of using nanoparticles drugs there are a better bioavailability and a rapid installation of the therapeutic effect in oral administration because the reduction in the particle size increases its rate of dissolution (Bai et al., 2011; Bawarski et al., 2008; Moddaresi et al., 2010). Besides, thanks to its small size, nanoparticles can cross the endothelium and epithelium (e.g., intestinal tract or the liver), the blood brain barrier (Fundaro et al., 2000) and even penetrate microcapillaries (Ehrenberg et al., 2008; Panyam et al., 2003). By having a much smaller diameter than capillaries have, aggregate formation, emboli or thrombi are avoided (Goldberg et al., 2006).

Despite the size, the hydrophobicity of the nanoparticles administered intravenously influences there in vivo destination and their clarification by the mononuclear phagocyte system (Olivier, 2005). The coating of nanoparticles with hydrophilic polymers or surfactants or the nanoparticle formulation using biodegradable copolymers with hydrophilic properties, e.g., polyethylene glycol (PEG) prolongs the circulation of the nanoparticles in the bloodstream, increasing the probability of success in freeing directed to specific locations (Bawarski, 2008; Olivier et al., 2005).

There are other factors that influence the rate of drug release. If the drug is placed into a matrix, solubility, diffusion and biodegradation of the matrix particles govern the release process (Moddaresi et al., 2010). In the case of nanospheres, in which the drug is uniformly distributed, its release occurs by diffusion through the degradable matrix so the speed of this degradation influences the drug diffusion. The quick delivery is mainly attributed to weak binding or adsorption of the drug to the surface of the nanoparticles (Goldberg et al., 2006). In contrast, if the nanoparticle is covered by a polymer, the release is controlled by diffusion of the drug through the polymer membrane, which acts as a barrier for drug release. Moreover, the release rate can also be affected by ionic interactions between the drug and formulation additives. When the caged drug interacts with additives can form a less water-soluble complex, which can also delay the release (Moddaresi et al., 2010).

Selective drug delivery can be active or passive. Active delivery requires therapeutic agent and is achieved by conjugating this or a support system to a tissue or a specific cell ligand. In contrast, the passive targeting is achieved by incorporating the therapeutic agent in a macromolecule or nanoparticle that passively reaches the target organ (Lamprecht et al., 2001; Moghimi et al., 2001). Nowadays, there are various types of nanoparticles that have been developed as drug delivery systems, such as: liposomes, micelles, dendrimers, nanotubes and polymer conjugates. These differ in their composition and structure, as detailed below. The choice of system depends on the way they are joined with the drug and the type of drug that is used in the treatment.

3.1.3.1 Polymeric Nanoparticles

In the case of polymeric nanoparticles, the drugs are conjugated to the side chain of a linear polymer with a linker (Rawat et al., 2006). We can distinguished two groups: the first ones are synthetized with natural polymers such as albumin, chitosan and heparin, the second ones are formed by synthetic polymers such as N-(2-hydroxypropyl)-methacrylamide copolymer (HPMA), polystyrene-maleic anhydride copolymer, polyethylene glycol (PEG), and poly L-glutamic acid (PGA), which was the first biodegradable polymer used for this purpose.

3.1.3.2 Liposomes

Liposomes are artificial spherical vesicles composed of a double lipid layer encapsulating an aqueous interior. These devices can fuse together cell membranes and easily penetrate the target cell. Thus, liposomes can incorporate into its structure hydrophilic molecules but also very hydrophobic drugs that can be encapsulated within the bilayer of phospholipid or at their interface. Although liposomes vary greatly in size, most are less than 400 nm (Fahmy et al., 2007).

They have a broad spectrum of use because of its application by intravenous, oral or cutaneous manner. Moreover the lipid nanoparticles protect drugs against degradation caused by water and can also be used as a formulation for providing prolonged release of active principles slightly soluble in water (Bawarski et al., 2008).

Liposomes are being used to treat diseases such as HIV, in which drugs such as Stavudine and Zidovudine are encapsulated in nanoparticles between 120 and 200 nanometers in size. Since both drugs have a short half-life, the cover of liposome keeps them active for longer periods (Mamo et al., 2010).

3.1.3.3 Micelles

Micelles are generally spherical colloidal particles with a diameter less than 50 nm. These nanoparticles are composed of a hydrophobic interior, where hydrophobic drugs can be collected, and a hydrophilic exterior that stabilized the hydrophobic core region and allows the polymer to be water-soluble. So, drugs can be trapped within the micelles or adhere covalently to the micelle surface. They can circulate for prolonged periods in the blood, as the hydrophilic shell limits opsonin adsorption and the small size allows the micelle to evade the immune system of the patient and the filtration of interendothelial cells in the spleen (Lu and Park, 2013). Therefore, they can be used for gradual drug release. The outer hydrophilic layer of the micelle is designed to be thermodynamically stable and biocompatible (Gaucher et al., 2005).

They have been used in prolonged administration of drugs by functionalizing the micelle surface with peptides and ligands. In this sense, they have resulted useful in administration of anticancer drugs in blood even for brain tumors, as these nanoparticles are capable of crossing the blood brain barrier (Orive et al., 2010). Micelles also provide a safe alternative for the administration of drugs poorly soluble in water because the hydrophobic core allows the solubilization of drugs with poor aqueous solubility (Lu and Park, 2013). For example, Paclitaxel (Lee et al., 2007) and Amphotericin-B (Dangi et al., 1998) formulation with micelles has enhanced the solubilization of these hydrophobic drugs.

3.1.3.4 Dendrimers

Dendrimers are synthetic three-dimensional nanomolecules composed of multiple branched monomers that emerge radially from the central core. The drugs are carried in the free ends of the branches. Dendrimers used in the administration of drugs are generally 10 to 100 nm in diameter (Wiener, E. C. et al., 1994).

The main advantage of these polymeric nanoparticles is the capacity to be simultaneously conjugated with several molecules, becoming a multifunctional drug delivery system (Cho et al., 2008; Svenson, and Tomalia, 2005). Pharmaceutical applications of dendrimers include formulations of nonsteroidal antiinflammatory, antibiotic, antiviral drugs and anticancer agents (Cheng et al., 2008).

3.1.3.5 Carbon Nanotubes

Carbon nanotubes are cylindrical molecules composed of hexagonal units of carbon atoms. They are hollow cylinders formed by rolling one or more layers of graphene sheets (Dresselhaus et al., 2004). These have two cylindrical structure forms: a monolayer carbon nanotube, composed of a cylindrical graphene layer terminated at both ends by a hemispherical arrangement of the carbon network in which the cylinder closing results from the inclusion of pentagonal and hexagonal C-C structures during the growth process; and multilayer carbon nanotubes, comprising several concentric cylinders of graphene meshes, each one comprising a monolayer nanotube. Multilayer nanotubes generally have an outer diameter of 2.5 to 100 nm, while the monolayer nanotube diameter varies from 0.6 to 2.4 nm (Joselevich, 2004). The sidewalls of graphene and the strong interactions p[1,5] confer hydrophobicity to the nanotubes, decreasing their solubility in aqueous medium. This is an obstacle to its use as a pharmaceutical excipient. To solve this problem, the surface of the nanotube may be modified by 4 different methods: with the use of surfactants; by solvents; by functionalization of the walls of the nanotubes; and by using biomolecules (Ham et al., 2005). For example, the solubility of these nanotubes and their dispersion in polar solvents have been improved by introducing polar functional groups at the surface of the nanotube to eliminate Van der Waals forces (Dyke and Tour, 2004) or with the use of unspecific bonds of an amphiphilic α-helical peptide with carbon nanotubes (Dieckmann et al., 2003).The applications of carbon nanotubes as excipients in pharmaceutical formulations are of great interest because of its ability to interact with macromolecules, such as proteins and DNA (Foldvari et al., 2008). In addition, nanotubes have shown excellent characteristics to perform the controlled and targeted drug administration (Heilmann, 1983). There are three modes of interaction between carbon nanotubes and the active components of pharmaceutical formulations. The first one is as porous absorbent to catch the active components within a nanotube mesh. The second mode is the functionalization of carbon nanotubes by adhering the compound of interest to its outer walls. Finally, the third approach involves the use of nanotubes as nanocatheters (Foldvari et al., 2008).

3.1.3.6 Quantum Dots

Quantum dots are colloidal semiconductor nanocrystals having a diameter of 2 to 10 nm. They can be synthesized from several kinds of semiconductor materials, being the most used: cadmium selenide (CdSe), cadmium telluride (CdTe), indium phosphide (InP), and indium arsenide (InAs). They are mainly used as contrast agents in images of biological tissues, but they have surface enough to attach therapeutic agents and therefore they could also be applied in the administration of drugs (Kaji et al., 2007).

However, the biocompatibility and physiological stability of quantum dots hinder its biomedical application: without the proper ligand, nanocrystals can form aggregates, not have adequate circulation time in the body, get caught in nonspecific links and be toxic. It is also necessary to obtain increasingly ecological quantum dots, as current applications rely heavily on nanocrystals of cadmium which is considered, along with mercury and lead, as highly toxic (Kaji et al., 2007).

3.2 CURRENT APPLICATIONS OF NANOMEDICINE

The characteristics of each drug delivery system have allowed the active substance of the drug to stay longer in the body without undergoing a premature degradation. They also let the drug be attached to a surface by chemical and physical interactions, be encapsulated and delivered on corresponding organ, tissue and cell. Nowadays, nanomedicine applies drug delivery systems to treat many diseases, including cancer, neurodegenerative disorders, diabetes and infectious disease.

3.2.1 DRUG DELIVERY FOCUSED ON CANCER

3.2.1.1 Why Is Important Drug Delivery Therapy for Cancer?

Nowadays, cancer causes the 14.6% of all human diseases. The uncontrolled division cycles of tumor cells, which lends them a kind of "immortality," is the main reason why cancer is especially difficult to treat. In addition, the specific symptoms developed in each patient and their different response for the main treatments have brought to find more personalized therapies in last years (Bernard et al., 2014).

The abnormal division of tumor cells originates several problems that make the disease deadly. Their limitless replication potential and their capacity to avoid apoptosis lead to the induction of angiogenesis and the invasion to other locations of the body though the invasion of capillary walls and basal membranes, resulting in metastasis. In addition, cancer cells are mostly insensitive to growth inhibitory signals (Hanahan and Weinberg, 2011; Silva et al., 2014).

Since its discovery, chemotherapy has been considered the most effective therapy to treat every kind of cancer. Nowadays, it is used, above all, in advanced states of the disease. However, the poor specificity of chemotherapeutic agents against tumors induces toxicity in health tissues. These side effects are also more significant when it is necessary to increase the doses of the drug in order to provide a suitable concentration in the tumor zone (Biswas and Torchilin, 2014; Gu et al., 2007; Silva et al., 2014). For this reason, nanotechnology-based on delivery systems represents a potential alternative to make a significant impact in cancer therapy.

During the last two decades, nanoparticle drug delivery systems have been studied for cancer therapy. Some of them have already been approved by US Food and Drug Administration (FDA) for their administration to patients and others are still being investigated in preclinic studies (Conde et al., 2012). The main advantages of using nanoparticular targeted drug delivery systems is their high specificity against tumor cells that reduces considerably toxic side effects of chemotherapy agents (e.g., nephrotoxicity, neurotoxicity, cardiotoxicity, etc.). In addition, they also allow reducing the amount of administered drug because its confinement in the nanocarrier improves the drug stability by reducing its degradation, which also contributes to reduce toxic effects in health tissues (Biswas and Torchilin, 2014). The drug delivery systems for anticancer applications are diverse: liposomes, polymeric nanoparticles, micelles, dendrimers, inorganic/metallic nanoparticles or bacterial nanoparticles, among othersm (Alexis, 2010; Biswas and Torchilin, 2014).

3.2.1.2 Passive and Active Drug Delivery Targeting in Cancer

Over the last year, several drug delivery systems have been designed for cancer therapy-based on passive and active approaches. The microenvironment generated by tumor tissues and cells has been used by passive delivery method through the enhanced permeability and retention (EPR) effect (Gu et al., 2007; Silva et al., 2014). It is characterized by a higher permeability to macromolecules, which results in an enhanced fluid retention in the tumor interstitial space. It leads to an increased concentration of nanocarriers in tumor tissues, which could be upto 100 times higher than in healthy ones. Thanks to this tumor signalization, the side effects of chemotherapeutic agents can be diminished because of their low penetration in health tissues (Biswas and Torchilin, 2014; Maeda, 1986; Maeda et al., 2000; Matsumura and Silva et al., 2014). The size and the superficial properties of the nanocarriers play a crucial role in the passive targeting. The accumulation of the vehicles in the tumor interstitium due to EPR, depends on the size of open gap junctions and channels of the inter and transendothelial, which have been reported to be between 400–600 nm. Thus, nanocarriers in the range of 100–300 nm are the perfect candidates to cross these pathways (Gu et al., 2007). The surface charge and steric stabilization are also important for their correct operation. For example, it has been recently shown that positively charged nanoparticles with a size rage of 50–100 nm can penetrate throughout the large tumors (Hu-Lieskovan et al., 2005), meaning that surface charge could affect cellular association and penetration in the tumors. In addition, steric stabilization must not be forgotten because it is indispensable to avoid early clearance executed by the reticule-endothelial system (RES). PEGylation (a surface functionalization with polyethylene glycol (PEG)) is the main technique to improve it. It can increase the circulation of the drug trough circulatory system, due to its capacity of avoiding the clearance system (Biswas and Torchilin, 2014). However, despite of the efforts made to reduce clearance in passive targeting; the inability to achieve the necessary concentration of drug in the tumor zone made it a difficult approach for drug delivery. Precisely because of that, active targeting drug delivery has converted into the center of attention in last years. Active tumor targeting is-based on the conjugation of drug delivery nanocarriers with molecules that can recognize and bind to specific ligand that are only present in tumor cells or only respond to some specific stimuli (Gu et al., 2007). This strategy reduces even more the toxic side effects produced by cancer drugs in normal tissues due to its high specific region of action. However, active targeting cannot be applied when metastasis phase is reached, being in this case the passive approach more recommendable.

3.2.1.3 Active Drug Delivery Targeting in Cancer Therapy

The diversity of available treatments for cancer and the fact that nowadays, there is no existence of a total cure for it, resides in the complexity of the process that are triggered in the organism and the capacity of the tumor cells to avoid drug agents. Also for this reason, there are several ways to treat cancer. Active drug delivery targeting has been applied by different strategies in cancer therapy, such as nanocarriers cross-linked with molecules able to bind to specific molecular markers of surface tumor cells: nanoparticles capable of recognizing characteristics and proteins overexpressed in the tumor microenvironment or stimuli-response nanocarriers (Biswas and Torchilin, 2014; Fleige and Quadir, 2012; Ganta et al., 2008; Gu et al., 2007; Silva et al., 2014). The main problem that active targeting has to face up is the correct selection of very specific molecular markers of tumor cells. A correct strategy for this is the active targeting-based on the microenvironment created by tumor cells, as in passive approach. In this way, nanocarriers can bind to receptors of the endothelial cells or extracellular matrix that are exposed in the closeness of the tumor zone creating this characteristic "microenvironment." As it was exposed before, tumors can expand to the rest of the body through a phenomenon known as angiogenesis, consisting in the formation of new blood vessel thereon preexisting ones. For this reason, preventing the recruitment of blood vessel can cause the tumor cell death and therefore the inhibition of metastasis process (Biswas and Torchilin, 2014; Gu et al., 2007). The main targets present on the tumor endothelial cells are vascular endothelial growth factor (VEGF), $\alpha v\beta 3$ integrin, vascular cell adhesion molecules (VCAM) and matrix metalloproteases (MMPs) (Biswas and Torchilin, 2014). All these receptors are overexpressed in tumor cells, lending to them their particular "immortality." For example, VEGF is the key mediator of angiogenesis and is up-regulated by oncogene expression, different growth factors and hypoxia. VEGF binds to two VEGF receptors (1 and 2 receptors) and the up-regulation of these receptors in endothelial tumor cells result in a larger interaction of VEGF with its receptors, which enhances angiogenesis (Carmeliet, 2005). Drug delivery targeting against VEGF can decrease angiogenesis at two levels: targeting VEGF to inhibit ligand binding to VEGF receptors and targeting the receptors to decrease the binding of VEFG (Biswas and Torchilin, 2014; Carmeliet, 2005). Another typical process that is triggered by tumor cells in angiogenesis process is the inflammation promoted by VCAM which is also overexpressed. They are immunoglobulin-like transmembrane glycoproteins whose molecular signaling has not been found in normal human endothelium, so their targeting allows modifying the behavior of endothelial tumor cells (Biswas and Torchilin, 2014). Migration of endothelial cells can be also inhibited by using integrins as targets. Integrins are endothelial surface receptors for various extracellular matrix proteins with a specific amino acid sequence (arginine-glycine-aspartic acid) (Biswas and Torchilin, 2014; Gosk et al., 2008). $\alpha v\beta 3$ integrin is overexpressed in some tumors and their inhibition by using active targeted nanocarriers leads to apoptosis and regression of metastatic tumor, being a very effective therapy in more advanced cancers (Gosk et al., 2008; Hood et al., 2002). Finally, MMPs can be also targeted for the same purposes than the other ones. They are zinc-dependent endopeptidases which are involved in angiogenesis, tumor invasion and metastasis. For example, membrane type 1 MMP (MT1-MMP) plays an essential role in the formation of new blood vessels by the recruitment of cell components and the degradation of extracellular matrix (Kondo et al., 2004). On the other hand, internal surface receptors or protein markers from tumor cells can be also suitable targets because of their generally overexpression in contrast to normal cells. In this approach, the nanocarrier confined with the drug has to cross vessels and interstitial spaces until the tumor cells, being possible the partial loss of the drug or its clearance, and therefore a lost in the effectiveness can take place (Biswas and Torchilin, 2014; Gu et al., 2007). Nowadays, several molecules are used for the specific surface functionalization, as for example: transferrin receptors, folate receptors, the epidermal growth factor receptor (EGFR), different monoclonal antibodies and aptamer targeting molecules. Transferrins, for example, are serum glycoproteins involved in iron homeostasis and cell growth regulation, which are overexpressed until 100-fold higher in tumor cells than in normal ones (Biswas and Torchilin, 2014). Folate receptors have been wider used in ovarian cancer therapy. Folic acid is a vitamin that plays a crucial role in the synthesis of nucleotide bases and their receptors are overexpressed in many tumors, especially in ovarian cancer (Biswas and Torchilin, 2014; Gu et al., 2007). The last performed study-based on folate receptors consisted in a folate conjugated ternary polymer for the dual delivery of drug and siRNA. Various attempts have been also developed with folate-polymer coated liposomes, which were loaded with doxorubicin (the most used drug in cancer therapy) and the results were well promising, increasing the up taking and the effectiveness of the drug (Biswas and Torchilin, 2014; Gu et al., 2007; Watanabe et al., 2012). The epidermal growth factor receptor has been also used as target for the same reason as the others: its overexpression and its role in cell proliferation of the tumors, angiogenesis and metastasis (Biswas and Torchilin, 2014). In this case, monoclonal antibody against this receptor has been used to surface functionalization of nanocarriers, as for example liposomes grafted with the Fab' portion of anti-HER2, a human epidermal receptor up-regulated in breast cancer (Park et al., 2002). Since the beginning of drug delivery and nanotechnology science, monoclonal antibodies (mAb) have been widely used for targeting molecules. This is clearly shown in the fact that the main drugs, which are already approved for clinical used, are-based on them. In the last years, many studies have been focused on the improvements of these targets, by using human mAb or adaptations from them that can reduce immunological response during their application. As a result, several drugs have been obtained like rituximab (Rituxan®), trastuzumab (Herceptin®), cetuximab (Erbitux®), and bevacizumab (Avastin®). However, in most cases symptoms and evolution of cancer are different in each patient, being necessary a particular treatment in each kind of tumor even in each person depending on their age, gender or metabolism. In order to improve this therapy, mAb has been conjugated with chemotherapeutic agents to increase the response in several patients. Nevertheless, although this last approach has overcome some downsides of mAb (like drug loading capacity and early excretion) there are still many challenges and limitations. For example, monoclonal antibodies require many engineering efforts at molecular level, which is expensive and complex. In addition, they can increase the size of the nanocarrier depending on the number of mAb grafted (Gu et al., 2007). Aptamer targeting molecules are a novel class of molecule which present several advantages over antibodies. They are DNA or RNA oligonucleotides that can interact with different ligands and fold by intramolecular interaction into unique confirmations. The main advantage over mAb is its simple method of synthesis, denominated systemic evolution of ligands by exponential enrichment (SELEX) which is easy to scale up. The synthetized aptamers are small and present low inmunotoxicity. They also are expected to present better tumor penetration and distribution than mAb and many of them are already being studied in preclinical trials (Biswas and Torchilin, 2014; Cho et al., 2008; Gu et al., 2007). The last drug delivery approach for cancer therapy that will be exposed in this chapter is-based on stimuli-response. In this case, the nanocarrier plays an active role in the drug delivery, being able to respond to a specific stimulus that will trigger the release of the active compound. The main advantage of this approach is the capacity of control, temporary and spatially, the release of the drug. In addition, the stimuli are specific for each disease, which present different biochemical characteristics (overexpressed enzymes and proteins, pH, temperature, redox potential, etc.) (Ganta et al., 2008; Fleige and Quadir, 2012). The intracellular and extracellular pH is really affected in cancer disease. In solid tumors, the pH is considerable more acidic (6.5) than in blood (7.4). Consequently, the use of nanocarriers functionalized with ionizable groups that respond to pH gradients found in cell pathways has been proposed as good vehicles for the transport of the drug at the specific zone of action. In addition, Glutathione (GSH) levels are also higher in tumor cells until 100–1000 fold which makes the intracellular space strongly reductive. This concentration gradient has been also used as a stimulus to the release of the drug in tumor cells by using disulfide cross-linked nanocarriers (Ganta et al., 2008; Fleige and Quadir, 2012). External stimuli are also another approach for drug delivery. In this case, magnetic field, ultrasound, light and heat have been used to guide nanocarriers to certain zone of application. Magnetic fields have been applied in targeted delivery of iron oxide nanoparticles in order to accumulate the loaded drug in the tumor cells. Recently, ultrasounds are the center of attention because of the uniform distribution that can be achieved with local sonication after the injection of targeted encapsulated drugs. The temperature is another variable that can be exploited by using hyperthermic stimuli in the tumor area, keeping the drug encapsulated in the nanocarrier until the application of the temperature, thus preventing the contact with the health tissues (Ganta et al., 2008; Fleige and Quadir, 2012). Photo-responsive nanocarriers are-based on the cleavage of a linker by light irradiation of a certain wavelength from outside the body. In this method, the main downside is the necessity that the patient stay in darkness during a period in order to limit the release of the drug to a particular time and space (Ganta et al., 2008; Fleige and Quadir, 2012).

3.2.1.4 Drug Delivery Systems Used in Cancer Therapy

The drug delivery systems used in cancer therapy are able to carry their loaded active drugs to cancer cells by using the EPR effect, the tumor microenvironment and the overexpressed receptors in tumor cells. Nanocarriers can be synthetized using multiple materials such as polymers, lipids, viruses and organometallic compounds (Cho et al., 2008). The structures of polymeric-based-nanocarriers are assorted. Depending on the method of preparation it can be distinguished: structure of capsule (polymeric nanoparticles), amphiphilic core/shell (polymeric micelles) or hyperbranched macromolecules (dendrimers). Besides, lipid-based-nanocarriers (liposomes) have been widely used over the years as drug containers and drug delivery systems because of its easy method of preparation (Cho et al., 2008). On the other hand, metallic nanoparticles have been used in medicine since XVI century, due to its high biocompatibility and low toxicity. The versatility of the method of synthesis permits to control many different properties like size, shape, structure, composition, assembly, encapsulation and tunable optical properties. Thanks to this, unique nanoparticles for a specific tumor can be obtained (Silva et al., 2014). Moreover, these nanoparticles can be functionalized with a wider range of ligands that make them even more specific for active targeting (Biswas and Torchilin, 2014; Gu et al., 2007; Sanna et al., 2014; Silva et al., 2014).

Gold NPs are de most interesting metallic nanoparticles. This is due to their tunable optical properties that can be controlled and modulated for the treatment and also for the diagnosis of this disease (Cho et al., 2008). Gold nanoparticles have multiple applications in cancer therapy, not only for drug delivery but also in photothermal therapy and radiotherapy, whose clinical trials are being developed. Concerning drug delivery, the main reasons why gold nanoparticles have been selected as good candidates have been already exposed: biocompatibility, low toxicity and versatility. However, these advantages could turn into disadvantages if the method of synthesis is not enough controlled and the nanoparticle does not present the appropriated characteristic to avoid immunologic response and renal clearance. The main toxic problems could derive from size, shape, surface charge and chemistry. It is known that Gold NP with smaller size in the range of biomolecules can evade cellular barriers achieving its medical function. However, if the nanoparticle presents a bigger size (in range of 10 nm more), the effectiveness of the therapy may be considerably reduced and not affect the tumor cells. In addition, surface modification can activate oxidative stress pathways and cause damage in DNA, as it has been shown in some studies. This toxicity can be partially reduced with PEGlycation of gold NPs (Brigger et al., 2002; Cobley et al., 2011; Knop et al., 2010). Despite of these toxicity problems, gold NPs are generally considered as safe systems for therapeutic use, being necessary a case-by-case study for each application in cancer therapy due to the specificity of the disease (Cho et al., 2008). Over the last years, several viruses and bacteriophages have been developed for biomedical and nanotechnology applications, which include tissue targeting and drug delivery using the capsid surface to bind multiple ligands or antibodies. Besides, some natural viruses as canine parvovirus present the affinity for receptors (e.g., transferrins, as it will be exposed above (Biswas and Torchilin, P. 2014; Silva et al., 2014) that are up-regulated on tumor cells (Cho et al., 2008). Other nanoparticles used in cancer therapy are the carbon nanotubes. They induced some toxicity due to its insolubility characteristics, but its functionalization with chemical agents has solved practically this problem and now they can be bind to several bioactive molecules. This advanced has supposed a fundamental advantage in the treatment of cancer and some drugs bound to carbon nanotubes are currently being studied in clinical trials (Cho et al., 2008).

3.2.1.5 Nanocarriers in Clinical Development for Cancer Therapy

Nowadays, there are several examples of nanocarriers designed for drug delivery which have already been approved for their used, while many of them are still in preclinical phases. Amongst all the bioactive compounds used to load these nanoparticulars, Doxorubicin has been widely used in passive and active drug delivery systems. In fact, the first encapsulated liposome, known as Doxil, approved by US Food and Drug Administration (FDA) in 1995 for clinical used contained Doxorubicin (Biswas and Torchilin et al., 2014; Sanna et al., 2014; Silva et al., 2014). Doxil has been used to treat many cancers, loaded in a variety of nanocarriers.

Paclitaxel is another bioactive compound which has been incorporated to different nanocarriers. Abraxane is an albumin-based nanosystem loaded with Paclitaxel that also was approved for clinical used in 2005. This nanocarrier has shown to be very effective in metastatic breast cancer (Biswas and Torchilin, 2014; Sanna et al., 2014).

3.2.2 *DRUG DELIVERY FOCUS ON ALZHEIMER DISEASE*

3.2.2.1 Why Is Important Drug Delivery for Alzheimer?

Alzheimer is, with Parkinson, one of the two most common neurodegenerative diseases that nowadays affect millions of people over the world. The symptoms of the disease are characterized by the progressive loss of memory, abilities and skills in patients, combined with other neurological alterations that are developed with the advanced of the disease like violent behaviors, depression, or delirium (Banks, 2012; John et al., 2015; Yan et al., 2014). Although, the origin of Alzheimer disease (AD) is still unknown, it is though that the 70% of the cases have a genetic base. In addition, several risk factors can contribute to its apparition, such as obesity, diabetes mellitus, hypertension, renal disease, smoking, traumatic brain injury or depression (Banks, 2012; John et al., 2015). The accumulation of extracellular amyloid plaques that are composed of unusually folded amyloid beta (Aβ) peptides and tau proteins is the main hallmark of AD that leads to neurodegeneration. Aβ peptides are a part of one larger precursor protein that is produced by neurons, astrocytes and other glial cells (especially under stress) and penetrates trough membrane of the neurons, being a crucial growth factor for them. Different proteins, enzymes and transporters participate in the production of Aβ, thus they can play a crucial role in the treatment of AD (Banks, 2012; John et al., 2015; Yan et al., 2014). The most daunting challenge that Alzheimer therapy has to confront is to cross the blood-brain barrier (BBB), which confers to the central nervous system (CNS) an immune-privileged space (Banks, 2012; John et al., 2015; Yan et al., 2014). The BBB is not only a physical barrier that control the molecules and elements that cross to the CNS, but also it can play a role of communication, nutrition and homeostasis. It is composed of different receptors and transporter proteins that can change its answer depending on the requests of the brain. For this reason, the BBB has to be understood as a regulatory interface (Banks, 2012). At this point, drug delivery therapy in Alzheimer disease has to overcome the BBB in order to reach a suitable concentration of drug into the CNS.

3.2.2.2 Drug Delivery Strategies in Alzheimer Disease

Several approaches have been developed in order to deal with AD. The significance of BBB in neurological diseases, converts itself into a therapeutic target trough to different strategies: (i) its disruption and (ii) the targeting against the several transporter elements and proteins present therein. The inhibition of Aβ could be also a good approach for Alzheimer treatment (Banks, 2012; John et al., 2015; Mourtas et al., 2014; Sarvaiya et al., 2015; Timbie et al., 2015; Yan et al., 2014). On the other hand, genetic therapy cannot be forgotten due to the inevitable genetic contribution of the disease (Banks, 2012; Sarvaiya et al., 2015; Timbie et al., 2015). Finally, the use of novel materials as carbon nanotubes and graphene has been also suggested for drug delivery in Alzheimer treatment. In addition, they have also shown a very good response in neuroregeneration due to their excellent physical properties and their ability to interface with neurons and neuronal circuits (John et al., 2015).

3.2.2.2.1 BBB Strategies to Deliver Drugs to the Brain

The main downside of intravenous Alzheimer drugs is the side effects that can produce to periphery organs due to its low absorption in the CNS. For this reason, other ways of administration, as for example intranasal route, have been developed in the last years. Intranasal administration is a noninvasive technique that can avoid side effects, but it is limited by the poor absorption across the nasal epithelium, which reduces its effectiveness and localization of the drugs (Banks, 2012; Sarvaiya et al., 2015). The dysfunction of BBB has been suggested as a good strategy to improve the drug delivery to the CNS. However, it has to be carefully controlled due to the neurotoxicity of many external substances that would enter to the brain (Banks, 2012; Timbie et al., 2015). The latest approach for the controlled disruption of BBB is-based in focused ultrasound (FUS). FUS therapy is applied in combination with contrast agent microbubbles that can be loaded with different drugs and targeted to specific zones of the brain. The ultrasound is applied trough the tissue causing the local disruption of the BBB. This approach presents some advantages as the possibility to be performed on awaken patients being not necessary the general anesthetic and a real-time monitoring of the treatment (Timbie et al., 2015). Although BBB disruption approach seems to be effective, in chronic disease as Alzheimer, the current methods are not worth due to its high cost/benefit ratio (Banks, 2012). On the other hand, targeting approach against elements present in the BBB is an alternative that has resulted in one of the only two drugs approved for Alzheimer treatment, Donezepil, which is able to cross the BBB by an organic cation transporter for choline (Banks, 2012; Kang et al., 2005; Kim et al., 2010). The main advantage of this strategy is the very specific localization of transporters and proteins, which permit to target drugs to specific zones of the brain (Banks, 2012). Surface functionalized nanocarriers, like antibodies, liposomes or biodegradable polymeric nanoparticles (chitosan or PLGA), act as "Trojan horse," crossing the BBB through endocytosis or saturated transporters, and delivering drugs in the desired zone of the brain (Sarvaiya et al., 2015).

3.2.2.2.2 Inhibition of Aβ

Regarding to the inhibition of Aβ, the main proposed target has been the β secretase, referred to as β-site amyloid precursor protein (APP) cleaving enzyme 1 (BACE1), which initiates the Aβ production. BACE1 is a key therapeutic target for Alzheimer treatment and several BACE1 inhibitors have entered in clinical trials (Yan et al., 2014). The inhibition can be carried out by gene suppression or with targeted nanocarriers, as for example antibodies used as therapeutic agents. In addition, another proposal is that antibodies recruit Aβ in the circulation (Banks, 2012; DeMattos et al., 2002), preventing its accumulation in the blood and crossing to the CNS.

3.2.2.2.3 Gene Delivery

As it was exposed below, the genetic character of Alzheimer disease has led to a gene treatment delivery that has been focus on the mutations related to Aβ production. More than 200 autosomal dominant mutations have been identified in APP and all these mutations increase the production of Aβ, and therefore, its accumulation as amyloid plaques (Yan et al., 2014).

Chitosan and other biodegradable polymeric nanocarriers have been proposed as good candidates for SiRNA (small interfering RNA) gene therapy for AD by using strands with 20–25 base pairlenght. It consists in the suppression of disease-linked genes (BACE1, APP, PS1 (presenilin 1), PS2) which are overexpressed due to mutations in DNA or epigenetic modifications as methylation (Sarvaiya et al., 2015).

3.2.3 DRUG DELIVERY FOCUSED ON DIABETES

Diabetes Mellitus (DM) is one of the most common metabolic disorders in the world. According to recent estimation, 382 million people worldwide have diabetes and this number is expected to rise to 592 million by 2035 (Guariguata et al., 2014). DM is a chronic disease characterized by decreased glucose tolerance due to a relative deficiency of the human body to produce insulin (INS) or a lack of sensitivity to the endogenous hormone (Reynolds and Hunt, 1981). INS is a 51-amino-acid protein secreted by the β-cells of the islets of Langerhans in the pancreas. This hormone regulates blood glucose levels by the stimulation of the glucose uptake from the systemic circulation (Dabelea, 2009) and the suppression of hepatic gluconeogenesis. If the disease is untreated, insufficient INS or decreased INS sensitivity results in hyperglycemia and long-term exposure of tissues to elevated ambient glucose concentrations is associated with the development of complications, including retinopathy, kidney and coronary heart disease, nerve degeneration, nephropathy and increased susceptibility to infection (Nathan, 1993; Sarwar et al., 2010; Wan et al., 2015). In contrast, INS overtreatment causes hypoglycaemia, which could lead to seizures, unconsciousness or death (Schulman et al., 2014).

The two most common forms of the disease are Insulin Dependent Diabetes Mellitus (or Type 1 diabetes) and Noninsulin Dependent Diabetes Mellitus (or Type 2 diabetes).

Type 1 diabetes accounts 10% of the patients (Dabelea, 2009) and results from cellular mediated autoimmune destruction of the pancreatic cells and subsequent hypoinsulinaemia and hyperglycaemia (Lieberman and Dilorenzo, 2003). The disease can affect people of all ages but usually occurs in children or young adults because the rate of β-cells destruction has a faster deterioration in young people than in adults (Diagnosis and Classification of Diabetes Mellitus, 2010). Type 2 diabetes accounts for 90–95% of all diabetes. These patients develop INS resistance, that is, their response to INS produced by β-cells is diminished, which leads to hyperglycaemia (Donath and Shoelson, 2011). Major metabolic syndromes like obesity, INS resistance, and dyslipidaemia have led to an epidemic of type 2 diabetes (Moller, 2001). Type 2 is the most common form of diabetes and is the fourth leading cause of death in developed countries (Mckinlay and Marceau, 2000).

3.2.3.1 Current Treatment and Its Disadvantages

The control of glucose level in blood is essential for long-term outcomes for DM patients (Pickup, 2012). The goal of management for the disease is the maintenance of these levels within healthy ranges (70–140 mg/dl) (Standards of Medical Care in Diabetes, 2013). The therapeutics for type 1 diabetes include INS injections (long-acting insulin to provide a basal level of INS and bolus injections of fast-acting INS at mealtimes) (Berenson et al., 2011; Tamborlane et al., 2008) to compensate for b-cell defects, dipeptidyl peptidase-4 (DPP-4) inhibition by Sitagliptin, and increased islet survival (Kim et al., 2008, 2009) and islet cell regeneration through islet neogenesis associated protein (INGAP) peptide therapy (Rafaeloff et al., 1997).

Initial recommendation for treating type 2 diabetes focuses on delaying disease progression through exercise and regulation of meals (Standards of Medical Care in Diabetes, 2013). Besides, treatment for type 2 diabetes includes oral hypoglycemic drugs, such as: sulfonylureas and repaglinide, that enhance INS secretion from the b-cells; troglitazone, that promotes the use of glucose by cells; metformin, that induces INS mechanism in liver tissue; and miglitol and acarbose that enact delayed carbohydrate absorption from food intake (Buse et al., 1999). However, in some cases other macromolecular diabetic such as Glucagon-like peptide (GLP) analogs like Exenatide and Liraglutide (Buse et al., 2009) must be injected subcutaneously due to the harsh environment of the gastrointestinal tract. The other major medications strategies constitutes combinational therapy of INS with sulfonylureas which reduced the daily requirement of INS (Riddle, 1996),

INS and metformin combination therapy (Golay, 1995), and troglitazone-INS in combination to reduce INS requirement (Buse et al., 1998).

The drugs used for the treatment of DM poses limitations in the sense that they produce significant patient annoyance such as weight gain due to INS therapy (Golay, 1995); daily self-injecting INS and the frequent self-monitoring of blood glucose by finger sticks, which is unpleasant for patients and implies pain; water retention that can lead to edema;, etc. On the other hand and despite the different therapies and combinations available, it remains difficult to maintain ideal glucose levels in the vast majority of patients (Pickup, 2012). It is estimated that almost 50% of patients do not achieve their target glucose blood levels throughout the day (Resnick et al., 2006), because INS is injected into the subcutaneous space and not into the portal blood where INS is secreted from the pancreas and lack of patient compliance (American Diabetes Association: Standards of Medical Care in Diabetes, 2014). Besides, several hypoglycemic agents are reported to have adverse effects, including lactic acidosis, diarrhea and vomit (Type 2 Diabetes and Metformin, 2014).

To provide clinical improvements, scientists are working in enable alternative routes of INS administration (Owens, 2002), optimize INS pharmacokinetics (Berenson et al., 2011) and develop new therapeutic entities (Mehanna, 2013).

3.2.3.2 Nanotechnology and Insulin Delivery

One of the aims of nanomedicine applied to DM is to minimize the frequency of injections by the use of long-acting nanoparticulate formulations of antidiabetic drugs (Peng et al., 2012).

3.2.3.2.1 Oral Insulin Delivery

The oral route is the most preferred form of chronic drug administration. However, physical and biochemical barriers of the gastrointestinal tract decrease its effectiveness (Diab et al., 2012).

The application of nanotechnologies in drug delivery is expected to achieve that drugs arrive intact to the target site, to improve drug water solubility, to increase the intestinal permeability of drugs and to improve treatment adherence by reducing dosing frequency (Diab et al., 2012). Focusing on this aim, different types of delivery systems and functional excipients have been explored.

3.2.3.2.2 Gold Nanoparticles

Gold nanoparticles (AuNPs) have been widely studied as INS delivery system. The surface of AuNPs can be modified in order to attach therapeutics onto AuNPs (Lee et al., 2011). AuNPs can be synthesized by various methods to achieve the desired size, shape, and surface functionality (Yeh et al., 2012). This way, production of nanoparticles can be achieved through chemical, physical, and biological methods.

In the chemical synthesis, many reducing and stabilizing agents such as hydrogen tetrachloroaurate ($HAuCl_4$) and citrate (Turkevich et al., 1951) have been proposed to create AuNPs. In this sense, chondroitin sulfate (CS), a biocompatible polymer, has been tested to study the effectiveness, stability and toxicity of CS-capped AuNPs for the oral delivery of INS. CS-capped AuNPs were prepared for INS delivery by dissolving CS in distilled water (DW) and mixing with $HAuCl_4 \cdot 3H_2O$ also solubilized in DW. To get AuNPs dispersion, this mixture was heated at 80°C for 5 h and cooled to room temperature and, then, INS dissolved in PBS was added into the AuNPs dispersion to prepare the AuNPs/INS dispersion and incubated. This way, INS was embedded in the AuNPs structure, resulting in a device of 120 nm of diameter that had maintained stability during tested period and negligible cytotoxicity. A higher efficacy of the CS-AuNPs/INS in delivering INS after oral administration was observed in diabetic rats when comparing to INS solution-treated control group, so is expected that developed CS-capped AuNPs can be used as efficient oral delivery systems of INS (Cho et al., 2014).

On the other hand, biological methods for nanoparticle synthesis using plants have also been suggested as possible ecological alternatives to chemical and physical methods (Daisy et al., 2012).

In this sense, gold nanoparticles synthesized from the extracts of *Cassia fistula* have been evaluated (Lee et al., 2011) as the stem barks of hexane extracts from this plant possess hypoglycemic and hypocholesterolemic effects (Nirmala et al., 2008). The extract was obtained in distilled water at 60°C for 15 min and filtered (Daisy et al., 2012). Gold nanoparticles were previously bioreduced to $HAuCl_4$-chloroaurate using an ultraviolet-visible spectrophotometer to monitor the process and a Fourier transform infrared spectroscopy was performed to check if functional groups were present in the sample. Aqueous $HAuCl_4$ was added to double distilled water containing the bark powder of the plant to perform the phytochemical mediated synthesis of the gold nanoparticles. The color change into ruby red indicated the formation of green gold nanoparticles and these NPs were characterized using a scanning electron microscopy (Daisy et al., 2012). The NPs biosynthetized from *C. fistula* were tested on experimental rats in which diabetes had been induced by Streptozotocin. Administration was performed via gastric intubation. Animals treated with Gold NPs for 30 days achieved an average of 168.47 mg/dL blood sugar levels. Although the glucose-lowering function of the experimental treatment was no as effective as that of INS, it reduced serum blood glucose concentrations, induced favorable changes in body weight, improved transaminase activity, achieved reversed renal, indicating that phytochemically synthesized gold nanoparticles are effective hypoglycaemic agents in the treatment of diabetes mellitus and in its associated complications (Daisy et al., 2012).

3.2.3.2.3 ***Polymeric Nanoparticles***

On the other hand, solid lipid nanoparticles (SLNs) have resulted successful in enhacing the absorption of some antidiabetic drugs, such as berberine (Xue et al., 2013) and INS (Zhang et al., 2012). These SLNs can be modified with lipid materials such as glycerol tripalmitate:soybean phospholipids (Xue et al., 2013) or stearic acid–octaarginine (Zhang et al., 2012) in order to protect drugs from enzymes and therefore these devices can be used as carriers for oral administration of therapeutics. These polymeric NPs were effective in lowering the blood glucose level in animal models and improved the bioavailability of the entrapped drug (Xue et al., 2013; Zhang et al., 2012)

In recent years, poly(lactide-coglycolide) (PLGA) NPs have also widely explored. The encapsulation of insulin in folate-(FA) coupled polyethylene glycol (PEG)ylated polylactide-*co*glycolide (PLGA) nanoparticles (FA-PEG-PLGA NPs) enhances its the oral absorption and hypoglycaemic activity. A double-emulsion solvent evaporation method resulted in NPs of 260 nm size that presented an encapsulation efficiency of 87.0 ± 1.92%. The study in diabetic rats demonstrated that Insulin NPs maintained a continual blood glucose level for 24 h, with a two-fold increase in the oral bioavailability as compared to subcutaneously administered standard insulin solution, all this suggesting that the once-daily administration would be sufficient to control diabetes, at least 24 h (Jain et al., 2012).

3.2.3.2.4 Micelles

Additionally, a novel formulation of pancreatic polypeptide (PP) in sterically stabilized phospholipid micelles (SSM) has demonstrated significant antidiabetic activity in a rodent model of pancreatogenic diabetes. This is a potentially fatal disease that occur secondary to pancreatic disorders. PP has significant antidiabetic efficacy but its therapeutic application is limited due to its short plasma half-life (Banerjee et al., 2013). In this study, empty micelles were prepared by dispersing 1,2-Distearoyl-sn-glycero-3-phosphatidylethanolamine-N [methoxy (polyethyleneglycol), 2000] sodium salt in PBS and allowing it to equilibrate for 1 h in the dark at 25°C. Afterwards, a solution of PP in PBS was added to these preformed micelles and the resulting dispersion became equilibrated in the dark for 2 h at 25°C to form PP-SSM. The intravenous administration of PP-SSM in rodent with PD revealed that this therapeutic significantly improved glucose tolerance, INS sensitivity and hepatic glycogen content compared to free peptide in buffer. Therefore, this study demonstrates the importance of micellar nanocarriers in protecting enzyme-labile peptides in vivo and delivering them to target site to enhance their therapeutic efficacy. This novel antidiabetic nanomedicine also showed the same significant decrease in blood glucose levels than that of metmorfin and even better than the decrease observed with INS therapy. Besides, PP-SSM was as effective as metformin in restoring INS sensitivity to normal levels (Banerjee et al., 2013).

3.2.3.2.5 Liposomes

Finally, liposomes have shown promising potential in oral INS delivery because of their facilitated absorption and their ability to protect the drug they carry from the harsh gastrointestinal (GI) environment. Although conventional phospholipid/cholesterol liposomes can be damaged by gastric acid or GI enzymes, these vesicles can be modified to elongate their GI survival. In this sense, the incorporation of bile salts to the liposome (Niu et al., 2011) coating the liposomes with polymers (Wu et al., 2004) and the design of multilayered carriers (Katayama et al., 2003) are some of the methods that have been assayed to improve oral bioavailability of these NPs. Biotinylation by incorporating biotin-conjugated 1,2-distearoyl-sn-glycero-3-phosphatidyl ethanolamine (DSPE) into the liposome membranes and preparation of INS-loaded biotinylated liposomes (BLPs) by using a reversed-phase evaporation method resulted in a particle size of ~150 nm that kept relatively high INS payload and stability (Zhang et al., 2014). This device demonstrated a significant hypoglycemic effect in diabetic rats and presented a bioavailability that was approximately double when comparing to that of conventional liposomes. This fact may be due to the facilitated absorption of the BLPs through receptor-mediated endocytosis, as biotin is a vitamin and there are many vitamins receptors expressed in the membranes of enterocytes (Hamman et al., 2007). The BLPs remained in a stable state after administration when in vitro survivability was tested in a simulated digestive media, indicating that the insertion of biotin-DSPE increased the strength of the liposome bilayer. Finally, this studied confirmed that the hypoglycemic profiles correlated well with the blood INS profiles, confirming the absorption of INS in its active form. The mild efficacy of the BLPs lasted for a prolonged time period, which was superior to subcutaneous INS administration (Zhang et al., 2014).

However, and despite the studies performed to show the potential benefits of all these formulations, no oral insulin formulation is commercially available nowadays, and there have been very few clinical trial reports with human data (Matteucci et al., 2015).

3.2.3.2.6 Nasal Administration

Nanocomplexes (NCs) comprising polysaccharides such as chitosan (CS) and alginate (ALG) can be used as mucosal drug-delivery device as they provide nontoxic, organic solvent-free, homogenous NCs with efficient limited release of entrapped drugs (Agnihotri et al., 2004). The inhalation route has the advantage of rapid absorption of the drug due to the large surface area and the proximity of the air and blood compartments. However, there are several factors to take into account when using this via because the efficiency of delivery for an inhaled drug depends on aerodynamic particle size (to avoid being cleared by alveolar macrophages), the fraction of dose delivered from the device, the fraction deposited in the alveolar region, the ventilatory parameters and the bioavailable fraction that is absorbed (Mastrandrea, 2010; McElroy et al., 2013).

In the search of intranasally drug delivery devices, the ability of insulinotropic repaglinide (REP) loaded onto chitosan/alginate NCs for controlling blood glucose level has been tested on diabetic rats (Elmowafy et al., 2014). In this study, spray-dried REP-loaded NCs in dry microparticles were prepared: NCs were made by the ionic gelation of CS with ALG; optimum size and entrapment efficiency (EE) REP-loaded CS/ALG NC dispersions were selected and mixed with Leucine and then spray-dried. These formulations were well tolerated according to the cytotoxicity and histopathological analysis and the antihyperglycemic activity of the nasally administered device was significantly sustained over 24 h, suggesting NC mucosal uptake. Indeed, these dry powders achieved better glycemic control compared with the conventional oral tablets, suggesting the usefulness of this controlled delivery system in the management of diabetes (Elmowafy et al., 2014).

3.3 CONCLUSION AND FUTURE EXPECTATIONS

The administration and drug formulations have been revolutionized with the development of nanotechnologies. Nanomedicine promises to solve the problems of drug delivery to specific cells and facilitate the movement of such drugs through different barriers in the body (e.g., the blood-brain). The challenge consists on characterizing the drug delivery systems to ensure that the therapeutic molecules they carry reach specific sites to act on the selected organ or cell exclusively.

Infectious diseases are one of the major goals of modern medicine. Nanoparticles have also been developed to create vaccines in the form of simple nasal drops. Thus, nasal administration vaccines could be used in poor or rural areas where there is a lack of healthcare personnel to administrate the injectable forms. In this sense, scientists are working on a vaccine against TB in aerosol (Stylianou et al., 2014) and a skin patch against West Nile virus and Chikungunya virus (Prow et al., 2010).

Several efforts have also been made in the improvement of traditional materials used in medicine such as prosthesis, catheters or sutures. Among all medical devices, sutures are one of the most used due to their crucial role in surgery for wounds closure. Because of the complexity of post-surgery treatments, the application of drug delivery technology in sutures has been proposed recently (Catanzano et al., 2014; Lee et al., 2013; Obermeier et al., 2014; Serrano et al., 2015). Traditionally, the treatment, which consisted of a combination of antiinflammatory, pain relief and antibacterial drugs, have been administered via oral or injected (Lee et al., 2013). However, thanks to the implementation of drug delivery sutures, the complexity and highly expense (due to long hospitalizations of the patients would be reduced (Obermeier et al., 2014). The first commercialized drug delivery suture was a Polyglactin 910 one with triclosan-based on the capacity of this to inhibit the bacteria attachment. In addition, US FDA approved it in 2002, obtaining therefore an antimicrobial suture available for clinical use (Lee et al., 2013; Obermeier et al., 2014). Nevertheless, the increasing amount of bacteria that have developed resistance against triclosan and antibiotics during the last years, has led to a wider use of antiseptic with an amply spectrum of action (Obermeier et al., 2014). In addition, drug delivery sutures have been developed with well-known antiinflammatory or pain relief drugs, like Diclofenac or ibuprofen in combination with biopolymers such as PLGA, resulting in useful advices with a dual function: closing the surgery wound and providing sustained localized delivery of these drugs (Catanzano et al., 2014; Lee et al., 2013).

For the moment, medicine will have to wait until a real application on nano-robots (nanobots). These molecular machines are expected to travel through the bloodstream and to have the ability to act on the DNA to repair it (essential for the cure of genetic diseases), modify proteins or even destroy whole cells, in the case of tumors. However, some experts have dared to explore the field of nanobots. This is the case of Robert Freitas, a researcher at the Institute for Molecular Manufacturing in California, who has created a kind of artificial red blood cell named respirocyte. This spherical robot mimics natural hemoglobin found inside the erythrocytes, but with the ability to release upto 236 times more oxygen per unit of volume than that of a natural red blood cell. The respirocytes will incorporate chemical and pressure sensors so they will be able to receive acoustic signals from the doctor, who will use ultrasound transmitter apparatus to give orders to modify the behavior of the nanobot while it is inside the patient's body.

KEYWORDS

- **alzheimer disease**
- **cancer**
- **delivery systems**
- **drugs**
- **insulin**
- **medicine**
- **nanoparticles**
- **nanotechnology**
- **polymers**

REFERENCES

Agnihotri, S. A.; Mallikarjuna, N. N.; Aminabhavi, T. M. Recent advances on chitosan-based micro and nanoparticles in drug delivery. *J. Control Release.* **2004**, *100*, 5–28.

Alexis, F.; Pridgen, E. M.; Langer, R.; Farokhzad, O. C. Nanoparticle Technologies for Cancer Therapy. *Handb. Exp. Pharmacol.* **2010**, 55–86.

Allhoff, F.; Lin, P.; Moore, D. What Is Nanotechnology and Why Does It Matter?: From Science to Ethics. John Wiley & Sons. **2010**, 3–5.

American Diabetes Association. Standards of medical care in diabetes. *Diabetes Care.* **2014**, *37*(1), 14–80.
Bai, D.; Xia, X.; Yow, C. M. N.; Chu, E. S. M.; Xu, C. Hypocrellin B-encapsulated nanoparticle-mediated rev-caspase-3 gene transfection and photodynamic therapy on tumor cells. *Euro J. Pharmacol.* **2011**, *650*, 496–500.
Banerjee, A.; Onyuksel, H. A novel peptide nanomedicine for treatment of pancreatogenic diabetes. *Nanomedicine.* **2013**, *9*(6), 722–728.
Banks, W. A. Drug delivery to the brain in Alzheimer's disease: Consideration of the blood–brain barrier. *Adv. Drug Deliv. Rev.* **2012**, *64*, 629–639.
Bawarski, W. E.; Pharm, D.; Chidlowsky, E.; Bharali, D. J.; Mousa, S. A. Emerging nanopharmaceuticals. *Nanomed Nanotech Biol Med.* **2008**, *4*, 273–282.
Berenson, D. F.; Weiss, A. R.; Wan, Z. L.; Weiss, M. A. Insulin analogs for the treatment of diabetes mellitus: therapeutic applications of protein engineering. *Ann. NY Acad. Sci.* **2011**, *1243*, E40–E54.
Bernard, W.; Stewart, C. P. Wild. World Cancer Report 2014. International Agency for Research Cancer. World Health Organization, **2014**.
Biswas, S.; Torchilin, V. P. Nanopreparations for organelle-specific delivery in cancer. *Adv. Drug Deliv. Rev.* **2014**, *66*, 26–41.
Brigger, I.; Dubernet, C.; Couvreû, P. Nanoparticles in cancer therapy and diagnosis. A*dv. Drug Deliv. Rev.* **2002**, *54*, 631–651.
Buse, J. B. Overview of current therapeutic options in type 2 diabetes. Rationale for combining oral agents with insulin therapy. *Diabetes Care.* **1999**, *22*(3), C65–70.
Buse, J. B.; Gumbiner, B.; Mathias, N. P.; Nelson, D. M. Troglitazone use in insulin-treated type 2 diabetic patients. *Diabetes Care.* **1998**, *21*(9), 1455–1461.
Buse, J. B.; Rosenstock, J.; Sesti, G.; Schmidt, W. E.; Montanya, E.; Brett, J. H.; Zychma, M.; Blonde, L.; LEAD-6 Study Group. Liraglutide once a day versus exenatide twice a day for type 2 diabetes: a26-week randomized, parallel-group, multinational, open-labeltrial (LEAD-6). *Lancet.* **2009**, *374*(9683), 39–47.
Buzea, C.; Pacheco, I.; Robbie, K. Nanomaterials and nanoparticles: Sources and toxicity. *Biointerphases.* **2007**, *2*, 17–71.
Carmeliet, P. VEGF as a key mediator of angiogenesis in cancer. *Oncology.* **2005**, *69*(3), 4–10.
Catanzano, O.; Acierno, S.; Russo, P.; Gervasio, M.; Del Basso De Caro, M.; Bolognese, A.; Sammartino, G.; Califano, L.; Marenzi, G.; Calignano, A.; Acierno, D.; Quaglia, F. Melt-spun bioactive sutures containing nanohybrids for local delivery of antiinflammatory drugs. *Mater Sci Eng C Mater Biol Appl.* **2014**, *43*, 300–309.
Cheng, Y.; Wang, J.; Rao, T.; He, X.; Xu, T. Pharmaceutical applications of dendrimers: promising nanocarriers for drug delivery. *Front. Biosci.* **2008**, *13*, 1447–1471.
Cho, H. J.; Oh, J.; Choo, M. K.; Ha, J. I.; Park, Y.; Maeng, H. J. Chondroitin sulfate-capped gold nanoparticles for the oral delivery of insulin. *Int. J. Biol. Macromol.* **2014**, *63*, 15–20.
Cho, K.; Wang, X.; Nie, S.; Chen, Z. G.; Shin, D. M. Therapeutic nanoparticles for drug delivery in cancer. *Clin Cancer Res.* **2008**, *14*(5), 1310–1316.
Cobley, C. M.; Chen, J.; Cho, E. C.; Wang, L. V.; Xia, Y. Gold nanostructures: a class of multifunctional materials for biomedical applications. *Chem. Soc. Rev. Chem. Soc. Rev.* **2011**, *40*, 44–56.

Conde, J.; Doria, G.; Baptista, P. Noble metal nanoparticles applications in cancer. *J Drug Deliv.* **2012**, 751075.

Dabelea, D. The accelerating epidemic of childhood diabetes. *Lancet.* **2009**, *373*, 1999–2000.

Daisy, P; Saipriya, K. Biochemical analysis of Cassia fistula aqueous extract and phytochemically synthesized gold nanoparticles as hypoglycemic treatment for diabetes mellitus. *Int. J. Nanomedicine.* **2012**, *7*, 1189–1202.

Dangi, J. S.; Vyas, S. P.; Dixit, V. K. The role of mixed micelles in drug delivery. I. Solubilization. *Drug Dev Ind Pharm.* **1998**, *24*(7), 681–684.

DeMattos, R. B.; Bales, K. R.; Cummins, D. J.; Paul, S. M.; Holtzman, D. M. Brain to plasma amyloid-beta efflux: a measure of brain amyloid burden in a mouse model of Alzheimer's disease. *Science.* **2002**, *295*, 2264–2267.

Diab, R.; Jaafar-Maalej, C.; Fessi, H.; Maincent, P. Engineered Nanoparticulate Drug Delivery Systems: The Next Frontier for Oral Administration. *AAPS J.* **2012**, *14*(4), 688–702.

Diagnosis and Classification of Diabetes Mellitus. *Diabetes Care.* **2010**, *33*(1), S62–S69.

Dieckmann, G. R.; Dalton, A. B.; Johnson, P. A.; Razal, J.; Chen, J.; Giordano, G. M.; Muñoz, E.; Musselman, I. H.; Baughman, R. H.; Draper, R. K. Controlled assembly of carbon nanotubes by designed amphiphilic Peptide helices. *J Am Chem Soc.* **2003**, *125*(7), 1770–1777.

Donath, M. Y.; Shoelson, S. E. Type 2 diabetes as an inflammatory disease. *Nat. Publ. Gr.* **2011**, *11*, 98–107.

Dresselhaus, M. S.; Dresselhaus, G.; Charlier, J. C.; Hernández, E. Electronic, thermal and mechanical properties of carbon nanotubes. *Philos. Trans. A. Math. Phys. Eng. Sci.* **2004**, *362*, 2065–2098.

Drexler, K. E. Engines of creation: the coming era of nanotechnology. *New York: Anchor Press/Doubleday*, **1986**.

Dyke, C. A.; Tour, J. M. Overcoming the Insolubility of Carbon Nanotubes Through High Degrees of Sidewall Functionalization. *Chem. Eur. J.* **2004**, *10*, 812–817

Ehrenberg, M. S.; Friedman, A. E.; Finkelstein, J. N.; Nter, G.; Rster, O.; Mcgrath, J. L. The influence of protein adsorption on nanoparticle association with cultured endothelial cells. *Biomaterials.* **2008**, *30*(4), 603–610.

Elmowafy, E.; Osman, R.; El-Shamy, A. H.; Awad, G. A. Nanocomplexes of an insulinotropic drug: optimization, microparticle formation, and antidiabetic activity in rats. *Int. J. Nanomedicine.* **2014**, *9*, 4449–4465

Fahmy, T. M.; Fong, P. M.; Park, J.; Constable, T.; Saltzman, W. M. Nanosystems for simultaneous imaging and drug delivery to T cells. *AAPS J.* **2007**, *9*, E171–180.

Feynman, R. P. There's plenty of room at the bottom. Eng Sci (CalTech) **1960**, *23*, 22–36.

Fleige, E.; Quadir, M. A.; Haag, R. Stimuli-responsive polymeric nanocarriers for the controlled transport of active compounds: Concepts and applications. *Adv. Drug Deliv. Rev.* **2012**, *64*, 866–884.

Foldvari, M.; Bagonluri, M. Carbon nanotubes as functional excipients for nanomedicines: I. pharmaceutical properties. *Nanomed Nanotech Biol Med.* **2008**, *4*, 173–182.

Foldvari, M.; Bagonluri, M. Carbon nanotubes as functional excipients for nanomedicines: II. Drug delivery and biocompatibility issues. *Nanomed Nanotech Biol Med.* **2008**, *4*, 183–200.

Freitas, R. A. Current Status of Nanomedicine and Medical Nanorobotics. *Journal of Computational and Theoretical Nanoscience.* **2005**, *2*, 1–25.

Fundaro, A.; Cavalli, R.; Bargoni, A.; Vighetto, D.; Zaraa, G. P.; Gasco, M. R. Non-stealth and stealth solid lipids nanoparticles (SLN) carrying doxorubicin: pharmacokinetics and tissue distribution after i.v. administration to rats. *Pharm Res.* **2000**, *42*, 337–343.

Ganta, S.; Devalapally, H.; Shahiwala, A.; Amiji, M. A review of stimuli-responsive nanocarriers for drug and gene delivery. *Journal of Controlled Release.* **2008**, *126*, 187–204.

Gaucher, G.; Dufresne, M. H.; Sant, V. P.; Kang, N.; Maysinger, D.; Leroux, J. C. Block copolymer micelles: preparation, characterization and application in drug delivery. *J Contr Rel.* **2005**, *109*, 169–188.

Golay, A.; Guillet-Dauphiné, N.; Fendel, A.; Juge, C.; Assal, J. P. The insulin-sparing effect of metformin in insulin-treated diabetic patients. *Diabetes. Metab. Rev.* **1995**, *11*(1), S63–67.

Goldberg, M.; Langer, R.; Jia, X. Nanostructured materials for applications in drug delivery and tissue engineering. *J Biomat Sci.* **2006**, *17*, 241–268.

Gosk, S.; Moos, T.; Gottstein, C.; Bendas, G. VCAM-1 directed immunoliposomes selectively target tumor vasculature in vivo. *Biochim. Biophys.* **2008**, 854–863.

Gu, F. X.; Karnik, R.; Wang, A. Z.; Alexis, F.; Levy-Nissenbaum, E.; Hong, S.; Langer, R. S.; Farokhzad, O. C. Targeted nanoparticles for cancer therapy. *Nano Today.* **2007**, *2*, 14–21.

Guariguata, L.; Whiting, D. R.; Hambleton, I.; Beagley, J.; Linnenkamp, U.; Shaw, J. E. Global estimates of diabetes prevalence for 2013 and projections for 2035. *Diabetes Res. Clin. Pract.* **2014**, *103*, 137–149.

Ham, H. T.; Choi, Y. S.; Chung, J. An explanation of dispersion states of single-walled carbon nanotubes in solvents and aqueous surfactant solutions using solubility parameters. *J. Colloid Interface Sci.* **2005**, *286*, 216–223.

Hamman, J. H.; Demana, P. H.; Olivier, E. I. Targeting Receptors, Transporters and Site of Absorption to Improve Oral Drug Delivery. *Drug Target Insights.* **2007**, *2*, 71–81.

Hanahan, D.; Weinberg, R. A. Hallmarks of Cancer: The Next Generation. *Cell.* **2011**, *44*, 646–674.

Heilmann K. Innovations in drug delivery systems. *Curr. Med. Res. Opin.* 1983, *8*(2), 3–9.

Hood, J. D.; Bednarski, M.; Frausto, R.; Guccione, S.; Reisfeld, R. A.; Xiang, R.; Cheresh, D. A. Tumor regression by targeted gene delivery to the neovasculature. Science. **2002**, *296*(5577), 2404–2407.

Hu-Lieskovan, S.; Heidel, J. D.; Bartlett, D. W.; Davis, M. E.; Triche, T. J. Sequence-Specific Knockdown of EWS-FLI1 by Targeted, Nonviral Delivery of Small Interfering RNA Inhibits Tumor Growth in a Murine Model of Metastatic Ewing's Sarcoma. *Cancer Res.* **2005**, *65*, 8984–8992.

Jain, S.; Rathi, V. V.; Jain, A. K.; Das, M.; Godugu, C. Folate-decorated PLGA nanoparticles as a rationally designed vehicle for the oral delivery of insulin. *Nanomedicine.* **2012**, *7*, 1311–1337.

John, A. A.; Subramanian, A. P.; Vellayappan, M. V.; Balaji, A.; Mohandas, H.; Jaganathan, S. K. Carbon nanotubes and graphene as emerging candidates in neuroregeneration and neurodrug delivery. *Int. J. Nanomedicine*. **2015**, *10*, 4267–4277.

Joselevich, E. Electronic structure and chemical reactivity of carbon nanotubes: a chemist's view. *Chem Phys Chem*. **2004**, *5*, 619–624.

Kaji, N.; Tokeshi, M.; Baba, Y. Quantum dots for single bio-molecule imaging. *Anal. Sci.* **2007**, *23*, 21–24.

Kang, Y. S.; Lee, K. E.; Lee, N. Y.; Terasaki, T. Donepezil, Tacrine and α-Phenyl-n-tert-Butyl Nitrone (PBN) Inhibit Choline Transport by Conditionally Immortalized Rat Brain Capillary Endothelial Cell Lines (TR – BBB). *Arch Pharm Res*. **2005**, *28*, 443–450.

Katayama, K.; Kato, Y.; Onishi, H.; Nagai, T.; Machida, Y. Double liposomes: hypoglycemic effects of liposomal insulin on normal rats. *Drug Dev. Ind. Pharm.* **2003**, *29*, 725–731.

Kim, M. H.; Maeng, H. J.; Yu, K. H.; Lee, K. R.; Tsuruo, T.; Kim, D. D.; Shim, C. K.; Chung, S. J. Evidence of carrier-mediated transport in the penetration of donepezil into the rat brain. *J Pharm Sci.* **2010**, *99*(3), 1548–1566.

Kim, S. J.; Nian, C.; Doudet, D. J.; McIntosh, C. H. Dipeptidyl peptidase IV inhibition with MK0431 improves islet graft survival in diabetic NOD mice partially via T-cell modulation. *Diabetes*. **2009**, 58, 641–651.

Kim, S. J.; Nian, C.; Doudet, D. J.; Mcintosh, C. H. Inhibition of Dipeptidyl Peptidase IV With Sitagliptin (MK0431) Prolongs Islet Graft Survival in Streptozotocin-Induced Diabetic Mice. *Diabetes*. **2008**, *57*(5), 1331–1339.

Knop, K.; Hoogenboom, R.; Fischer, D.; Schubert, U. S. Poly(ethylene glycol) in Drug Delivery: Pros and Cons as Well as Potential Alternatives. *Angew. Chem. Int. Ed.* **2010**, *49*, 6288–6308.

Kondo, M.; Asai, T.; Katanasaka, Y.; Sadzuka, Y.; Tsukada, H.; Ogino, K.; Taki, T.; Baba, K.; Oku, N. Anti-neovascular therapy by liposomal drug targeted to membrane type-1 matrix metalloproteinase. *Int J Cancer.* **2004**, *108*(2), 301–306.

Lamprecht, A.; Ubrich, N.; Yamamoto, H.; Schäfer, U.; Takeuchi, H.; Maincent, P.; Kawashima, Y.; Lehr, C. M. Biodegradable Nanoparticles for Targeted Drug Delivery in Treatment of Inflammatory Bowel Disease. *J Pharm Exper Therap.* **2001**, *299*, 775–781

Lee, J. E.; Park, S.; Park, M.; Kim, M. H.; Park, C. G.; Lee, S. H.; Choi, S. Y.; Kim, B. H.; Park, H. J.; Park, J. H.; Heo, C. Y.; Choy, Y. B. Surgical suture assembled with polymeric drug-delivery sheet for sustained, local pain relief. *Acta Biomater.* **2013**, *9*(9), 8318–8327.

Lee, K.; Lee, H.; Lee, K. W.; Park, T. G. Optical imaging of intracellular reactive oxygen species for the assessment of the cytotoxicity of nanoparticles. *Biomaterials*. **2011**, *32*, 2556–2565.

Lee, S. C.; Huh, K. M.; Lee, J.; Cho, Y. W.; Galinsky, R. E.; Park, K. Hydrotropic polymeric micelles for enhanced paclitaxel solubility: in vitro and in vivo characterization. *Biomacromolecules*. **2007**, *8*(1), 202–208.

Lieberman, S. M.; Dilorenzo, T. P. A comprehensive guide to antibody and T-cell responses in type 1 diabetes. *Tissue Antigens*. **2003**, *62*, 359–377.

Lu, Y.; Park, K. Polymeric micelles and alternative nanonized delivery vehicles for poorly soluble drugs. *Int. J. Pharm.* **2013**, *453*, 198–214.

Lubick, N. Silver socks have cloudy lining. *Environ Sci Technol.* **2008**, *42*(11), 3910.

Maeda, H.; Wu, J.; Sawa, T.; Matsumura, Y.; Hori, K. Tumor vascular permeability and the EPR effect in macromolecular therapeutics: a review. *J. Control. Release.* **2000**, *65*, 271–284.

Mamo, T.; Moseman, E. A.; Kolishetti, N.; Salvador-Morales, C.; Shi, J.; Kuritzkes, D. R.; Langer, R.; von Andrian, U.; Farokhzad, O. C. Emerging Nanotechnology Approaches for HIV/AIDS Treatment and Prevention. *Nanomedicine (Lond).* **2010**, *5*(2), 269–285.

Mastrandrea, L. D. Vascular Health and Risk Management Dovepress inhaled insulin: overview of a novel route of insulin administration. *Vasc. Health Risk Manag.* **2010**, *6*, 47–58.

Matsumura, Y.; Maeda, H. A New Concept for Macromolecular Therapeutics in Cancer Chemotherapy: Mechanism of Tumoritropic Accumulation of Proteins and the Antitumor Agent Smancs1. *CANCER Res.* **1986**, *46*, 6387–6392.

Matteucci, E.; Giampietro, O.; Covolan, V.; Giustarini, D.; Fanti, P.; Rossi, R. insulin administration: present strategies and future directions for a noninvasive (possibly more physiological) delivery. *Drug Des. Devel. Ther.* **2015**, *9*, 3109–3118.

McElroy, M. C.; Kirton, C.; Gliddon, D.; Wolff, R. K. Inhaled biopharmaceutical drug development: nonclinical considerations and case studies. *Inhal. Toxicol.* **2013**, *25*, 219–232.

Mckinlay, J.; Marceau, L. US public health and the twenty-first century: Diabetes mellitus. *Lancet.* **2000**, *356*, 757–761.

Mehanna, A. Antidiabetic agents: past, present and future. *Future Med. Chem.* **2013**, *5*, 411–430.

Moddaresi, M.; Brown, M. B.; Zhao, Y.; Tamburic, S.; Jones, S. A. The role of vehicle nanoparticle interactions in topical drug delivery. *Int. J. Pharm.* **2010**, *400*, 176–182.

Moghimi, S. M.; Hunter, A. C.; Murray, J. C. Long-Circulating and Target-Specific Nanoparticles: Theory to Practice. *Pharm Rev.* **2001**, *53*, 283–318.

Moller, D. E. New drug targets for type 2 diabetes and the metabolic syndrome. *Nature.* **2001**, *414*, 821–827.

Mourtas, S.; Lazar, A. N.; Markoutsa, E.; Duyckaerts, C.; Antimisiaris, S. G. Multifunctional nanoliposomes with curcumin-lipid derivative and brain targeting functionality with potential applications for Alzheimer disease. *Eur J Med Chem.* **2014**, *80*, 175–183.

Nathan, D. M. Long-Term Complications of Diabetes Mellitus. *N. Engl. J. Med.* **1993**, *328*, 1676–1685.

Navalakhe, R. M.; Nandedkar, T. D. Application of nanotechnology in biomedicine. *Indian J. Exp. Biol.* **2007**, *45*, 160–165.

Nie, S.; Xing, Y.; Kim, G. J.; Simons, J. W. Nanotechnology applications in cancer. *Annu Rev Biomed Eng.* **2007**, *9*, 257–288.

Nirmala, A.; Eliza, J.; Rajalakshmi, M.; Edel, P.; Daisy, P. Effects of hexane extract of Cassia fistula barks on blood glucose and lipid profile in streptozotocin diabetic rats. *Int J Pharmacol.* **2008**, *4*(4), 292–296.

Niu, M.; Lu, Y.; Hovgaard, L.; Wu, W. Liposomes containing glycocholate as potential oral insulin delivery systems: preparation, in vitro characterization, and improved protection against enzymatic degradation. *Int. J. Nanomedicine.* **2011**, *6*, 1155–1166.

Obermeier, A.; Schneider, J.; Wehner, S.; Matl, F. D.; Schieker, M.; von Eisenhart-Rothe, R.; Stemberger, A.; Burgkart, R. Novel High Efficient Coatings for Anti-Microbial Surgical Sutures Using Chlorhexidine in Fatty Acid Slow-Release Carrier Systems. *PLoS One.* **2014**, *9*(7), 101426.

Olivier, J. C. Drug transport to brain with targeted nanoparticles. *NeuroRx.* **2005**, *2*(1), 108–119.

Orive, G.; Ali, O. A.; Anitua, E.; Pedraz, J. L.; Emerich, D. F. Biomaterial-based technologies for brain anticancer therapeutics and imaging. *BBA – Rev. Cancer.* **2010**, *1806*(1), 96–107.

Owens, D. R. New horizons–alternative routes for insulin therapy. *Nat. Rev. Drug Discov.* **2002**, *1*, 529–540.

Panyam, J.; Labhasetwar, V. Biodegradable nanoparticles for drug and gene delivery to cells and tissue. *Adv. Drug Deliv. Rev.* **2003**, *55*, 329–47.

Park, J. W.; Hong, K.; Kirpotin, D. B.; Colbern, G.; Shalaby, R.; Baselga, J.; Shao, Y.; Nielsen, U. B.; Marks, J. D.; Moore, D.; Papahadjopoulos, D.; Benz, C. C. Anti-HER2 immunoliposomes: enhanced efficacy attributable to targeted delivery. *Clin Cancer Res.* **2002**, *8*(4), 1172–1181.

Peng, Q.; Zhang, Z. R.; Gong, T.; Chen, G. Q.; Sun X. A rapid-acting, long-acting insulin formulation-based on a phospholipid complex loaded PHBHHx nanoparticles. *Biomaterials.* **2012**, *33*, 1583–1588

Pickup, J. C. Insulin-pump therapy for type 1 diabetes mellitus. *N. Engl. J. Med.* **2012**, *366*, 1616–1624.

Pickup, J. C. Management of diabetes mellitus: is the pump mightier than the pen? *Nat. Rev. Endocrinol.* **2012**, *8*, 425–433.

Prow, T. W.; Chen, X.; Prow, N. A.; Fernando, G. J.; Tan, C. S.; Raphael, A. P.; Chang, D.; Ruutu, M. P.; Jenkins, D. W.; Pyke, A.; Crichton, M. L.; Raphaelli, K.; Goh, L. Y.; Frazer, I. H.; Roberts, M. S.; Gardner, J.; Khromykh, A. A.; Suhrbier, A.; Hall, R. A.; Kendall, M. A. Nanopatch-Targeted Skin Vaccination against West Nile Virus and Chikungunya Virus in Mice. *Queensland.* **2010**, *4108*, 1776–1784.

Rafaeloff, R.; Pittenger, G. L.; Barlow, S. W.; Qin, X. F.; Yan, B.; Rosenberg, L.; Duguid, W. P.; Vinik, A. I. Cloning and Sequencing of the Pancreatic Islet Neogenesis Associated Protein (INGAP) Gene and Its Expression in Islet Neogenesis in Hamsters. *J. Clin. Invest.* **1997**, *99*, 2100–2109.

Rawat, M.; Singh, D.; Saraf, S. Nanocarriers: promising vehicle for bioactive drugs. *Biol. Pharm. Bull.* **2006**, *29*, 1790–1798.

Resnick, H. E.; Foster, G. L.; Bardsley, J.; Ratner, R. E. Achievement of American Diabetes Association clinical practice recommendations among US adults with diabetes, 1999–2002: the National Health and Nutrition Examination Survey. *Diabetes Care.* **2006**, *29*, 531–537.

Reynolds, C.; Hunt, J. Diabetes update. *Can. Fam. Physician.* **1981**, *27*, 1255–1261.

Riddle, M. C. Combined therapy with a sulfonylurea plus evening insulin: safe, reliable, and becoming routine. *Horm. Metab. Res.* **1996**, *28*, 430–433.

Sahoo, S. K.; Parveen, S.; Panda, J. J. The present and future of nanotechnology in human health care. *Nanomedicine.* 2007, *3*(1), 20–31.

Sanchez, F. Nanotechnology in concrete – A review. *Constr. Build. Mater.* **2010**, *24*, 2060–2071.

Sanna, V.; Pala, N.; Sechi, M. Targeted therapy using nanotechnology: focus on cancer. *Int. J. Nanomedicine.* **2014**, *9*, 467–483.
Sarvaiya, J.; Agrawal, Y. K. Chitosan as a suitable nanocarrier material for anti Alzheimer drug delivery. *Int J Biol Macromol.* **2015**, *72*, 454–465.
Sarwar, N.; Gao, P.; et al. Emerging Risk Factors Collaboration, Diabetes mellitus, fasting blood glucose concentration, and risk of vascular disease: a collaborative meta-analysis of 102 prospective studies. *Lancet.* **2010**, *375*, 2215–2222.
Schulman, R.; Moshier, E.; Rho, L.; Casey, M.; Godbold, J.; Mechanick, J. Association of Glycemic Control Parameters with Clinical Outcomes in Chronic Critical Illness. *Endocr. Pract.* **2014**, *20*, 884–893.
Serrano, C.; García-Fernández, L.; Fernández-Blázquez, J. P.; Barbeck, M.; Ghanaati, S.; Unger, R.; Kirkpatrick, J.; Arzt, E.; Funk, L.; Turón, P.; del Campo, A. Nanostructured medical sutures with antibacterial properties. *Biomaterials.* **2015**, *52*, 291–300.
Shrivastava, S.; Dash, D. Applying Nanotechnology to Human Health: Revolution in Biomedical Sciences. *J. Nanotechnol. Hindawi Publishing Corporation.* **2009**, 14.
Silva, J.; Fernandes, A. R.; Baptista, P. V. Application of Nanotechnology in Drug Delivery. **2014**.
Sondi, I.; Salopek-Sondi, B. Silver nanoparticles as antimicrobial agent: a case study on E. coli as a model for Gram-negative bacteria. *J Colloid Interface Sci.* **2004**, *275*(1), 177–182.
Standards of Medical Care in Diabetes-2013. *Diabetes Care.* **2013**, *36*(1), 11–66.
Stylianou, E.; Diogo, G. R.; Pepponi, I.; van Dolleweerd, C.; Arias, M. A.; Locht, C.; Rider, C. C.; Sibley, L.; Cutting, S. M.; Loxley, A.; Ma, J. K.; Reljic, R. Mucosal delivery of antigen-coated nanoparticles to lungs confers protective immunity against tuberculosis infection in mice. *Eur J Immunol.* **2014**, *44*(2), 440–449.
Svenson, S.; Tomalia, D. A. Dendrimers in biomedical applications—reflections on the field. *Adv. Drug Deliv. Rev.* **2005**, *57*(15), 2106–2129.
Tamborlane, W. V.; Beck, R. W.; Bode, B. W.; Buckingham, B.; Chase, H. P.; Clemons, R.; et al. Juvenile Diabetes Research Foundation Continuous Glucose Monitoring Study Group, Continuous Glucose Monitoring and Intensive Treatment of Type 1 Diabetes. *N Engl J Med.* **2008**, *359*(14), 1464–1476.
Timbie, K. F.; Mead, B. P.; Price, R. J. Drug and gene delivery across the blood–brain barrier with focused ultrasound. *J Control Release.* **2015**, *3659*(15), 30101–30102.
Turkevich, J.; Stevenson, P. C.; Hillier, J. Discussions of the Faraday Society. **1951**, *11*, 55–75.
Wan, T. T.; Li, X. F.; Sun, Y. M.; Li, Y. B.; Su, Y. Recent advances in understanding the biochemical and molecular mechanism of diabetic retinopathy. *Biomed. Pharmacother.* **2015**, 74, 145–147.
Watanabe, K.; Kaneko, M.; Maitani, Y. Functional coating of liposomes using a folate–polymer conjugate to target folate receptors. *Int. J. Nanomedicine.* **2012**, *7*, 3679–3688.
Wiener, E. C.; Brechbiel, M. W.; Brothers, H.; Magin, R. L.; Gansow, O. A.; Tomalia, D. A.; Lauterbur, P. C. Dendrimer-based metal chelates: a new class of magnetic resonance imaging contrast agents. *Magn. Reson. Med.* **1994**, *31*(1), 1–8.

Wissing, S. A.; Müller, R. H. Cosmetic applications for solid lipid nanoparticles (SLN). *Int. J. Pharm.* **2003**, *254*, 65–68.

Wu, Z. H.; Ping, Q. N.; Wei, Y.; Lai, J. M. Hypoglycemic efficacy of chitosan-coated insulin liposomes after oral administration in mice. *Acta Pharmacol Sin.* **2004**, *25*, 966–972.

Xue, M.; Yang, M. X.; Zhang, W.; Li, X. M.; Gao, D. H.; Ou, Z. M.; et al. Characterization, pharmacokinetics, and hypoglycemic effect of berberine loaded solid lipid nanoparticles. *Int. J. Nanomedicine.* **2013**, *8*, 4677–4687.

Yan, R.; Vassar, R. Targeting the β secretase BACE1 for Alzheimer's disease therapy. *Lancet Neurol.* **2014**, *13*, 319–329.

Yeh, Y. C.; Creran, B.; Rotillo, V. M. Gold Nanoparticles: Preparation, Properties, and Applications in Bionanotechnology. *Nanoscale.* **2012**, *4*, 1871–1880.

Zhang, X.; Qi, J.; Lu, Y.; He, W.; Li, X.; Wu, W. Biotinylated liposomes as potential carriers for the oral delivery of insulin. *Nanomedicine Nanotechnology, Biol. Med.* **2014**, *10*, 167–176.

Zhang, Z. H.; Zhang, Y. L.; Zhou, J. P.; Lv, H. X. Solid lipid nanoparticles modified with stearic acid–octaarginine for oral administration of insulin. *Int. J. Nanomedicine.* **2012**, *7*, 3333–3339.

CHAPTER 4

POLYMERIC MATRIX SYSTEMS FOR DRUG DELIVERY

SNEŽANA ILIĆ-STOJANOVIĆ,[1,*] LJUBIŠA NIKOLIĆ,[1]
VESNA NIKOLIĆ,[1] DUŠICA ILIĆ,[1] IVAN S. RISTIĆ,[2] and ANA TAČIĆ[1]

[1]*University of Niš, Faculty of Technology, Bulevar oslobodjenja 124, 16000 Leskovac, Republic of Serbia, *E-mail: ilic.s.snezana@gmail.com*

[2]*University of Novi Sad, Faculty of Technology, Bulevar Cara Lazara 1, 21000 Novi Sad, Republic of Serbia*

CONTENTS

ABSTRACT

This chapter provides an overview of the drug delivery systems, DDS, based on natural and/or synthetic polymers as carriers for the active substances, proteins or cells. The polymers used in the DDS must possess properties that make them suitable for interaction with the human organism. The modified rate or the place of the active substances release can be achieved by addition of excipients and/or by special technological procedures. The matrix systems represent the dispersion of active ingredients in an inert matrix and they can be manufactured in the form of tablets, nanoparticles and microparticles. The DDS based on biodegradable natural polymers (proteins and polysaccharides) and synthetic polymers [poly(lactic acid), poly(D,L-lactide-coglycolide) and poly(ε-caprolactone)] have been studied. The DDS based on synthetic non-biodegradable polymers [poly(methylmethacrylate), poly(2-hydroxyethyl methacrylate), poly(acrylic acid), poly(*N*-isopropylacrylamide), poly(acrylamide) and poloxamer] were also considered.

4.1 INTRODUCTION

Systems for the delivery of drugs (DDS – drug delivery system) are defined as formulations or carriers that allow loading of therapeutic agent into the body, improve its efficiency and safety by controlling the rate, time and/or place of release in the body. DDS include the administration of a therapeutic agent; release the active ingredients (drug) from it and the transport of active substances through biological membranes for the treatment of diseases to the site of action. The pharmacokinetic and pharmacodynamic properties of the drug, toxicity, immunogenicity, biorecognition ability and thus increase efficiency are modified by using DDS systems.

In order to reduce the drug degradation, to prevent the side effects and increase bioavailability, different systems for the targeted drugs delivery, such as hydrogels, polymers, microparticles, microcapsules, lipoproteins, liposomes, nanoparticles, micelles and dendrimers were examined (Shaik et al., 2012). The pharmaceutical technology aims a development of process for active ingredients incorporation in drug dosage forms, which is acceptable for the application and allows the active ingredient to release following application in accordance with the aims of therapy (Milić and Petrović, 2003).

4.1.1 THE PHARMACEUTICAL FORMS OF MODIFIED RELEASE OF THE ACTIVE SUBSTANCE

The pharmaceutical forms with modified release were the drug dosage forms that are prepared with the addition of special auxiliary substances (excipients) and/or special technological procedures, where rate or place of active substance release from the drug dosage forms is modified (Milić and Petrović, 2003). The modified release formulations have been designed to achieve one of following criteria's:

- reduction of drug fluctuations in the plasma,
- a decrease in frequently giving and better acceptability for the patient,
- improving the pharmacological activity of drug.

Based on mechanism of drug release, they are divided on:

- preparations in which the control mechanism is the dissolution (encapsulated and matrix systems) and
- preparations in which the control mechanism is the diffusion (membrane-reservoir and matrix systems with dispersed drug).

Based on the release rate of active substances, preparations with modified release were divided into (Milić and Petrović, 2003):

- preparations with the rapid release of the active ingredient (fast action);
- preparations with slow/controlled release of active ingredient (with delayed and/or extended action, depot-preparations, retard preparations, preparations with prolonged action), which can be further divided into the preparations with:
    - a time-controlled release of active ingredients,
    - with active substance release kinetics at approximately zero order,
    - two- or multiphase release of the active ingredient (extended release tablets which possess an initial rapid release, and extended release tablet exhibiting a fast final release).

4.1.1.1 The Influence of Different Factors on the Release Rate of the Active Ingredient

Factors that affect the release rate of active ingredients are related to the structure, chemical and physical characteristics of the matrix and the active substance. In most formulations, drug release is controlled by diffusion, but also factors such as the polymer molecular weights, type of polymer, the crystallinity and method of drug incorporation also significantly influence release of drugs (Freiberg and Zhu, 2004). One of the objectives to be achieved by advanced pharmaceutical forms is drug release at a constant rate. The release of active substance from the modern pharmaceutical form is mainly composed of two processes:

- an initial release of the active substance from the carrier surface;
- release of the active substance at a constant rate which is dependent on diffusion or the biodegradation of the matrix (Ghaderi et al., 1996; Mogi et al., 2000).

In some formulations relatively constant release of the active ingredient after the initial release is achieved. In certain formulations, the release follows zero order kinetics with no significant effects of the initial release of the active ingredient. In many modern pharmaceutical forms are represented considerably more complex release of the active ingredient, depending on the desired application (Berkland et al., 2002; Kakish et al., 2002; Makino et al., 2000; Narayani and Rao, 1996; Yang et al., 2000).

4.1.2 THE MATRIX SYSTEMS FOR THE DELIVERY OF ACTIVE INGREDIENT

The matrix systems represent a dispersion of the active ingredient in an inert matrix, obtained by various techniques, and can be prepared in the forms of tablets, microparticles and others (Milić and Petrović, 2003). The homogeneous matrix system is a dispersion of an active ingredient and matrix in which it is the physically or chemically bonded. The heterogeneous matrix system constitutes an active ingredient and a matrix which does not have a high degree of microhomogeneity, and may also contain additives (fillers, plasticizers, etc.) in order to increase the porosity.

DDS can be obtained from natural or synthetic polymers, which can be biodegradable or non-biodegradable, and they are designed for the release of pharmacologically active substances, proteins, or cells. The polymers used in the drug delivery systems have possess the properties that make them suitable for interaction with the human organism. The properties of synthetic polymers can be adjusted in the synthesis process, changing the composition or manufacture method. Biodegradable polymers are especially interesting for use in DDS, because when they used, do not require removal from the body or any additional intervention, due to their degradation on non-toxic components. Their degradation products or metabolites are the usual products which can be easily metabolized and excreted from the body (Coelho et al., 2010). Another possible method of making DDS with satisfactory characteristic is the combination of natural and synthetic polymers. The ultimate goal is to obtain a polymeric material which exhibits the best properties of natural polymers (biodegradability and compatibility) and synthetic polymers (mechanical characteristics). The production technologies of modern pharmaceutical forms are highly specialized and provide modified release of the active ingredient, which significantly contributes to the quality and safety of the therapy (Rizkalla et al., 2006). DDS may maintain therapeutic effect over an extended period of time, reducing side effects (Orive et al., 2005) and providing a higher efficiency of the active ingredient (Steffens et al., 2002). The applied dose of the active ingredient passes through more physiological barrier before it reaches the place of action. Unstable drugs, usually peptides, proteins and enzymes, may lose their activity under the influence of environmental factors in the body (Brannon-Peppas, 1997). Polymeric carriers can protect the active ingredient from degradation on its path to the target site (Gander et al., 1996). Traditional systems for the drug delivery were characterized by short time effects, which requires frequently drug taking and resulting in large fluctuations of drug concentrations in plasma (Dong et al., 2005).

4.1.2.1 Advanced Systems for the Delivery of Drugs

Design and development of modern drug delivery system are the main goal of the research in pharmacy, which are mostly focused on advanced and intelligent systems for the drug transfer compared to the simple conventional drug formulations. The main specificity of intelligent polymers are their extreme response to minor changes in environmental conditions, such as temperature, pH, UV radiation, magnetic fields and/or light intensity. These external stimuli cause changes in the structural characteristics of intelligent polymers, which possess significant potential for different types of applications in the biotechnology and biomedicine.

Hydrogels are polymeric materials that contain a large number of hydrophobic groups that are able to absorb large quantities of water in their three-dimensional network, wherein its structure remains intact. In the swollen state, hydrogels are soft and rubbery, show good biocompatibility, thermal and mechanical stability, allowing a wide application in chemistry, medicine and pharmacy (Chen et al., 2013). Hydrogels resemble living tissue more than any other type of synthetic biomaterials. The high water content and soft consistency of the hydrogel in the swollen state reduces irritation and friction with the surrounding tissue and the like living tissue, allowing the diffusion of metabolism products. The pH-sensitive hydrogels are mostly used for controlled drug release. Cationic gels achieve maximum swelling capacity in acidic media, which allows the drug release in the stomach. Anionic hydrogels achieve maximum swelling capacity in neutral and weakly alkaline media and release drugs in the intestines. The hydrogels may be used as a solid form in which they are synthesized (contact lenses, cylinders, plates), compressed powders, microparticles (bioadhesive wound treatment), coating (for implants and catheters), membranes (for transdermal administration) and fluid (which pass into gel at the place of application).

Nanogels enabling the controlled release of active ingredients due to the slow degradation, biocompatibility, ability to respond to external stimuli and the ability for target drug delivery. Nanogels possess properties of gels and colloids, so in recent times are very actual. The structure of these systems in their complexity could greatly mimic the structure of living cells, where different parts of the nanoparticles interact with each other, exchanging matter, receive energy, perform mechanical work, changing the physical and mechanical properties, and all this in response to a specific external stimulus. These particles are composed from a hybrid structure of the polymer-particles type, that possess core-layer structure (Figure 4.1).

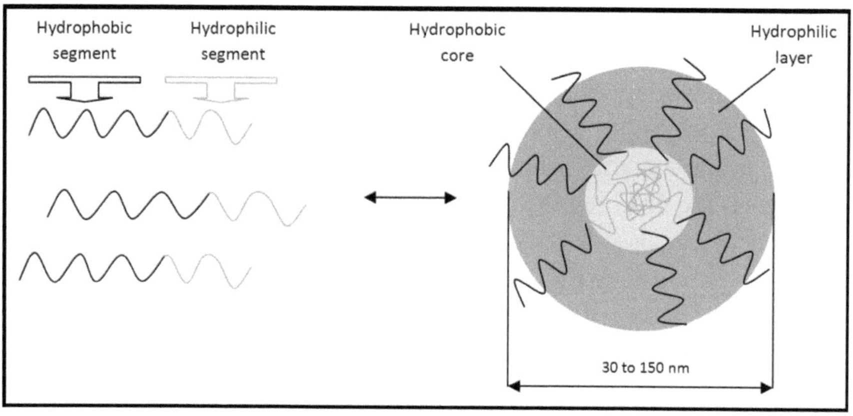

FIGURE 4.1 The modification of the particle surface with hydrophilic segments.

The core may be made of metal, metal oxides, nonmetal oxides, salts, polymers, liquid or gas (hollow structure). The coating is a polymer whose surface characteristics may be modified in different ways, so that it can respond to the external stimuli. The drug molecules can be bonded to the particle surface, or encapsulated within the core. Surface modification, synthesis and selection of nanoparticles are dependent on the drug nature, release period and its stability (Hans and Lowman, 2002). The organism recognizes hydrophobic substances as foreign objects, absorbs them and eliminates *via* phagocytic system. If it is necessary to the drug long-term release, the hydrophobic surface may be modified in order to prevent phagocytosis and the drug became "invisible" to the immune system due to created steric barrier (Storm et al., 1995). The particles can be modified with hydrophilic coating of poly(ethylene glycol), polysorbates, or poly(vinylpyrrolidone) (Torchilin and Trubetskoy, 1995).

By modifying the surface of certain polymers is also possible to achieve localized drug effect due to increased drug concentrations in certain tissues (Shenoy and Amiji, 2005) or allow the passage of drug through the blood-brain barrier (Borchard et al., 1994). Specific strategy for surfaces modification is the ligand binding to the nanoparticles surface that enables the recognition of the receptor on the specific cell surface. For this purpose, lectins which are involved in many aspects of cellular recognition and adhesion on the cells surface were the most examined (Rodrigues et al., 2003). In contact with polymeric nanoparticles, it enables efficient transport through the intestinal mucosa.

The behavior of polymeric or hybrid nanoparticles as drug carriers includes receiving external signals (physical or chemical), chemical reaction or the material properties change and transfer of those changes, resulting in the drug release at the target site. Receiving a chemical and biochemical signals is based on physical or chemical interactions between the polymer and signaling factors (Yurke et al., 2005).

The chemical structure of polymer is changing in interaction with environment, which is followed by changes in conformation, phase transitions, changes in the optical, magnetic and surface properties of the polymer. Compared with other materials, polymeric materials exhibit the ability to modify various DDS properties, such as biodegradability, biocompatibility, reproducibility, chemical, topological and dimensional stability. Lately, a large number of DDS in the form of microspheres, films, tablets or implants, based on appropriate polymer properties are actual. All polymers used for drug delivery are classified according to:

- chemical nature (polyesters, polyanhydrides, polyamino acids);
- stability of the backbone chain (biodegradable and non-biodegradable); and
- solubility in the water (hydrophobic and hydrophilic).

The efficiency of the drug encapsulation within a given complex polymeric systems is one of the most important requirements for polymeric DDS. In addition the chemical encapsulation, which takes place by complex mechanisms of the polymer crosslinking, also, the physical encapsulation of the drug is very important. Frequently, the physical drug encapsulation is more attractive than the chemical. The drug distribution within the body is another very important parameter for polymeric DDS. It depends on the value of critical micelle concentration (CMC), surface charge and the target site of drug delivery. The small size of micelles and low CMC enable high stability of micelles and increase the time of circulation, which leads to the accumulation of micelles at the target site and controlled drug release. The surface charge is an important factor for the micelle biodistribution. The presence of a poly(ethylene glycol) in the micelle coating increases the time of drug half-life in the plasma and reduces its interaction with blood components. A large number of companies concerned with the development of multiple technologies carrier in order to prolong the lives of patients are achieving a competitive advantage and increasing participation of their products on the market (Milić and Petrović, 2003). A lot of the clinical drugs have been successfully encapsulated whereby the bioavailability, the efficiency and the controlled release have been increased (Budhian et al., 2005; Cheng et al., 2008; Gomez-Gaete et al., 2007). Nanodrugs for diseases such as tumors of different etiology (Mu and Feng, 2003), AIDS (Coester et al., 2000), diabetes (Damge et al., 2007), malaria (Date et al., 2007), prion diseases (Calvo et al., 2001), tuberculosis (Ahmad et al., 2006) are in various stages of testing, wherein some of them are commercialized (Kim and Lee, 2001). An overview of some commercial products of drugs with polymer carriers is presented in the Table 4.1.

4.2 DRUG DELIVERY SYSTEMS BASED ON BIODEGRADABLE POLYMERS

Polymer degradation represent the process of decomposing a polymer into smaller fragments, which results in a change of its characteristics, such as molecular weight, color, shape and resistance to stretching. These changes occur during processing, storage and use of polymer under the influence of temperature (thermodegradation), light (photolysis), water (hydrolysis) and enzymes (biodegradation). Polymer degradation depends on its characteristics (chemical composition, structure, molecular weight and morphology), the processing procedure, sterilization, the environmental conditions (pH, ionic strength, temperature, light) and the degradation method (Shaik et al., 2012).

Biodegradable polymers are often used as matrices for active substances, primarily due to their low toxicity. They also provide controlled release of the drug at the site of action during a certain period of time. Drug nanoencapsulation increases the efficiency and tolerance to the drug and represents better alternative to conventional drugs (Fassas et al., 2003; Kreutera et al., 1997; Raghuvanshi et al., 2002; Safra et al., 2000; Schroeder et al., 1998). Important properties of polymeric nanoparticles as drug carriers are particle size and particle size distribution, surface morphology, charge and adhesion, surface erosion,

TABLE 4.1 Commercial Drugs as DDS

Drug name	Polymer + Active Ingredient	DDS	Drug delivery applications	Producer
Enantone® Monats-Depot	PLGA + Leuprolide acetate	Microcapsules for depot suspension	Treatment of prostate cancer	Takeda GmbH
Lupron Depot®	PLGA + Leuprolide acetate	Microspheres for depot suspension	Therapy of prostate cancer, endometriosis and/or uterine fibroids	For AbbVie Inc. by Takeda Ltd.
Sandostatin® LAR Depot	PLGA + Ostreotide acetate	Microspheres for injectable suspension	Long-term treatment of the profuse watery diarrhea associated with VIP-secreting tumors	Novartis
Nutropin® Depot	PLGA + recombinant human growth hormone	Microspheres for injectable suspension	Long-term treatment of growth failure due to a lack of adequate endogenous GH secretion	Novartis
Signifor® LAR	PLGA + Pasireotide pamoate	Microspheres for injectable suspension	Treatment of patients with acromegaly	Novartis
Vivitrol®	PLGA + Naltrexone	Extended-release injectable suspension	Treatment of alcohol dependence and prevention of relapse to opioid dependence	Alkermes, Inc.
Profact® Depot	PLGA + Buserelin acetate	Implant	Treatment of advanced prostate cancer	Sanofi-Aventis, Deutschland
Genexol-PM	PEG/PLA + Paclitaxel	Micellar nanoparticles	Treatment of breast, lung and ovarian cancer	Samyang Biopharm
Abraxane®	Albumin + Paclitaxel	Albumin-bound particles for injectable suspension	Treatment of advanced pancreatic cancer (with gemcitabine)	Celgene
Bellafill® Artefill®	PMMA beads + collagen + lidocaine	Microspheres in collagen gel	Correction of facial wrinkles	Suneva Medical, Inc.
Optipac Knee Refobacin®	PMMA + gentamycin	Bone cement with gentamicin	Uni knee, hybrid knee, two-step total knee or shoulder	Biomet Orthopaedics, Switzerland
Optipac Refobacin® Revision	PMMA + gentamycin + clindamycin	Bone cement with gentamycin and clindamycin	Revision Surgery	Biomet Orthopaedics, Switzerland

diffusion, encapsulation efficiency, stability, drug release kinetics and hemodynamics (Feng, 2004). In order to meet increasing number of requests and achieve better therapeutic efficiency of the drugs, significant efforts in developing new and improving existing biodegradable polymers are invested (Ristić et al., 2011, 2013a). Biodegradable polymers used in medicine and pharmacy should be biocompatible and their degradation products should be nontoxic.

Ideal biodegradable polymer for use in medicine and pharmacy should have the following properties:

- does not cause inflammation/toxic reaction,
- can be easily sterilized,
- after completes its action, it should be degraded by appropriate metabolic pathway and eliminated from the organism (Ristić, 2011).

In this chapter, DDS based on proteins (collagen, gelatin and albumin) and polysaccharides (chitosan, dextran and alginate) as natural polymers, as well as DDS based on polyester (poly(lactic acid), PLA, poly(D,L-lactide-coglicolide), PLGA, and poly(ε-caprolactone), PCL, as synthetic polymers, were studied.

4.2.1 DRUG DELIVERY SYSTEMS BASED ON NATURAL POLYMERS

4.2.1.1 Proteins

Proteins are high molecular weight compounds composed of amino acids linked by peptide bonds. As the main structural components of the human organism proteins are suitable for DDS development and collagen, gelatin and albumin are commonly used.

Collagen is the most abundant protein in the human organism as a major component of skin, bone and cartilage. This natural polymer is non-toxic, biodegradable and poorly immunogenic. The collagen microspheres are studied as carriers for glucocorticoids (hydrocortisone). For delivery of high molecular weight drugs, collagen is used in the form of pellets. It has been shown that the release profile of such drugs from the collagen carriers is two-step process which depends on the molecular weight of the drug and the size of pellet. The collagen matrices were used for sustained release of growth factors (Coelho et al., 2010). Prabhakaran et al. (2009) shown that collagen matrix could be used for realizing of hydroxyapatite. The collagen matrix degrades over time, while hydroxyapatite fills the remaining space without changing the characteristics and morphology of injured tissue, compared to healthy tissue.

Gelatin is a polyampholyte which contains cationic and anionic groups, in addition to hydrophilic groups. It is obtained by partial hydrolysis of collagen and in aqueous solutions below 35°C, gelatin undergoes a sol-gel transformation. Due to its high water solubility and poor mechanical properties, cross-linking of gelatin is preferred. It is known that swelling, mechanical and thermal properties depend on cross-linking degree of gelatin (Oppenhiem, 1981). Gelatin is widely used in the food industry and medicine because it is nonimmunogenic, nontoxic, biodegradable, bioactive, biocompatible and low cost. It is suitable for the preparation of nano formulation of various drugs for cancer, HIV infection and malaria treatment. It provides various benefits, such as modified release, localized effects of the drug and prevention of adverse drug effects (Lu et al., 2004). Cross-linked gelatin microspheres are used for delivery of colchicine in the cancer treatment. Gelatin microspheres loaded with doxorubicin, 5-fluorouracil, bleomycin and mytomicin C are also prepared (Wallace et al., 2000). The cross-linked hydrogels containing gelatine (as natural polymer) and synthetic polymers can be used for target drug delivery and, alternatively, as implants.

Albumin is the most abundant protein in human plasma, with molecular weight of about 66 000 $gmol^{-1}$. It is hydrosoluble, biodegradable, nontoxic and nonimmunogenic, which makes it suitable for pharmaceutical use. Albumin microspheres are suitable carriers for anticancer drugs, because albumin is a source of nitrogen and energy for cancer cells after taking it over by endocytosis and decomposing in lysosomes. In this way, the anticancer drug can be delivered in the targeted cancer cells, which reduces drug systematic toxicity (Coelho et al., 2010).

4.2.1.2 Polysaccharides

Polysaccharides are high molecular weight compounds composed of repeated monosaccharide units. They contain hydroxyl groups, which can react directly with carboxylic groups of active substances and form an ester. The ester linkage can be easily hydrolyzed and release active substance at the site of action. The presence of reactive lateral groups allows modification of polysaccharides structure increasing their possible applications. In DDS development, chitosan, alginate and dextran are widely used polysaccharides (Vilar et al., 2012).

Chitosan is a modified natural polysaccharide obtained by partial *N*-deacetylation of chitin, which is found naturally in the shell of crustaceans. It is insoluble in water and organic solvents, but it can be dissolved in acidic mediums. Biodegradability, low toxicity and good biocompatibility make chitosan suitable for pharmaceutical applications (Coelho et al., 2010). Chitosan is a suitable carrier for glycyrrhizin because, in the system with PEG and tripolyphosphate anions as surface modifiers, it optimizes the process of drug release in human organism (Wu, 2005). Also, it is possible to successfully encapsulate insulin and cyclosporin A in chitosan matrix, whose selectivity is significantly changed and allows direct localized effect of the drug in target tissues (De Campos et al., 2001; Sarmento et al., 2007). Microspheres and nanoparticles of chitosan are widely used as drug carriers (Tiyaboonchai, 2003) for proteins, anticancer and ophthalmic drugs, as well as carriers for genes in gene therapy (Coelho et al., 2010).

Alginic acid is a cationic polysaccharide extracted from brown algae. It is a block copolymer composed of two uronic acid units: β-D-mannuronic acid and α-L-glucuronic acid, with molecular weight higher than 500,000 $gmol^{-1}$. It is usually used in the form of sodium salt (Coelho et al., 2010). Sodium alginate readily forms gel in contact with divalent cations (Ca^{2+}) at room temperature. In addition, alginic gel is biocompatible and nonimmunogenic, but cannot be degraded by the mammals' enzymes (Coelho et al., 2010). Hydrogels of sodium alginate can be used for sustained localized release of low molecular weight drugs and macromolecules. The release of active substance can be adjusted by the covalent cross-linking of alginate. In the case of alginate microspheres loaded with growth factors, the problem of immediate release is solved by cross-linking of alginate with ethylenediamine (Coelho et al., 2010).

Dextran is a polysaccharide composed of D-glucopyranose units linked by α-1,6-glycosidic bond. The side branches can be linked by α-1,2-, α-1,3- and α-1,4-glycosidic bonds. Dextran is a suitable material for application in biomedicine due to its biodegradability, biocompatibility, nonimmunogenicity and nonantigenicity. It is used as a carrier for delivery of proteins, lipophilic drugs, genes, as well as for preparation of self-nano-emulsifying systems for delivery of ibuprofen and naproxen. These systems are stable over a wide pH range (4–11) and enable sustained release of hydrophobic drugs (Coelho et al., 2010).

4.2.2 DRUG DELIVERY SYSTEMS BASED ON SYNTHETIC POLYMERS

4.2.2.1 Poly(lactic acid)

Lactide is the common name for a cyclic ester of lactic acid, which occurs in the form of two optically active and one optically nonactive stereoisomer (Figure 4.2a and 4.2b).

Thus, three different lactides can be formed: L(-)-lactide (S,S), D(+)-lactide (R,R) and optically inactive meso-lactide (R,S). A racemic mixture of L- and D-lactide is generally known as D,L-lactide. For polymer synthesis, L- and D,L-lactide are used (Figure 4.2c). Poly(lactic acid) and poly(lactide) are mainly referred by name formed of prefix *poly* and the name of monomer from which they derived. Thus, poly(lactic acid) is a polymer obtained from lactic acid and poly(lactide) is obtained from lactide. Sometimes, poly(lactide) is referred as poly(lactic acid), because both forms have the same constitutional repeated units, H-[OCH(CH_3)CO]$_n$-OH. These polymers are biodegradable (aliphatic

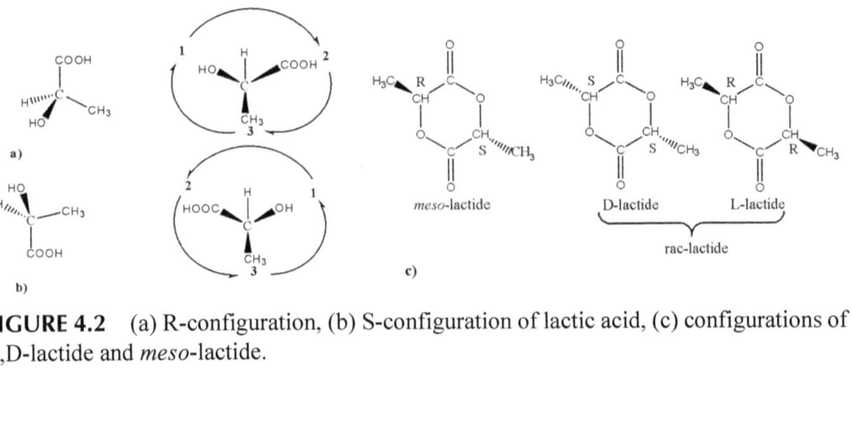

FIGURE 4.2 (a) R-configuration, (b) S-configuration of lactic acid, (c) configurations of L,D-lactide and *meso*-lactide.

polyester base is particularly sensitive to hydrolysis under the water and heat influence) and biocompatible, because their hydrolysis in physiological mediums gives lactic acid, nontoxic component, which can be eliminated from the body through the Krebs cycle as H_2O and CO_2 (Amecke et al., 1995). Lactic acid can be obtained by fermentation from renewable sources, such as corn, sugar, whey and potato starch (Mobley, 1994). The lactide monomer can be obtained by fermentation of carbohydrates using microorganisms, such as species of the *Lactobacillus* genus. A number of studies has confirmed that synthesis of poly(lactide) catalyzed by specific enzymes (Reeve et al., 1994), such as protease K, is possible.

Polymers synthesized from lactic acid have found numerous applications in medicine as fixators for fractures, surgical suture, implants, systems for controlled release of drugs (Gilding and Reed, 1979; Leenslag et al., 1987) and in tissue engineering (Lunt, 1998).

For the application as drug carrier, poly(lactide) should be defined molecular weight. Thus, catalyst, which can reduce the polymer chain length, and small molecular weight compounds, that can significantly change the properties of PLA, are used in poly(lactide) synthesis (Ristić et al., 2013b). Pyridine, in the form of 4-(dimethylamino) pyridine (DMAP), is used as a catalyst for the regulation of poly(D,L-lactide) chains lengths. The high molecular weight poly(lactide) depolymerize with primary alcohols at 38°C in the solution and at 185°C in the mass. Poly(lactide) with desired molecular weight can be obtained by transesterification.

The traditional method of poly(lactide) synthesis performs by ring-opening polymerization, using Sn(II) 2-ethylhexanoate as initiator. This process takes over 30 h at 120°C and required high vacuum. The application of microwaves in the PLA synthesis, by the same polymerization mechanism and using the same initiator, significantly reduces the time of polymerization (30 min at 100°C) with evident energy saving (Nikolić et al., 2007; 2010). The effects of the polymerization duration and the monomer/initiator mole ratio on the average molecular weight of polymer and the polydispersity index have examined. It has been shown that average molecular weight and the polydispersity index increase with delay of reaction time and increasing of monomer/initiator mole ratio.

Microspheres of PLA synthesized by microwave heating are prepared and used as carriers for controlled release of allicin and its transforments, ajoene and viniyldithiine (Ilić et al., 2012). The release of allicin transforments from PLA microspheres was examined in simulated conditions of digestive tract (pH = 3 and pH = 8). It has been shown that the release is more efficient in the acidic environment.

Oridonin is a diterpenoid compound used in the treatment of esophagus and liver cancers. The therapeutic use of this drug is limited by the low water solubility and low therapeutic index. By oridonin incorporation into the PLA nanoparticles, prolonged drug retention in the blood and drug accumulation in target tissues after intravenous administration, were achieved (Xing et al., 2007). Modification of the properties of PLA nanoparticles, with hydrophilic PEG, enables longer drug retention in the circulation and thus, sustained action. The release of progesterone from PLA-PEG-PLA nanoparticles depends on the PEG content, the PEG molecular weight and the total molecular weight of the copolymer (Matsumoto et al., 1999).

Poly(L-lactide) (PLLA) can be used as a substitute for collagen fibers in the bone matrix. It has been shown that hydroxyapatite (HAp) as biomaterial could be applied in reparation of bone defects. HAp/PLLA bicomposite was tested as an artificial substitute for bone matrix. It provides adequate microenvironment for the development and activity of bone marrow cells and hematopoiesis (Vasiljević et al., 2009).

4.2.2.2 Poly(D,L-lactide-coglycolide)

Poly(D,L-lactide-coglycolide) is one of the most commonly used polymers for nanodrug development. By hydrolysis of PLGA, lactic and glycolic acid are produced. They are nontoxic, biodegradable and can be included in metabolic processes. Therefore, PLGA is one of the most important biodegradable polymers for the preparation of system for controlled release of drugs. PLGA nanoparticles can be prepared by methods such as emulsification-diffusion, evaporation-extraction, interfacial polymer deposition and nanoprecipitation (Sah and Sah, 2015).

Microspheres and nanoparticles of PLGA have been used in the controlled delivery of proteins, vaccines, genes, antigens and growth factors.

They are also suitable for encapsulation of anticancer drugs (Coelho et al., 2010), as cisplatin and carboplatin. Cisplatin is an anticancer drug, extremely toxic to healthy cells (Moreno et al., 2010). Encapsulation of cisplatin in PLGA significantly improves drug selectivity and enhanced the effect towards tumor cells which increases efficiency and reduces side effects of the drug. PLGA microspheres loaded with carboplatin meet the requirements set for intracerebral implementation and can be used in animal studies (Chen et al., 1997). PLGA nanoparticles are suitable for encapsulation of estradiol, which is used for the relief of menopausal symptoms, as a part of a replacement therapy. Estradiol is well absorbed after oral administration, but it has poor bioavailability due to first-pass metabolism through the liver. Encapsulation with PLGA can increase oral bioavailability of estradiol, reduce dosage frequency, minimize the dose-dependent side effects and increase patient compliance (Mittal et al., 2007).

4.2.2.3 Poly(ε-caprolactone)

Poly(ε-caprolactone) is a semicrystalline polymer with low melting point (55–60°C) and glass transition temperature around −60°C. PCL is decomposed by hydrolysis of the ester bond under physiological conditions, without producing acid products. PCL alone or in combination with other polymers can be used for encapsulation of various active substances, because it is permeable for small molecules. It is especially interesting as material for the long-term release of active substances, due to its slower degradation compared to PLA.

The insulin encapsulation in nanoparticles made of PCL and polycationic polymer Eudragit RS enables the preservation of the insulin activity after oral administration. Also, absorption of active substance is facilitated due to mucoadhesive action of Eudragit RS (Damge et al., 2007). Amphotericin B is a polyenoic antifungal agent and drug for treatment of leishmaniosis. Encapsulation of amphotericin B in the PCL nanospheres can reduce drug toxicity and enhance activity, by increasing the available drug concentration at the site of action (Espuelas et al., 2002). Gelatin-coated PCL formulation for sustained ibuprofen release shows enhanced adhesion properties and can be applied in wound healing (Coelho et al., 2010).

4.3 DRUG DELIVERY SYSTEMS BASED ON NONBIODEGRADABLE POLYMERS

In the case of non-biodegradable matrix systems, the release of the therapeutic agent is carried out by diffusion of water into the matrix and/or diffusion of therapeutic agent through the matrix channels. There is a risk that the therapeutic agent is decomposed or inactivated within non-biodegradable matrix before releasing. This is particularly pronounced in the case of macromolecules and polypeptides, since these compounds are generally unstable in a buffer (Wallace et al., 2000). Most of the studies of synthetic non-biodegradable polymers in drug delivery systems are focused on testing of poly(methylmethacrylate), PMMA; poly(2-hydroxyethyl methacrylate), PHEMA; poly(acrylic acid), PAA; poly(*N*-isopropylacrylamide), PNIPAM; poly(acrylamide), PAM and poloxamers (PEO-PPO-PEO triblock copolymers) (Figure 4.3).

4.3.1 *POLY(METHYLMETHACRYLATE)*

Poly(methylmethacrylate) is a biocompatible and stable polymer. It is transparent, does not absorb water and can be processed by conventional methods. PMMA is widely used in manufacturing of contact

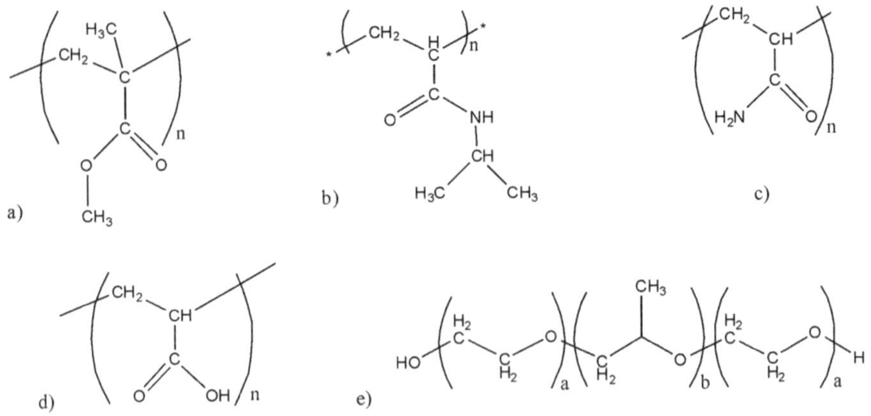

FIGURE 4.3 Chemical structure of non-biodegradable polymers: (a) poly(methylmethacrylate), (b) poly(*N*-isopropylacrylamide), (c) poly(acrylamide), (d) poly(acrylic acid), and (e) poloxamers.

lenses, artificial teeth and dentures, as well as bone cement in orthopedic surgery (Anderson, 1997; Craig and Peyton, 1997; Liso et al., 1997; Philips, 1991). The presence of residual monomers in polymers is important because it can affect the quality and application of polymers in medicine, pharmacy and dentistry. In papers (Kostić et al., 2009, 2011), the effect of the residual methyl methacrylate (MMA) monomer reduction on the dental materials quality-based on PMMA was examined. The improvement of mechanical properties of dental materials was investigated by introducing itaconate in a polymerization of MMA (Spasojević et al., 2011). Džunuzović et al. (2010) were synthesized and characterized methyl methacrylate nanocomposites obtained by *in situ* polymerization via free radical, using titanium dioxide as inorganic filler. One of the most important materials is a copolymer of methyl methacrylate and methacrylic acid known as Eudragit® (Kakehi et al., 2000; Porsch et al., 2000; Ristić et al., 2009). In *in vivo* tests in animals for the treatment of inflammatory bowel diseases, Eudragit® microspheres with ellagic acid were used. They allow delivery and release of active substance in the caecum and ascending colon (Jeong et al., 2001).

The macroporous polymers are promising materials for different purposes. Polymers with the open pores, may be used as adsorbents for the preparation of ion exchange resins or as the inert carriers for the catalytic particles, enzyme or microorganism cells. Numerous methods for the synthesis of the low cross-linked polymeric material involving the initial prepolymerization process in the emulsion with high water phase content are described in the patent literature (Adamski et al., 1999; Barby and Haq, 1985; Bass and Brownscombe, 1993; Beshouri, 1993; Brownscombe et al., 1993; DesMarais et al., 1994; Edwards et al., 1988; Gregory et al., 1989; Haq, 1985; Mork et al., 2000; Young et al., 1994). Vinyl monomers with one double bond were used as the starting compounds for polymerization. During the polymerization, in case of monomers with two or more double bonds, one double bond is used for formation of main polymer chains and other ones are used for crosslinking between the polymer chains. Using described method, insoluble cross-linked polymers can be obtained. The polymer swelling capacity, elasticity or solid properties depend on the crosslink density and the monomer structures. The final step is the rinsing of resulting polymer, the opening of the pores, and drying. Macroporous copolymer-based on methyl methacrylate obtained by suspension copolymerization with glycidyl methacrylate could be used for the selective sorption of the metal ions (Jovanović et al., 1994, 1999, 2000). The synthesis by the sol-gel polymerization (in CO_2 at supercritical conditions) is well-known, wherein the methyl methacrylate provides the hardness of the final copolymer (Cooper et al., 2000; Cooper, 2000). The possible applications of porous poly(methyl methacrylate) as implant in the cardiovascular surgery (MacGregor, 1986) and as enzyme carrier were described (Russell and Beckman, 1996).

The macroporous, cross-linked copolymer of methyl methacrylate and acrylamide, with a free volume of the open pores from 50 to 90%, which is insoluble and non-swellable in organic solvents were synthesized (Nikolić et al., 2002, 2004; Nikolić, 2003). Obtained macroporous copolymer of poly(methyl methacrylate-*co*-acrylamide), PMMA-AA, has a porous structure in the solid state, in contrast to the ordinary cross-linked copolymers which become porous only after swelling in a solvent. The synthesis comprises the emulsion pre-polymerization and then the copolymerization sol-gel process. The final cross-linking by methylol derivatives of melamine or glycoluril was carried out. If in the synthesis the pores remain closed, foamed material is obtained, which may be used as a light construction material or as a good insulator. If the obtained copolymer has open pores, it could be used as the starting material for the preparation of ion exchange resins, as adsorbent, or the inert carrier of the catalyst particles, and particularly suitable for the immobilization of microorganisms or enzymes within the pores. The advantage of this method, compared to the existing methods for synthesis of porous cross-linked polymers with a low density, is the crosslinking between the polymer chains, which is carried out by chemical condensation of amido, amino, and hydroxyl groups at high temperatures. Cross-linked porous copolymer PMMA-AA was applied to immobilize *Saccharomyces cerevisiae* cells and then used for alcoholic fermentation in a bioreactor with vibrating stirrer. The results showed that the effect of axial mixing has a positive impact on the substrate conversion degree and the product yields (Nikolić, 2003). Scanning electron micrographs of the cross-linked porous copolymer and the cross-linked porous copolymers with immobilized cells of *Saccharomyces cerevisiae* are shown in Figures 4.4a and 4.4b, respectively (Nikolić, 2003; Nikolić et al., 2004).

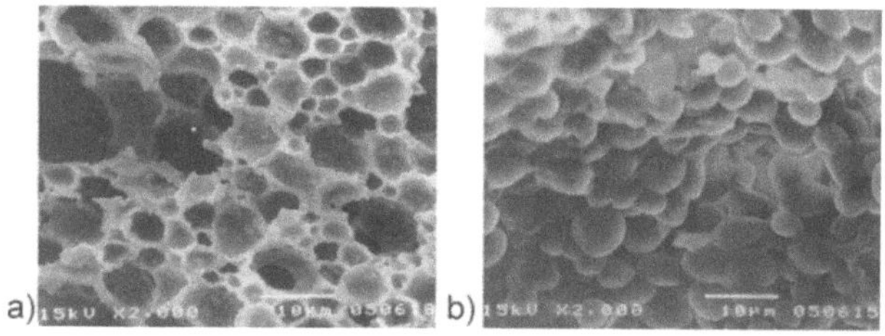

FIGURE 4.4 (a) SEM micrograph of cross-linked porous copolymer PMMA-AA, magnified x 2000, bar = 10 μm, (b) SEM micrograph of cross-linked porous copolymer PMMA-AA with immobilized yeast *Saccharomyces cerevisiae* cells, magnified x 2000, bar = 10 μm (From Nikolić et al., *J. Appl. Polym. Sci. 91*, 387, 2004a. With permission from John Wiley.)

In the paper of Todorovic et al. (2012) the synthesis of porous PMMA by suspension polymerization and the synthesis of porous PMMA-AA copolymer by the emulsion polymerization were carried out in the presence of inert compounds. The morphology of the formed pores in PMMA and PMMA-AA was determined by SEM microscopy (Figures 4.5a and 4.5b, respectively). It was observed that the copolymer has a higher porosity. The content of residual monomers in PMMA

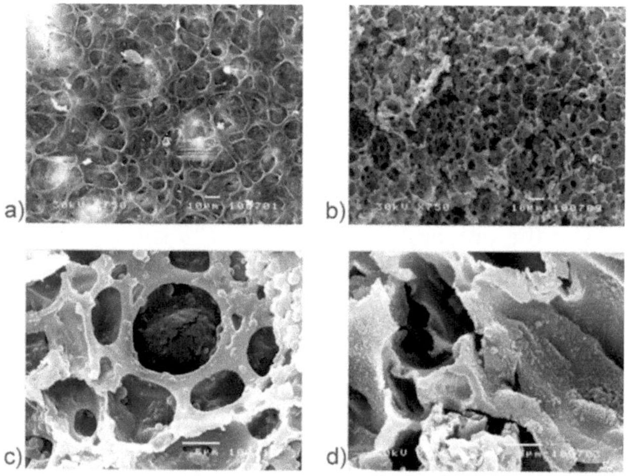

FIGURE 4.5 SEM micrographs of: (a) PMMA, magnification x750, bar = 10 μm, (b) PMMA-AA; magnification x 750 magnification, bar = 10 μm, (c) PMMA with the immobilized enzyme, magnification x 3500, bar = 5 μm, (d) PMMA-AA with the immobilized enzyme, magnification x 3500, bar = 5 μm (From Todorović et al., *Advanced Technol. 1*(2), *11*, 2012. With permission from Advanced Technologies.)

and PMMA-AA were minimized after boiling in the water. The polymers were used for the immobilization of enzymes, and it is observed that on the PMMA surface was less immobilized α-amylase than on the PMMA-AA surface. The morphologies of PMMA and PMMA-AA after immobilization of the enzyme α-amylase were shown in Figures 4.5c and 4.5d, respectively.

4.3.2 POLY(2-HYDROXYETHYL METHACRYLATE)

Poly(2-hydroxyethyl methacrylate) is a biostable polymer, which is able to form a hydrogel and with a broad potential application. Hydrogel-based on 2-hydroxyethyl methacrylate with ethylene dimethacrylate was first synthesized in 1956 and has been applied in ophthalmology (Wichterle and Lim, 1961). PHEMA hydrogels are important drug carriers in cancer and neurological disorders treatments (Coelho et al., 2010). Copolymers of hydroxyethyl methacrylate, styrene and N-isopropylacrylamide were used in the systems for the drug and enzymes delivery (Samra et al., 2003). Hydrogel-based on hydroxyethyl methacrylate and ethylene glycol was applied in rhinoplasty (Voldřich et al., 1975).

4.3.3 POLY(ACRYLIC ACID)

Poly(acrylic acid) is transparent, inexpensive, safe and easy to use anionic polymer. On the markets, there are a number of gels-based on PAA, which differ in molecular weight, types and degree of crosslinking of the polymer. For the preparation of gel 0.5–1% solution of PAA is used. PAA hydrogels are applied as biomaterials in the field of biosensors and activator (Kim et al., 2005). There is a possibility of use in dermatology such as adhesives (Onuki et al., 2008), drug carriers, protective surface equipment in dentistry, for the synthesis of materials used for soft contact lenses producing (Craig and Tamburic, 1997). Carbopol® Ultrez 10 is easily dispersing acrylic acid polymer. Similar to the other Carbomers, it additionally stabilizes the composition affecting the gel consistency. Anti-irritant potential and antimicrobial activity of the synthesized allicin and its transforments (ajoenes and vinyldithiins) incorporated in Carbopol®

Ultrez 10 were studied by Savic et al. (2011). The examined activities of incorporated agents were preserved, wherein transforments showed stronger activity compared to allicin.

4.3.4 POLY(N-ISOPROPYLACRYLAMIDE)

Poly(*N*-isopropylacrylamide) is the temperature-sensitive polymer, which possess hydrophilic and hydrophobic groups. The lower critical solution temperature (LCST) at which PNIPAM passing through the phase transition is 32°C. At temperatures below the LCST, PNIPAM manifests hydrophilic nature, establishing the hydrogen bonds between water molecules and *hydrogel* side groups, and gel swells. Above LCST, PNIPAM's become more hydrophobic because hydrogen bonds with water molecules break up, the polymer chains eliminate water and polymer network collapse. For the hydrogel synthesis, the copolymerization with certain amount of anionic monomer – acrylic, methacrylic, maleic or itaconic acid, are performed, in order to obtain pH-sensitive hydrogels (Zhao et al., 2006). By varying the copolymer composition is possible to modify the LCST value to physiological temperature (36–38°C), which is essential for pharmaceutical application and controlled drug release (Alexander et al., 2014; Andersson et al., 1997; Geever et al., 2006). Potential applications of poly(*N*-isopropylacrylamide-co-2-hydroxypropyl methacrylate) hydrogels, p(NIPAM-HPMet), as carriers for modified/controlled drug release depending on temperature change were investigated in vitro (Ilić-Stojanović et al., 2011, 2012a, 2012b, 2013, 2014a, 2014b; Ilić-Stojanović, 2013; Nizam El-Din, 2011). Caffeine, paracetamol, ibuprofen, phenacetin, piroxicam and naproxen, as model drugs, were incorporated in p(NIPAM-HPMet) hydrogels and drug release from hydrogels were investigated by suitable analytical methods.

4.3.5 POLY(ACRYLAMIDE)

Poly(acrylamide) is obtained by polymerization of acrylamide which forms linear chains or cross-linked structures, usually by using *N,N'*-methylenebisacrylamide as a crosslinker. PAM absorbs a large amount of water creating a soft gel which is used for the manufacture of soft contact lenses. In the linear form, PAM is used as a thickening or suspending agent. Over 20 years PAM hydrogels have been used extensively to correct defects and deformities of soft tissue as well as in posttraumatic facial reconstruction (Pinchuk et al., 1996). Many studies have shown that the PAM is nontoxic, without allergic, teratogenic, embryotoxic and mutagenic effects, and it is practically non-biodegradable (McCollister et al., 1964; Stephen, 1991). Implants made of poly(acrylamide) gels do not undergo the macroscopic structural changing after 18 months (Niechajev, 2000). Products-based on poly(acrylamide) are new generation of fillers for soft tissue with 2.5% of poly(acrylamide) and 97.5% water, and they are approved as a medical device for an esthetic facial correction and medical indications (de Cássia Novaes and Berg, 2003). The controlled-release of drugs may be achieved by incorporation of drugs into the PAA microspheres, which are pH- and thermo-sensitive (Fang and Kawaguchi, 2002; Kim et al., 1994, 2001).

4.3.6 POLOXAMERS

Poloxamers, PEO-PPO-PEO triblock copolymers, belong to the group of symmetrical, amphiphilic, nonionic, macromolecular surface active agents. They are built of hydrophobic polypropylene oxide (PPO) central part and external hydrophilic polyethylene oxide (PEO) part, with general molecular formula $HO(C_2H_4O)_a(C_3H_6O)_b(C_2H_4O)_a$ and the structure shown in Figure 4.3e (Pepić, 2005; Patel et al., 2009). All of poloxamers have a similar chemical structure, but different contents and molecular weight of hydrophilic PEO (a) and hydrophobic PPO (b) blocks. Poloxamers represent the auxiliary substances in the preparation of pharmaceutical products. In the poloxamers nomenclature exists simple rules, so the name (generic or trade) shows the main properties of molecular materials. After the generic name of poloxamer follows a three-digit number; the first two digits multiplied by 100 gives the approximate relative molecular weight of PPO part, while the third digit multiplied by 10 corresponds to the mass percentage of PEO in the molecule. The tradename includes the letters that indicate the aggregate state of the copolymer: L (liquid), P (paste),

F (flakes-solid) and two-digit or three-digit number in which the last digit indicates the average mass fraction PEO in percentages. The remaining digit or first two provide an average molecular weight of the central part of the PPO expressed in Da after multiplication by 300. The ratio of hydrophilic and lipophilic units in the copolymer molecule represents a hydrophilic-lipophilic balance (HLB) (Pepić, 2005). PEO-PPO-PEO block copolymers with different number of hydrophilic and hydrophobic units possess different HLB value.

4.3.6.1 Micelles of Poloxamer

Due to the amphiphilic character, micelles of poloxamer possess surfactant properties, including the ability to interact with the hydrophobic surfaces and biological membranes. The micelle core consists of hydrophobic PPO blocks, which are separated from the aqueous medium by a shell of hydrophilic PEO chains. Various therapeutic and diagnostic agents are incorporated in PPO core. The shell of PEO chain ensures that micelles remain in the dispersed state and reduces the undesirable interactions of drug with cells and proteins (Batrakova and Kabanov, 2008). The dissolved drugs can modify aggregation and micelle stability. A small amount of *o*-xylene increases the tendency of ambiphiles to form aggregates, which applies to the PEO-PPO-PEO triblock copolymers, while urea increases and the phenol decreases CMC value due to the interaction with the PEO chains (Jiang et al., 2001). Naproxen and indomethacin does not affecting CMC, but lead to a slight reduction in the micelles size and reduction of the aggregation number (Sharma and Bhatia, 2004). Short-chain alcohols, e.g., ethanol, prevent the formation of micelles in the water, while the hydrophobic aliphatic alcohols, such as butanol, favor the aggregation (Pandya et al., 1993; Armstrong et al., 1996). The temperature required for the beginning of the micelles formation reduces by increase of the polymer concentration (Alexandridis and Hatton, 1995; Alexandridis et al., 1995).

In preparation of Pluronic P123 micelles in different HCl solutions, the value of the critical micellar temperature increases with acid addition (Yang et al., 2006). This is probably consequence of an increased interaction between the alkyl group and the protonated water molecules.

The maximum solubilization of amphiphilic molecules depends on the ability to form micelles (Dong et al., 2004). Generally, solubilization increases the diameter of the micelles core and reduces the CMC. Micelle diameter increases due to incorporation of molecules and increase the aggregation number. Physical and chemical micelles stabilization performs to solve a problem of aggregates dissociation at low concentrations and inability to retain a hydrophobic drug in the core. Stability is critical when the derivative is more hydrophilic (Chiappetta and Sosnik, 2007). Several methods of stabilization the poloxamer micelles were examinated. Directly radically cross-linking of micelles core and reactive monomer (e.g., styrene) reduces the capacity of micelles due to a high-density core. A small concentration of vegetable oil added in diluted poloxamer micelle solution reduces degradation due to dilution, but does not affect the charging capacity. The increasing stability of PEO-PPO-PEO block copolymer micelles was achieved by adding a second nonionic surfactant. A new strategy for the micelles stabilization is achieved by cross-linking micelle layers, thereby increasing stability and permeability (Rodriguez-Hernandez et al., 2005).

By incorporating hydrophobic drugs into micelles, their solubility increases, drug decomposition (hydrolytic, metabolic) in the pharmaceutical form and/or in the body slows, sustained and/or controlled-release achieves, drug bioavailability and the pharmacokinetic properties improves. (Pepić, 2005). By incorporating 9-nitro-camptothecin lactone in the poloxamer micelles solubility was greater 13 times for P123, 8 times for F127, and 5 times for F68, with improved stability (Pepić, 2005; Saha et al., 2013). Solubility of ziprasidone increased 48 times in a 5% aqueous solution of poloxamer 188 at 25°C, compared to the pure drug. The dissolution rate of ziprasidone solid dispersion is higher than the pure drug, wherein the best ratio for preparation drug-carrier dispersion is 1:2 (Prasanthi et al., 2011).

4.3.6.2 The Poloxamer Gels

PEO-PPO-PEO solutions with high concentration (~20 wt% of poloxamer 407) reversibly increase the viscosity at room, or physiological temperatures, compared to lower temperatures (~4°C), which is explained by thermoreversible gellation. The phase transition of the micellar solution in the gel is an important advantage of PEO-PPO-PEO copolymers in different applications. Aqueous solution of P407 with concentration greater than 20% has a lower phase transition temperature (<25°C) and becomes a gel at room temperature, which limited the drug delivery application. In order to develop thermosensitive gel with phase transition at higher temperature in the solution P407 has been incorporated poloxamer P188. Carbopol 940 (C940) was added to increase firmness of thermosensitive P408/P188 gel. The obtained composite thermosensitive hydrogel-based on P407, P188 and C940 is a convenient system for the transdermal delivery of active substances (Chen et al., 2013).

Polymers used for drug delivery are considered to be biologically inert materials that protect the drug from degradation, extend drug exposure to the tissue and allow drug transport to the cell. However, today it is increasingly evident that synthetic polymers may dramatically change the specific cellular response. Besides its ability to build micelle, Pluronic block copolymers exhibit the ability to modify cell responses which can lead to increased sensitivity of cancer cells to the drugs and increased drug transport through the cell membrane (Batrakova and Kabanov, 2008). Poloxamer 188 reduces inflammation and tissue damage after injury. It is believed that protective mechanism of this polymer includes a surfactant effect on the oxidative stress and inflammatory responses (Hartikka et al., 2001). In vitro studies have demonstrated that poloxamer 407 increases the transfection efficiency of the adenovirus vectors in vascular smooth muscle cells (Gilbert et al., 1998). Poloxamer decomposition time *in vitro* after polymerization is 1–20 h, depending on the concentration. Poloxamer applications *in vivo*, in 22% cases, lead to complete occlusion, then a complete arteries cleavage after 10–90 min, without any complications (Block et al., 1995). Poloxamer 407 is used for making intraperitoneal barrier for preventing the formation of postsurgical adhesions in animal models. It is suitable for application in laparoscopic surgery due to reversible gelation properties at body temperature.

4.4 CONCLUSIONS

A large number of studies about biodegradable and non-biodegradable polymerss in drug delivery systems was considered. Design of drug delivery systems is essential for the development of modern pharmaceutical matrix systems with new or existing commercial drugs. There were numerous attempts to include the new technological approaches of different polymer design into practice and to determine the efficiency of their application in drug delivery. Future applications of intelligent polymers and matrix systems in drug delivery can provide the development of innovative materials and technologies which will enable to overcome the disadvantages of conventional therapy.

ACKNOWLEDGMENT

Financial support provided by the Ministry of Education, Science and Technological Development, the Republic of Serbia (Projects No. TR 34012 and III 45022) is gratefully acknowledged.

KEYWORDS

- **biodegradable polymers**
- **matrix systems**
- **nanoparticulates**
- **natural polymers**
- **non-biodegradable polymers**
- **synthetic polymers**

REFERENCES

Adamski, R. P.; Beshouri, S. M.; Chamupathi, V. G. Process to prepare low density porous cross-linked polymeric materials. U.S. Patent 5,912,276, Jun 15, 1999.

Ahmad, Z.; Pandey, R.; Sharma, S.; Khuller, G. K. Alginate nanoparticles as antituberculosis drug carriers: formulation development, pharmacokinetics and therapeutic potential. *Indian J. Chest Dis. Allied Sci.* **2006**, *48*(3), 171–176.

Alexander, A.; Khan, J.; Saraf, S.; Saraf, S. Polyethylene glycol (PEG)-Poly (*N*-isopropylacrylamide) (PNIPAAm) based thermosensitive injectable hydrogels for biomedical applications. *Eur. J. Pharm. Biopharm.* **2014**, *88*(3), 575–585.

Alexandridis, P.; Hatton, T. A. Poly(ethylene oxide)-poly(propylene oxide)-poly(ethylene oxide) block copolymer surfactants in aqueous solutions and at interfaces: thermodynamics, structure, dynamics, and modeling. *Colloids Surf. A*. **1995**, *96*(1), 1–46.

Alexandridis, P.; Nivaggioli, T.; Hatton, T. A. Temperature effects on structural properties of pluronic P104 and F108 PEO-PPO-PEO block copolymer solutions. *Langmuir* **1995**, *11*(5), 1468–1476.

Amecke, B.; Bendix, D.; Entenmann, G. Synthetic Resorbable Polymers Based on Glycolide, Lactides and Similar Monomers. In: *Encyclopedic Handbook of Biomaterials and Bioengineering, Part A: Materials;* Wise, L. D.; Trantolo, J. D.; Altobelli, E. D.; Yaszernski, J. M.; Gresser, D. J.; Schwarts, R. E.; Eds.; Marcel Dekker, Inc., NY, 1995; Vol. *2*(5), p 977.

Andersson M.; Axelsson A.; Zacchi G. Diffusion of glucose and insulin in a swelling N-isopropylacrylamide gel. *Int. J. Pharm*. **1997**, *157*(2), 199–208.

Armstrong, J.; Chowdhry, B.; Mitchell, J.; Beezer, A.; Leharne, S. Effect of cosolvents and cosolutes upon aggregation transitions in aqueous solutions of the poloxamer F87 (poloxamer P237): a high sensitivity differential scanning calorimetry study. *J. Phys. Chem*. **1996**, *100*(5), 1738–1745.

Barby, D.; Haq, Z. Low density porous cross-linked polymeric materials and their preparation and use as carriers for included liquids. U.S. Patent 4,522,953, Jun 11, 1985.

Bass, R. M.; Brownscombe, T. F. Process for preparing low density porous cross-linked polymeric materials. U.S. Patent 5,210,104, May 11, 1993.

Batrakova, E. V.; Kabanov, A. V. Pluronic block copolymers: evolution of drug delivery concept from inert nanocarriers to biological response modifiers. *J. Control. Release* **2008**, *130*(2), 98–106.

Berkland, C.; King, M.; Cox, A.; Kim, K.; Pack, D. W. Precise control of PLG microsphere size provides enhanced control of drug release rate. *J. Control. Release* **2002**, *82*, 137–147.

Beshouri, S. M. Process for preparing low density porous cross-linked polymeric materials. U.S. Patent 5,200,433, Apr 6, 1993.

Block, T. A.; Aarsvold, J. N.; Matthews, K. L.; Mintzer, R. A.; River, L. P.; Capelli-Schellpfeffer, M.; Wollmann, R.; Tripathi, S.; Chen, C. T.; Lee, R. C. The 1995 Lindberg Award. Nonthermally mediated muscle injury and necrosis in electrical trauma. *J. Burn Care Rehabil*. **1995**, *16*(6), 581–588.

Borchard, G.; Audus, K. L.; Shi, F.; Kreuter, J. Uptake of surfactant-coated poly(methyl methacrylate)- nanoparticles by bovine brain microvessel endothelial cell monolayers. *Int. J. Pharm*. **1994**, *110*, 29–35.

Brannon-Peppas, L. Polymers in Controlled Drug Delivery. *Med. Plast. Biomater*. **1997**, *4*, 34–44.

Brownscombe, T. F.; Gergen, W. P.; Bass, R. M.; Mores, M.; Wong, P. K. Process for preparing low density porous cross-linked polymeric materials. EP 0642535 A1, Mar 15, 1995.

Budhian, A.; Siegel, S. J.; Winey, K. I. Production of haloperidol-loaded PLGA nanoparticles for extended controlled drug release of haloperidol. *J. Microencapsul*. **2005**, *22*(7), 773–785.

Calvo, P.; Gouritin, B.; Brigger, I.; Lasmezas, C.; Deslys, J. P.; Williams, A.; Andreux, J. P.; Dormont, D.; Couvreur, P. PEGylated polycyanoacrylate nanoparticles as vector for drug delivery in prion diseases. *J. Neurosci. Methods* **2001**, *111*(2), 151–155.

Chen, J.; Zhou, R.; Li, L.; Li, B.; Zhang, X.; Su, J. Mechanical, rheological and release behaviors of a poloxamer 407/poloxamer 188/carbopol 940 thermosensitive composite hydrogel. *Molecules* **2013**, *18*(10), 12415–12425.

Chen, W.; He, J.; Olson, J. J.; Lu, D. R. Carboplatin-loaded PLGA microspheres for intracerebral implantation: in vivo characterization. *Drug Deliv.* **1997**, *4*(4), 301–311.

Cheng, Q.; Feng, J.; Chen, J.; Zhu, X, Li, F. Brain transport of neurotoxin-I with PLA nanoparticles through intranasal administration in rats: a microdialysis study. *Biopharm. Drug Dispos.* **2008**, *29*(8), 431–439.

Chiappetta, D. A.; Sosnik, A. Poly(ethylene oxide)-poly(propylene oxide) block copolymer micelles as drug delivery agents: Improved hydrosolubility, stability and bioavailability of drugs. *Eur J. Pharm. Biopharm.* **2007**, *66*(3), 303–317.

Coelho, J. F.; Ferreira, P. C.; Alves, P.; Cordeiro, R.; Fonseca, A. C.; Góis, J. R.; Gil, M. H. Drug delivery systems: Advanced technologies potentially applicable in personalized treatments. *EPMA J.* **2010**, *1*(1), 164–209.

Coester, C.; Kreuter, J.; von Briesen, H.; Langer, K. Preparation of avidin-labeled gelatin nanoparticles as carriers for biotinylated peptide nucleic acid (PNA). *Int. J. Pharm.* **2000**, *196*(2), 147–149.

Cooper, A. I. Polymer synthesis and processing using supercritical carbon dioxide. *J. Mater. Chem.* **2000**, *10*(2), 207–234.

Cooper, A. I.; Wood, C. D.; Holmes, A. B. Synthesis of well-defined macroporous polymer monoliths by sol-gel polymerization in supercritical CO_2. *Ind. Eng. Chem. Res.* **2000**, *39*(12), 4741–4744.

Craig, D. Q. M.; Tamburic, S. Dielectric analysis of bioadhesive gel systems. *Eur. J. Pharm. Biopharm.* **1997**, *44*(1), 61–77.

Craig, R. G.; Ward, M. L. *Restorative Dental Materials*, 10th ed., Mosby, St. Louis, MO, 1997.

Damge, C.; Maincent, P.; Ubrich, N. Oral delivery of insulin associated to polymeric nanoparticles in diabetic rats. *J. Control. Release* **2007**, *117*(2), 163–170.

Date, A. A.; Joshi, M. D.; Patravale, V. B. Parasitic diseases: Liposomes and polymeric nanoparticles versus lipid nanoparticles. *Adv. Drug Deliv. Rev.* **2007**, *59*(6), 505–521.

De Campos, A. M.; Sanchez, A.; Alonso, M. J. Chitosan nanoparticles: a new vehicle for the improvement of the delivery of drugs to the ocular surface. Application to cyclosporin A. *Int. J. Pharm.* **2001**, *224*(1–2), 159–168.

de Cássia Novaes, W.; Berg, A. Experiences with a new nonbiodegradable hydrogel (Aquamid): A pilot study. *Aesth. Plast. Surg.* **2003**, *27*(5), 376–380.

DesMarais, T. A.; Stone, K. J.; Thompson, H. A.; Young, G. A.; LaVon, G. D.; Dyer, J. C. Absorbent foam materials for aqueous body fluids and absorbent articles containing such materials. U.S. Patent 5,331,015, Jul 19, 1994.

Dong, C. M.; Guo, Y. Z.; Qiu, K. Y.; Gu, Z. W.; Feng, X. D. In vitro degradation and controlled release behavior of D, L-PLGA50 and PCL-b-D, L-PLGA50 copolymer microspheres. *J. Control. Release* **2005**, *107*(1), 53–64.

Dong, J.; Chowdhry, B. Z.; Leharne, S. A. Solubilization of polyaromatic hydrocarbons in aqueous solutions of poloxamine T803. *Colloid. Surface. A* **2004**, *246*(1), 91–98.

Džunuzović, E. S.; Marinović-Cincović, M. T.; Džunuzović, J. V.; Jeremić, K. B.; Nedeljković, J. M. Influence of the way of synthesis of poly(methyl methacrylate) in the presence of surface modified TiO_2 nanoparticles on the properties of obtained nanocomposites. *Hem. Ind.* **2010**, *64*(6), 473–489.

Edwards, C. J. C.; Gregory, D. P.; Sharples, M. Low density porous elastic cross-linked polymeric materials and their preparation. U.S. Patent 4,788,225, Nov 29, 1988.

Espuelas, M. S.; Legrand, P.; Loiseau, P. M.; Bories, C.; Barratt, G.; Irache, J, M. In Vitro Antileishmanial Activity of Amphotericin B Loaded in Poly(epsilon-Caprolactone) Nanospheres. *J. Drug Target.* **2002**, *10*(8), 593–599.

Fang, S. J.; Kawaguchi, H. A thermosensitive amphoteric microsphere and its potential application as a biological carrier. *Colloid Polym. Sci.* **2002**, *280*(11), 984–989.

Fassas, A.; Buffels, R.; Kaloyannidis, P.; Anagnostopoulos, A. Safety of high-dose liposomal daunorubicin (daunoxome) for refractory or relapsed acute myeloblastic leukemia. *Br. J. Haematol.* **2003**, *122*(1), 161–163.

Feng, S. S. Nanoparticles of biodegradable polymers for new-concept chemotherapy. *Expert Rev. Med. Devices* **2004**, *1*(1), 115–125.

Freiberg, S.; Zhu, X. X. Polymer microspheres for controlled drug release. *Int. J. Pharm.* **2004**, *282*(1–2), 1–18.

Gander, B.; Johansen, P.; Hô Nam-Trân, H.; Merkle, P. Thermodynamic approach to protein microencapsulation into poly(D, L-lactide) by spray drying. *Int. J. Pharm.* **1996**, *129*(1–2), 51–61.

Geever, L. M.; Devine, D. M.; Nugent, M. J. D.; Kennedy, J. E.; Lyons, J. G.; Hanley, A.; Higginbotham, C. L. Lower critical solution temperature control andswelling behavior of physically cross-linked thermosensitive copolymers-basedon *N*-isopropylacrylamide. *Eur. Polym. J.* **2006**, *42*, 2540–2548.

Ghaderi, R.; Struesson, C.; Carlfors, J. Effect of preparative parameters on the characteristics of poly(d, l-lactidecoglocolide) microspheres made by the double emulsion method. *Int. J. Pharm.* **1996**, *141*(1–2), 205–216.

Gilbert, P.; Jones, M. V.; Allison, D. G.; Heys, S.; Maira, T.; Wood, P. The use of poloxamer hydrogels for the assessment of biofilm susceptibility towards biocide treatments. *J Appl Microbiol.* **1998**, *85*(6), 985–990.

Gilding, D. K.; Reed, A. M. Biodegradable polymers for use in surgery- pol(ethylene oxide) poly(ethylene-terephthalate) (PEO-PET) copolymers. *Polymer* **1979**, *20*, 1459–1464.

Gomez-Gaete, C.; Tsapis, N.; Besnard, M.; Bochot, A.; Fattal, E. Encapsulation of dexamethasone into biodegradable polymeric nanoparticles. *Int. J. Pharm.* **2007**, *331*(2), 153–159.

Gregory, D. P.; Sharples, M.; Tucker, I. M.; Porous material, EP 0299762 A3, Jun 7, 1989.

Hans, M. L.; Lowman, A. M. Biodegradable nanoparticles for drug delivery and targeting. *Curr. Opin. Solid State Mater. Sci.* **2002**, *6*(4), 319–327.

Haq, Z. Porous cross-linked absorbent polymeric materials. U.S. Patent 4,536,521, Aug 20, 1985.

Hartikka, J.; Sukhu, L.; Buchner, C.; Hazard, D.; Bozoukova, V.; Margalith, M.; Nishioka, W. K.; Wheeler, C. J.; Manthorp, M.; Sawdey, M. Electroporation-facilitated delivery of plasmid DNA in skeletal muscle: plasmid dependence of muscle damage and effect of poloxamer 188. *Mol. Ther.* **2001**, *4*, 407–415.

Ilić-Stojanović, S. PhD. Dissertation, Faculty of Technology, University of Nis, **2013**.

Ilić-Stojanović, S. S.; Nikolić, Lj.B.; Nikolić, V. D.; Milić, J. R.; Stamenković, J.; Nikolić, G. M.; Petrović, S. D. Synthesis and characterization of thermosensitive hydrogels and the investigation of modified release of ibuprofen. *Hem. Ind.* **2013**, *67*(6), 901–912.

Ilić-Stojanović, S. Synthesis and characterization of negatively thermosensitive hydrogels, *LAP Lambert Academic Publishing*, Akademikeverlag GmbH & Co. KG: Saarbrücken, Germany, DE, **2015**.

Ilić-Stojanović, S.; Mladenović-Ranisavljević, I.; Nikolić, V.; Takić, Lj.; Stojiljković, D.; Stojiljković, S.; Nikolić, Lj. Thermo-responsive hydrogels for controlled release of paracetamol, 43th International October Conference Proceedings (Kladovo) Serbia, October 12–15, 2011, Marković, D.; Živković, D.; Nestorović S. Eds.; University of Belgrade – Technical Faculty in Bor: Bor, 707–710.

Ilić-Stojanović, S.; Nikolić Lj.; Nikolić V.; Ristić I.; Budinski-Simendić J.; Kapor A.; Nikolić G. The structure characterization of thermosensitive poly(*N*-isopropylacrylamide-*co*-2-hydroxypropyl methacrylate) hydrogel. *Polym. Int.* **2014a**, *63*, 973–981.

Ilić-Stojanović, S.; Nikolić, L. J.; Nikolić, V.; Milić, J.; Petrović, S. D.; Nikolić, G. M.; Kapor, A. Potential application of thermo-sensitive hydrogels for controlled release of phenacetin. *Hem. Ind.* **2012a**, *66*(6), 831–839.

Ilić-Stojanović, S.; Nikolić, L. J.; Nikolić, V.; Petrović, S.; Stanković, M. Process for synthesis of thermosensitive hydrogels and pharmaceutical applications. RS 53220 B, Aug 28, 2014b.

Ilić-Stojanović, S.; Nikolić, L. J.; Nikolić, V.; Stanković, M.; Stamenković, J.; Mladenović-Ranisavljević, I.; Petrović, S. D. Influence of monomer and crosslinker molar ratio on the swelling behavior of thermosensitive hydrogels. *CI&CEQ*. **2012b**, *18*(1), 1–9.

Ilić, D.; Ristić, S. I.; Nikolić, L. J.; Stanković. M.; Nikolić, G.; Stanojević, L. J.; Nikolić, V. Characterization and release kinetics of allylthiosufinate and its transforments from poly(d, l-lactide) microspheres. *J. Environ. Eng.* **2012**, *20*(1), 80–87.

Jeong, Y. I.; Prasad, Yv, R. P.; Ohno, T.; Yoshikawa, Y.; Shibata, N.; Kato, S.; Takeuchi, K.; Takada, K. Application of Eudragit P-4135F for the delivery of ellagic acid to the rat lower small intestine. *J. Pharm. Pharmacol.* **2001**, *53*(8), 1079–1085.

Jiang, L. Q.; Zheng, Y. Y.; Zhao, J. X. Effect of phenol on micellization of Pluronic block copolymer F127 and solubilization of anthracene in the micelle. *Fine Chemicals*. **2001**, *18*(12), 731–735.

Jovanović, S. M.; Nastasović, A.; Jovanović, N. N.; Jeremić, K.; Savić, Z. The influence of inert component composition on the porous structure of glycidyl methacrylate/ethylene glycol dimethacrylate copolymers. *Angew. Makromol. Chem.* **1994**, *219*(1), 161–168.

Jovanovic, S. M.; Nastasovic, A.; Jovanovic, N. N.; Novakovic, T.; Vukovic, Z.; Jeremic, K. Synthesis, properties and applications of cross-linked macroporous copolymers-based on methacrylates. *Hem. Ind.* **2000**, *54*(11), 471–479.

Jovanovic, S.; Nastasovic, A.; Novakovic, T.; Jovanovic, N. Influence of the inert component on the porosity parameters of GMA-*co*EGDMA. *Hem. Ind.* **1999**, *53*(11), 372–376.

Kakehi, T.; Yamashita, M.; Yasuda, H. Syntheses and adhesion properties of novel syndiotactic block copolymers of alkyl(meth)acrylate with methacrylic acid and its analogs. *React. Funct. Polym.* **2000**, *46*(1), 81–94.

Kakish, H. F.; Tashtoush, B.; Ibrahim, H. G.; Najib, N. M. A novel approach for the preparation of highly loaded polymeric controlled release dosage forms of diltiazem HCl and diclofenac sodium. *Eur. J. Pharm. Biopharm.* **2002**, *54*(1), 75–81.

Kim, E. J.; Cho, S. H.; Yuk, S. H. Polymeric microspheres composed of pH/temperature-sensitive polymer complex. *Biomaterials* **2001**, *22*(18), 2495–2499.

Kim, H.; Park, S. J.; Kim, S. I.; Kim, S. J. Electroactive polymer hydrogels composed of polyacrylic acid and poly(vinyl sulfonic acid) copolymer for application of biomaterial. *Synth. Met.* **2005**, *155*(3), 674–676.

Kim, S. Y.; Lee, Y. M. Taxol-loaded block copolymer nanospheres composed of methoxy poly(ethylene glycol) and poly(epsilon-caprolactone) as novel anticancer drug carriers. *Biomaterials* **2001**, *22*(13), 1697–1704.

Kim, Y. H.; Bae, Y. H.; Kim, S. W. pH/temperature sensitive polymers DOE macromolecular drug loading and release. *J. Control. Release* **1994**, *28*(1–3), 143–152.

Kostić, M.; Krunić, N.; Nikolić, L. J.; Nikolić, V.; Najman, S.; Kocić, J. Residual monomer content determination in some acrylic denture base materials and possibilities of its reduction. *Vojnosanit. pregl.* **2009**, *66*(3), 223–227.

Kostić, M.; Krunić, N.; Nikolić, L. J.; Nikolić, V.; Najman, S.; Kostić, I.; Rajković, J.; Manić, M.; Petković, D. Influence of residual monomer reduction on acrylic denture base resins quality. *Hem. Ind.* **2011**, *65*(2), 171–177.

Kreutera, J.; Petrov, V. E.; Kharkevich, D. A.; Alyautdin, R. N. Influence of the type of surfactant on the analgesic effects induced by the peptide dalargin after its delivery across the blood–brain barrier using surfactant-coated nanoparticles. *J. Control. Release* **1997**, *49*(1), 81–87.

Leenslag, J. W.; Penning, A. J.; Bos, R. M.; Rozema, F. R.; Boering, G. Resorbable materials of poly(L-lactide). *Biomaterials* **1987**, *8*, 311.

Liso, P. A.; Gaquez, B. V. Rebuelta, M.; Hernaez, M. L. Analysis of the leaching and toxicity of new amine activators for the curing of acrylic bone cements and composites. *Biomater.* **1997**, *18*, 15–20.

Lu, Z.; Yeh, T. K.; Tsai, M.; Au, J. L.; Wientjes, M. G. Paclitaxel-loaded gelatin nanoparticles for intravesical bladder cancer therapy. *Clin. Cancer Res.* **2004**, *10*, 7677–7684.

Lunt, J. Large-scale production, properties and commercial applications of polylactic acid polymers. *Polim. Degrad. Stab.* **1998**, *59*, 145–152.

MacGregor, D. C. Cardiovascular prosthetic devices and implants with porous systems. U.S. Patent 4,627,836, Dec 9, 1986.

Makino, K.; Mogi, T.; Ohtake, N.; Yoshida, M.; Ando, S.; Nakajima, T.; Ohshima, H. Pulsatile drug release from poly(lactide-*co*-glycolide) microspheres: how does the composition of the polymer matrixes affect the time interval between the initial burst and the pulsatile release of drugs? *Colloids Surf., B.* **2000**, *19*(2), 173–179.

Matsumoto, J.; Nakada, Y.; Sakurai, K.; Nakamura, T.; Takahashi, Y. Preparation of nanoparticles consisted of poly(L-lactide)-poly(ethylene glycol)-poly(L-lactide) and their evaluation in vitro. *Int. J. Pharm.* **1999**, *185*(1), 93–101.

McCollister, D. D.; Hake, C. L.; Sadek, S. E.; Rowe, V. K. Toxicologic investigations of polyacrylamides. *Toxicol. Appl. Pharmacol.* **1965**, *7*(5), 639–651.

Milić, J.; Petrović, S. D. The characteristics of novel dosage forms. *Hem. Ind.* **2003**, *57*(10), 424–436.

Mittal, G.; Sahana, D. K. V.; Bhardwaj, M. N. V.; Kumar, R. Estradiol loaded PLGA nanoparticles for oral administration: Effect of polymer molecular weight and copolymer composition on release behavior in vitro and in vivo. *J. Control. Release* **2007**, *119*(1), 77–85.

Mobley, D. P.; Ed. *Plastic from microbes: Microbial synthesis of polymers and polymer precursors*; Carl Hanser Verlag, Munich, DE, 1994.

Mogi, T.; Ohtake, N.; Yoshida, M.; Chimura, R.; Kamaga, Y.; Ando, S.; Tsukamoto, T.; Nakajima, T.; Uenodan, H.; Otsuka, M.; Matsuda. Y.; Ohshima, K.; Makino, K. Sustained release of 17 β-estradiol from poly(lactide-*co*glycocide) microspheres in vitro and in vivo. *Colloids Surf., B.* **2000**, *17*, 153–165.

Moreno, D.; Zalba, S.; Navarro, I.; Tros de Ilarduya, C.; Garrido, M. J. Pharmacodynamics of cisplatin-loaded PLGA nanoparticles administered to tumor-bearing mice. *Eur. J. Pharm. Biopharm.* **2010**, *74*(2), 265–274.

Mork, S. W.; Park, C. P.; Solc, J. H. Emulsionen mit grosser innerer phase und daraus hergestellte poröse materialien. EP 1,003,788 A1, May 31, 2000.

Mu, L.; Feng, S. S. A novel controlled release formulation for the anticancer drug paclitaxel (Taxol®): PLGA nanoparticles containing vitamin E TPGS. *J. Control. Release,* **2003**, *86*(1), 33–48.

Narayani, R.; Rao, K. P. Gelatin microsphere cocktails of different sizes for the controlled release of anticancer drugs. *Int. J. Pharm.* **1996**, *143*(2), 255–258.

Niechajev, I. Lip enhancement: Surgical alternatives and histologic aspects. *Plast. Reconstr. Surg.* **2000**, *105*(3), 1173–1183.

Nikolić, L. J.; Nikolić, V.; Skala, D.; Stamenković, J.; Veljković, V.; Lazić, M. Macroporous copolymer methylmethacrylate and acrylamide, procedure of synthessis and usage like support of microorganisms and enzymes. RS Pat. Appl. P-606/02, Aug 13, 2002.

Nikolić, L. J.; Nikolić, V.; Stanković, M.; Adnađević, B.; Jovanović, J.; Ristić, I. Poly(lactide) synthesis by microwaves. RS Pat. Appl. P-2007/0324, Jul 31, 2007.

Nikolić, L. J.; Skala, D.; Nikolić, V.; Stamenković, J.; Babić, D.; Ilić-Stojanović, S. Methyl methacrylate and acrylamide cross-linked macroporous copolymers. *J. Appl. Polym. Sci.* **2004**, *91*, 387–395.

Nikolić, Lj. PhD. Dissertation, Faculty of Technology, University of Nis, **2003.**

Nikolic, Lj.; Ristic, I.; Adnadjevic, B.; Nikolic, V.; Jovanovic, J.; Stankovic, M. Novel microwave-assisted synthesis of poly(D, L-lactide): The influence of monomer/initiator molar ratio on the product properties. *Sensors*, **2010**, *10*(5), 5063–5073.

Nizam El-Din, H. M. Characterization and caffeine release properties of *N*-isopropylacrylamide/hydroxypropyl methacrylate, copolymer hydrogel synthesized by gamma radiation. *J. Appl. Polym. Sci.* **2011**, *119*(1), 577–585.

Onuki, Y.; Nishikawa, M.; Morishita, M.; Takayama, K. Development of photocross-linked polyacrylic acid hydrogel as an adhesive for dermatological patches: Involvement of formulation factors in physical properties and pharmacological effects. *Int. J. Pharm.* **2008**, *349*(1–2), 47–52.

Oppenheim, R. C. Paclitaxel loaded gelatin nanoparticles for intravesical bladder cancer therapy. *Int. J. Pharm.* **1981**, *8*, 217.

Orive, G.; Hernández, R. M.; Gascón, A. R.; Pedraz, J. L. Micro and nano drug delivery systems in cancer therapy. *J. Can. Ther.* **2005**, *3*, 131–138.

Pandya, K.; Bahadur, P.; Nagar, T. N.; Bahadur, A. Micellar and solubilizing behavior of Pluronic L-64 in water. *Colloids Surf. A Physicochem. Eng. Aspects.* **1993**, *70*(3), 219–227.

Patel, H. R.; Patel, R. P.; Patel, M. M. Poloxamers: A pharmaceutical excipients with therapeutic behaviors. *Int. J. PharmTech. Res.* **2009**, *1*(2), 299–303.

Pepić, I. PEO-PPO-PEO triblock copolymers. *Farm. Glas.* **2005**, *61*(3), 131–152.

Philips, R. W. *Skinner's Science of Dental Materials*, 9th Ed., Saunders, Philadelphia, PA, 1991.

Pinchuk, M. P.; Kebuladze, I. M.; Stasenko, A. A. State of cellular and humoral immunity after soft tissue endoprosthesis implantation and correction of body forms using biogel Interfall. *Klin. khir.* **1997**, *(9–10)*, 36–37.

Porsch, B.; Hillang, I.; Karlsson, A.; Sundelöf, L. O. Ion-exclusion controlled size-exclusion chromatography of methacrylic acid–methyl methacrylate copolymers. *J. Chromatogr. A.* **2000**, *872*(1), 91–99.

Prabhakaran, M. P.; Venugopal, J.; Ramakrishna, S. Electrospun nanostructured scaffolds for bone tissue engineering. *Acta Biomater.* **2009**, *5*(8), 2884–2893.

Prasanthi, N. L.; Manikiran, S. S.; Sowmya, S.; Anusha, B.; Rao, N. R. Effect of poloxamer 188 on in vitro dissolution properties of antipsychotic solid dispersions. *Int. J. Pharm. Sci. Rev. Res.* **2011**, *10*(1), 15–19.

Raghuvanshi, R. S.; Katare, Y. K.; Lalwani, K.; Ali, M. M.; Singh, O.; Panda, A. K. Improved immune response from biodegradable polymer particles entrapping tetanus toxoid by use of different immunization protocol and adjuvants. *Int. J. Pharm.* **2002**, *245*(1–2), 109–121.

Reeve, M.; McCarthy, S.; Downey, M.; Gross, R. Polylactide stereochemistry: Effect on enzymic degradability. *Macromolecules*, **1994**, *27*(3), 825–831.

Ristić, I. S. PhD. Dissertation, Faculty of Technology, University of Novi Sad, **2011**.

Ristić, I. S.; Marinović-Cincović, M.; Cakić, S. M.; Tanasić, L. M.; Budinski-Simendić, J. K. Synthesis and properties of novel star-shaped polyesters-based on L-lactide and castor oil. *Polym. Bull.* **2013a**, *70*(6), 1723–1738.

Ristić, I. S.; Nikolić, Lj.B.; Nikolić, V. D.; Budinski-Simendić, J. K.; Zdravković, V. S. The influence of monomer molar ratio on the properties of copolymers-based on methyl methacrylate and methacrylic acid. *Hem. Ind.* **2009**, *63*(6), 611–619.

Ristić, I. S.; Radusin, T.; Pilić, B.; Cakić, S.; Budinski-Simendić, J. The influence of isosorbide on thermal properties of poly(L-lactide) synthesized by different methods. *Polym. Eng. Sci.* **2013b**, *53*(7), 1374–1382.

Ristić, I. S.; Tanasić, L.; Nikolić, Lj.B.; Cakić, S. M.; Ilić, O. Z.; Radičević, R. Z.; Budinski-Simendić, J. K. The properties of poly(l-Lactide) prepared by different synthesis procedure. *J. Environ. Eng.* **2011**, *19*(2), 419–430.

Rizkalla, N.; Range, C.; Lacasse, F. X.; Hildgen, P. Effect of various formulation parameters on the properties of polymeric nanoparticles prepared by multiple emulsion method. *J. Microencapsul.* **2006**, *23*, 39–57.

Rodrigues, J. S.; Santos-Magalhães, N. S.; Coelho, L. C.; Couvreur, P.; Ponchel, G.; Gref, R. Novel core(polyester)-shell(polysaccharide) nanoparticles: protein loading and surface modification with lectins. *J. Control. Release*, **2003**, *92*(1–2), 103–112.

Rodriguez-Hernandez, J.; Chécot, F.; Gnanou, Y.; Lecommandoux, S. Toward 'smart' nano-objects by self-assembly of block copolymers in solution. *Prog. Polym. Sci.* **2005**, *30*(7), 691–724.

Russell, A. J.; Beckman, E. J. Addition and condensation polymers containing proteins, enzymes. U.S. Patent 5,482,996, Jan 9, 1996.

Safra, T.; Muggia, F.; Jeffers, S.; Tsao-Wei, D. D.; Groshen, S.; Lyass, O.; Henderson, R.; Berry, G.; Gabizon, A. Pegylated liposomal doxorubicin (doxil): reduced clinical cardiotoxicity in patients reaching or exceeding cumulative doses of 500 mg/m^2. *Ann. Oncol.* **2000**, *11*(8), 1029–1033.

Sah, E.; Sah, H. Recent trends in preparation of poly(lactide-*co*glycolide) nanoparticles by mixing polymeric organic solution with antisolvent. *J. Nanomater*. **2015**, *16*(1), 1–22.

Saha, S. C.; Patel, D.; Rahman, S.; Savva, M. Physicochemical characterization, solubilization, and stabilization of 9-Nitrocamptothecin using pluronic block copolymers. *J. Pharm. Sci.* **2013**, *102*(10), 3653–3665.

Samra, B.; Galaev, I.; Mattiasson, B. Gels with a shape memory. U.S. Patent 6,538,089, March 25, 2003.

Sarmento, B.; Ribeiro, A.; Veiga, F.; Sampaio, P.; Neufeld, R.; Ferreira, D. Alginate/Chitosan nanoparticles are effective for oral insulin delivery. *Pharm. Res*. **2007**, *24*(12), 2198–2206.

Savić, V.; Ilić, D.; Nikolić, V.; Nikolić, L. J.; Stanković, M.; Stanojević, L. J.; Tasić-Kostov, M. *Book of abstract*, 5th Congress of Pharmacy of Macedonia with International Participation, (Ohrid) Macedonia, September 21–25, 2011; The Macedonian Pharmaceutical Association and the Faculty of Pharmacy: Skopje, 262.

Schroeder, U.; Sommerfeld, P.; Ulrich, S.; Sabel, B. A. Nanoparticle technology for delivery of drugs across the blood–brain barrier. *J. Pharm. Sci.* **1998**, *87*(11), 1305–1307.

Shaik, M. R.; Korsapati, M.; Panati, D. Polymers in Controlled Drug Delivery Systems. *Int. J. Pharm. Sci.* **2012**, *2*(4), 112–116.

Sharma, P. K.; Bhatia, S. R. Effect of antiinflammatories on Pluronic® F127: micellar assembly, gelation and partitioning. *Int. J. Pharm.* **2004**, *278*(2), 361–377.

Shenoy, D. B.; Amiji, M. M. Poly(ethylene oxide)-modified poly(epsilon caprolactone) nanoparticles for targeted delivery of tamoxifen in breast cancer. *Int. J. Pharm.* **2005**, *293*(1–2), 261–270.

Spasojević, M. P.; Zrilić, M.; Stamenković, S. D.; Veličković, J. S. The effect of the accelerated aging on the mechanical properties of the PMMA denture base materials modified with itaconates. *Hem. Ind.* **2011**, *65*(6), 707–715.

Steffens, G. C. M.; Nothdurft, L.; Buse, G.; Thissen, H.; Höcker, H.; Klee, D. High density binding of proteins and peptides to poly(D, L-lactide) grafted with polyacrilic acid. *Biomaterials*, **2002**, *23*(16), 3523–3531.

Stephen, S. H. Final report on the safety assessment of polyacrylamide. *J. Am. Coll. Toxicol.* **1991**, *10*, 193–203.

Storm, G.; Belliota, S. O.; Daemen, T.; Lasicc, D. D. Surface modification of nanoparticles to oppose uptake by the mononuclear phagocyte system. *Adv. Drug Deliv. Rev*. **1995**, *17*(1), 31–48.

Tiyaboonchai, W. Chitosan nanoparticles: a promising system for drug delivery. *Naresuan Univ. J.* **2003**, *11*(3), 51–66.

Todorović, Z. S.; Nikolić, L. J. B.; Nikolić, V. D.; Vuković, Z. M.; Mladenović-Ranisavljević, I. I.; Takić, Lj. M. Cross-linked macroporous polymers and copolymers, their synthesis and characterization. *Advanced technol.* **2012**, *1*(2), 11–19.

Torchilin, V. P.; Trubetskoy, V. S. Which polymers can make nanoparticulate drug carriers long-circulating? *Adv. Drug Deliv. Rev.* **1995**, *16*, 141–155.

Vasiljevic, P.; Najman, S.; Djordjevic, Lj.; Savic, V.; Vukelic, M.; Zivanov-Curlis, J.; Ignjatovic, N.; Uskokovic, D. Ectopic osteogenesis and hematopoiesis after implantantion of bone marrow cells seeded on HAP/PLLA scaffold. *Hem. Ind.* **2009**, *63*(4), 301–307.

Vilar, G.; Tulla-Puche, J.; Albericio, F. Polymers and drug delivery systems. *Curr. Drug Deliv.* **2012**, *9*(4), 367–394.

Voldřich, Z.; Tománek, Z; Vacìk, J.; Kopeček, J. Long-term experience with poly(glycol monomethacrylate) gel in plastic operations of the nose. *J. Biomed. Mater. Res., Part A*, **1975**, *9*(6), 675–685.

Wallace, D. G.; Reich, C. J.; Shargill, N. S.; Vega, F.; Osawa, A. E. Fragmented polymeric compositions and methods for their use. U.S. Patent 6,066,325, May 23, 2000.

Wichterle, O.; Lìm, D. Process for producing shaped articles from three-dimensional hydrophilic high polymers. U.S. Patent 2,976,576, Mar 28, 1961.

Wu, Y.; Yang, W.; Wang, C.; Hu, J.; Fu, S. Chitosan nanoparticles as a novel delivery system for ammonium glycyrrhizinate. *Int. J. Pharm.* **2005**, *295*(1–2), 235–245.

Xing, J.; Zhang, D.; Tan, T. Studies on the oridonin-loaded poly(D, L-lactic acid) nanoparticles in vitro and in vivo. *Int. J. Biol. Macromol.* **2007**, *40*(2), 153–158.

Yang, B.; Guo, C.; Chen, S.; Ma, J.; Wang, J.; Liang, X.; Zheng, L; Liu, H. Effect of Acid on the Aggregation of Poly(ethylene oxide)-Poly(propylene oxide)-Poly(ethylene oxide) Block Copolymers. *J. Phys. Chem. B.* **2006**, *110*(46), 23068–23074.

Yang, Y. Y.; Chung, T. S.; Bai, X. L.; Chan, W. K. Effect of preparation conditions on morphology and release profiles of biodegradable polymeric microspheres containing protein fabricated by double-emulsion method. *Chem. Eng. Sci.* **2000**, *55*, 2223–2236.

Young, G. A.; LaVon, G. D.; Taylor, G. W. High efficiency absorbent articles for incontinence management. U.S. Patent 5,318,554 A, Jun 7, 1994.

Yurke, B.; Lin, D. C.; Langrana, N. A. Use of DNA nanodevices in modulating the mechanical properties of polyacrylamide gels; Carbone, A.; Pierce, N. A. (Eds.); in DNA Computing, Springer: Berlin Heidelberg, DE, **2006**, pp. 417.

Zhao, Y.; Yang Y.; Yang X; Xu, H. Preparation and pH-sensitive swelling behavior of physically cross-linked polyampholyte gels. *J. Appl. Polym. Sci.* **2006**, *102*, 3857–3861.

Zillies, J.; Coester, C. Evaluating gelatin-based nanoparticles as a carrier system for double stranded oligonucleotides. *J. Pharm. Pharmaceut. Sci.* **2004**, *7*(4), 17–21.

CHAPTER 5

APPLICATIONS OF NANOBIOMATERIALS IN DRUG DELIVERY

YASER DAHMAN[1,*] and HAMIDEH HOSSEINABADI[2]

[1]*Department of Chemical Engineering, Ryerson University, Toronto, Ontario, M5B 2K3, Canada, *E-mail: ydahman@ryerson.ca*

[2]*Department of Mechanical and Industrial Engineering, Ryerson University, Toronto, Ontario, M5B 2K3, Canada*

CONTENTS

ABSTRACT

The need for affordable and effective treatments for different diseases and disorders has been the motivation for researchers to develop and study novel drug delivery techniques. In this regards, nanomaterials have been widely used for the development of nano-based drug delivery and cancer therapy. The nano-based drug delivery systems (NDDSs) offer several advantages over the conventional drug delivery systems, such as multiple targeting functionalization, *in vivo* imaging, multiple drug delivery to drug resistant cells, extended circulation half-life, and controlled drug release. The use of NDDSs for biomedical applications constitutes a new research field called "nanomedicine" which involves a multistep process from the design to synthesis and in vitro experiments. This paper reviews the basics of NDDSs and application of nanomaterials such as liposomes, dendrimers, nanoparticles and nanotubes in this field.

5.1 INTRODUCTION

Nanomaterials (NMs) size range is similar to viruses (20–450 nm), proteins (5–50 nm), or genes (2-nm wide and 10–100 nm long). Unique characteristics of NMs, such as the possibility of manipulating them with an external magnetic field and being observable instrumentally, makes them attractive for various biomedical applications (Brigger et al., 2002; Gong et al., 2006). Moreover, NMs offer other advantages including possibility of attaching different kinds of biomolecules to them through chemical or physical bonds, which makes them suitable for medical purposes. Examples of NMs used in biomedical applications are nanoparticles (NPs), quantum dots (QDs), nanotubes, and nanostructures composed of these individuals (Huang et al., 2007; Osterfield et al., 2008; Su et al., 2010).

The nanoscale size of NMs leads to having different physical and mechanical properties than that of the bulk materials. In order to take advantage of the unique properties that arise from the nanoscale dimensions of NMs, it is important to control and manipulate their size, shape, and surface functional groups, and organize them into special structures to create new products, devices, and technologies (Baranov et al., 2010; Carbone et al., 2006; Deka et al., 2009). Nanotechnology is the science of engineering the properties of NMs for the purpose of building microscopic structures that serve to achieve specific purposes. Two different approaches are used to manipulate and control the properties of these nanostructures, "top-down" or "bottom-up." In the "top-down" approach, nanostructures are made out of large sized materials through lithography or any other outside force that impose order on NMs, while in the "bottom-up" method, NMs are assembled into nano-sized structures (Kubik et al., 2005).

NMs are attractive for applications in medicine due to the role of nanoscale phenomena, such as enzyme action, cell cycle, cell signaling, and damage repair. NMs can be used to create tools for analyzing the nanoscale structures such as cells and tissues from the atomic and cellular levels and to synthesize biocompatible NMs for therapies, diagnostics, and replacements. NMs can be used to target drugs to specific proteins and nucleic acids associated with the disease and disorders. In addition, NMs are able to deliver several biomolecules to specific sites of action and, at the mean time, protect them from degradation, immune attack or other effects of the harsh environment inside the body (Cozzoli et al., 2005; Vo et al., 2006).

Nanomaterial-based drug delivery systems, NDDSs, offer extraordinary properties that can be beneficial in cancer therapy by effectively delivering biologically active agents into living cells. "Nanomedicine" is the field of research that deals with biomedical applications of NDDSs through optimizing the design, synthesis, in vitro testing and initial administration to cross the tissue endothelium barrier and introduction into the interstitial space of tissues, through the cell membrane into organelles of cells and even through the pronuclear membrane into the nucleus of the cells.

Liposomes are one of the NMs used to encapsulate anticancer drugs. Magnetic nanoparticles (NPs) have been studied in the analyzes of blood, urine, and other body fluids to speed up separation and improve selectivity are ongoing (Aguilar et al., 2010; Mahmoudin et al., 2011; Milhelm and G. F.; 2008; Wagner et al., 2006). NMs have been shown to effectively deliver vaccine antigens with enhanced antibody and cellular responses (Aguilar et al., 2012; Pusic et al., 2011; Yang et al., 2008). Chitosan, with size range of 40 nm to <1 μm is a novel nasal delivery system for vaccines and inorganic particles (Demers et al., 2000; Ueno et al., 2005; Yang et al., 2014). However, a disadvantage of using NMs in biomolecule delivery is that they are cleared from the body very rapidly (Cherukuri et al., 2004; Pusic et al., 2011). On the other hand, NMs larger than 10 nm will avoid single pass renal clearance and the presence of negative charge minimizes nonspecific interaction with proteins and cells to achieve pharmacokinetic (PK) manipulations. These NMs can be highly effective when readily taken up by antigen-presenting cells (Pusic et al., 2011).

NMs have been studied for their application as drug delivery systems, or as drug carriers. The advantages of using NMs for drug delivery include the possibility of controlling release of drug, preventing degradation of labile molecules such as proteins and peptides, and targeted delivery of drug to specific sites (Singh and Lillard, 2009). There are already a number of nano-enabled drugs sold in the market, that take advantage of small size of the particles, which leads to higher surface to volume ratio and provides enhanced bioavailability of poorly soluble drugs (Liversidge, 2011).

This chapter is focused on the application of NMs in drug delivery systems (DDSs), and is dedicated to reviewing some of the most recent studies on the nanocarriers used for treatment of cancer and other diseases/disorders.

5.2 NANO DRUG DELIVERY SYSTEMS

NMs posses some unique properties that are advantageous in drug delivery. The nano-sized particles allow for extravasation through the endothelium in inflammatory sites, tumors, epithelium (e.g., intestinal tract and liver), or penetrate micro capillaries (Singh and Lillard, 2009). Cell uptake of the drugs is increased by using NMs and also drug can be selectively delivered to specific targets (Dekker, 1999; Linhardt, 1989; Panyam et al., 2003). Since the smallest capillaries in the body have diameter of 5–6 μm, the size of particles entering the bloodstream should be significantly smaller than 5 μm, and have no tendency to aggregate, to prevent embolism caused by the particles. Most of the studied NMs for drug delivery applications are in the range of 10 to 200 nm, while particles larger than 200 nm are not favorable for this purpose (Kreuter, 1994).

AS NMs distribute in the bloodstream, interactions take place with plasma proteins, cells, and other blood components as reviewed by Moghimi et al. (2001). Accumulation of liposomes and other NMs in the liver and spleen could be related to the nature of proteins that adsorb onto the surface of systemically administered NMs (Kamps and Scherpof, 1998).

In order to use NMs as drug carriers, the drug may be dissolved, entrapped, adsorbed, attached, or encapsulated into or onto them. The NMs can be engineered with different drug loading and release properties for any therapeutic agent (Barratt, 2000; Couvreur et al., 2002; Pitt et al., 1981). NMs have been used in various forms as drug delivery system. Examples are nanocapsules, where a polymer membrane entraps the drug in its cavity (Singh and Lillard, 2009); or nanosphere, where the drug is physically or chemically dispersed uniformly inside or outside the structure; liposomes, dendrimers, magnetic NPs and hydrogels.

Conventional drug carriers, modify the drug biodistribution profile as it is mainly delivered to the mononuclear phagocyte system (MPS), also called reticuloendothelial system (RES) in the liver, spleen, lungs, and bone marrow (Moghimi et al., 2001). This is while NMs used in drug delivery may be recognized by the host immune system and cleared by phagocytes from circulation (Muller et al., 1996). In addition to the size, hydrophobicity of the NMs is important in determining the level of blood components that bind on the particle surface that may lead to an immune response (Brigger et al., 2002). As a result, reducing the osponization is a deterministic factor in prolonging the circulation of NMs in the body. In order to achieve this purpose, coating the NPs with hydrophilic polymers/surfactants or by formulating NPs with biodegradable copolymers with hydrophilic characteristics such as polyethylene glycol (PEG) and polyethylene oxide is used (Singh and Lillard, 2009). Studies have proved that presence of PEG on NP surfaces prevents osponization by complement and other serum factors, leading to reduced phagocytosis (Bhadra et al., 2002). Another example is the super-paramagnetic iron oxide (SPIO) NPs coated with dextran that are used as magnetic resonance imaging (MRI) contrast agents. The nanoscale size (less than 20 nm), surface crosslinking, and PEGylation of the SPIO NPs have been shown to prolong circulation times (Park et al., 2008). On the other hand, larger SPIO NPs (50–150 nm) that are not dextran coated are rapidly eliminated from circulation by the liver and spleen and serve to enhance the MR contrast in these organs (Park et al., 2008).

Using NMs in drug delivery systems offers advantages for therapeutics (Park et al., 2008). First, drugs and imaging agents delivered with nanocarriers distributed over smaller volumes, which leads to decrease associated side effects (Drummond et al., 1999). Second, the biodistribution of the drugs at specific target is enhanced and pharmacokinetics is improved, which results in increased efficacy (Au et al., 2001; Moghimi and Szebeni, 2003). Third, the concentration of drug in healthy tissues and thus its toxicity is minimized (Das et al., 2011). The nanocarriers have been engineered to target tumors and disease sites that have permeable vasculature allowing easy delivery of payload. Specific targeting and reduced clearance increases the therapeutic index that consequently lowers the dose required for efficacy (Das et al., 2011). Fourth, nanocarriers can be used to improve the solubility of hydrophobic therapeutics in aqueous medium. Fifth, nanocarriers can efficiently stabilize labile molecules and prevent them from degradation (Koo et al., 2005; Kristl et al., 2003).

Due to the unique properties of NMs, they offer promising improvements in efficacy of conventional drugs through nano-therapeutic drug delivery. This promise is based on the ability of NMs to cross the various obstacles between the administration of the drug and the site of drug delivery. These advantages make the nano drug delivery systems superior to the conventional ones.

5.3 FACTORS AFFECTING DRUG DELIVERY BY NANOMATERIALS

Using nanocarriers is beneficial due to increasing the bioavailability of the drug at the target site, reducing the frequency of drug administration, and reaching target sites that are inaccessible through conventional drug delivery systems. Biocompatibility is the most important factor to be considered when using NMs as drug carriers. There are some other factors affecting the efficacy of nanocarriers such as the drug loading and releasing characteristics of the NMs, their surface charge, and also size and distribution.

One of the most important factors that affect the loading mechanism of drug and modification procedure of NMs for targeted drug delivery is surface charge of NPs. Surface charge is measured in terms of the NPs' zeta potential, which indicates the electrical potential of particles. Zeta potential depends on the composition of the NMs and the medium where they are dispersed (Couvreur et al., 2002). Studies have shown that NMs exhibiting a zeta potential above ± 30 mV are stable in suspension as the surface charge prevents aggregation of the particles (Singh and Lillard, 2009). Drugs may be loaded through covalent conjugation, hydrophobic interaction (HI), charge–charge interaction, or encapsulation, which is decided upon by considering the characteristics of the drug and the targeting molecules, and also surface charge of the NPs. Changes in zeta potential can also be used as an indicator of whether a drug is loaded on the NMs. Dynamic light scattering (DLS) results can be used to find out the average zeta potential of the NPs.

Size and size distribution of the particles determine the chemical and physical properties of NMs. Metal NMs are mostly synthesized in organic solvents and the core/shell size of particles are characterized before converting them into water soluble particles (to make them biocompatible). Transmission electron microscopy (TEM) is mostly used to study core or core/shell size and size distribution. Once the NMs are converted into water-soluble forms through applying amphiphilic coatings, their size and size distribution can be studied using DLS or photon correlation spectroscopy. The hydrodynamic size of the NMs (in water-soluble form) is larger than the core/shell size and is critical in determining the *in vivo* distribution, toxicity and targeting ability of NMs. It also affects the drug loading/release characteristics of NMs.

Studies have proved the effect of nano scale size of NMs on increasing the cell uptake and mobility of the drug, which makes them superior to microparticles for drug delivery applications (Panyam and Labhasetwar, 2003). Following the opening of the endothelium tight junction (TJ), NMs can cross the blood–brain barrier (BBB) by hyperosmotic mannitol. This may provide a route for sustained delivery of therapeutic agents for difficult-to-treat diseases like brain tumors (Kroll et al., 1998). Zauner et al. (2001) showed that majority of cells take up NMs but not larger microparticles. $CaCo_2$ cells showed uptake of 100 nm NPs at 2.5 fold greater than 1 μm microparticles and a 6 fold greater uptake than 10 μm microparticles. This proves that particle size plays an important role in the particle distribution in cells.

5.4 METHODS/TECHNIQUES FOR DRUG LOADING

Drug loading is the process of incorporation of drug on or in a NM. NM drug delivery system should possess high drug loading capacity to minimize the need of repeating drug doses, minimize aggregation tendency and show dispersibility for efficient delivery of the drugs. Factors influencing the drug loading efficiency include drug solubility in the NMs, dispersion medium, size and composition of the NMs, drug molecular weight and solubility, drug–NM interaction, and the presence of surface functional groups (e.g., carboxyl, amine) on either the drugs or on the NMs (Aguilar et al., 2010; Govender et al., 1999; Panyam et al., 2004).

PEG is used in synthesizing NMs as it does not affect the drug loading and interactions (Peracchia et al., 1997). Studies have shown that the ionic interaction between the drug and matrix materials can be used to increase drug-loading capacity. The loading capacity is generally expressed in percent related to the NP. Iscan et al. (1999) reported 10–20% loading capacity for tetracaine and etomidate.

Drug loading method is determined considering the properties of the NM and the drug and can be pursued in several ways. Loading the drug on the NM surface can be done through charge–charge interaction, covalent bonding, or hydrophobic interaction. Another method is by loading the drug inside the NM core through encapsulation during the synthesis of nanocarriers or by hydrophobic interactions. When drug loading is done through adsorption or absorption methods, it is done after NP formation; by incubating the NM with a concentrated drug solution (Muller et al., 2000).

Solid lipid NM loading is generally done through encapsulation method. Encapsulation may be done at high or low temperatures, depending on the drug's temperature sensitivity and hydrophilicity. In hot homogenization technique, the lipid is melted at 108°C above its melting point. Then the drug is added to the melt and the resulting mixture is stirred to disperse in a hot aqueous surfactant solution. After that, the resulting pre-emulsion is homogenized with a piston-gap homogenizer and finally the hot oil is cooled down to room temperature in water nanoemulsion until the lipid recrystallizes into solid lipid NPs. In the case of glycerides being composed of short-chain fatty acids, e.g., Dynasan (Rejman et al., 2004) it might be necessary to cool the nanoemulsions to even lower temperatures to initiate recrystallization. On the other hand, in cold homogenization technique, after the mixing of the molten lipid and the drug, it is cooled to reach the solidification point. Following that, the lipid microparticles (~50 ± 100 mm) are ground and the resulting material is dispersed in a cold surfactant solution, which is then homogenized at or below room temperature to create solid lipid NPs. This procedure minimizes the loss of hydrophilic drugs to the water phase. One way to minimize the loss of hydrophilic compounds to the aqueous phase of the solid lipid NP dispersion is using low-solubility liquids for the drug like oils or PEG 600, instead of water. As this dispersion can be converted into soft gelatin capsules for oral drug delivery, this procedure is advantageous (Muller et al., 1993).

Cisplatin, an anticancer drug with side effects such as ototoxicity and nephrotoxicity, is an example of encapsulated drug in biomedical applications. In a study by Das et al., albumin-based cisplatin NPs were shown to be a sustained release carrier for cisplatin. Ultraviolet spectroscopy results indicated that the drug encapsulation varied from 30 to 80% for different ratios of cisplatin and protein. In vitro release kinetics results indicated that the NP-based formulation had biphasic release kinetics and was capable of sustained release compared with the free drug (80% release in 45 h) (Das et al., 2011).

In another study by Choi et al. (Choi et al., 2013), rattle structures NMs composed of a gold nanorod (AuNR) in a mesoporous silica ($mSiO_2$) nanocapsule were prepared. Rattle-structured nanomaterials are hollow-structured nanomaterials with a movable core. The mesoporous silica shell has advantageous properties such as low cytotoxicity, uniform size, high stability, easy functionalization, and high pore volumes (Slowing et al., 2007). Doxorubicin hydrochloride (DOX) was used as a model drug encapsulated in the nanostructured carrier. The UV-vis absorption spectra recorded before and after loading the DOX into the rattle-structured NM showed that the DOX loading efficiency reached upto 71.2% and the loading content was 59.7 mg DOX per mg of the rattle-structured NM. TEM images of corresponding to all the steps of synthesis of rattle-structures NM can be seen in Figure 5.1.

Another method of drug loading is on the surface of a NP. For this purpose, NP should have functional groups on its surface that allows formation of covalent or charge–charge bonding, or hydrophobic interaction between the drug and the NP. Common covalent coupling methods involve formation of a disulfide bond, cross-linking between two primary amines, reaction between a carboxylic acid and primary amine, reaction between maleimide and thiol, reaction between hydrazine and aldehyde, and reaction between a primary amine and free aldehyde (Nobs et al., 2004).

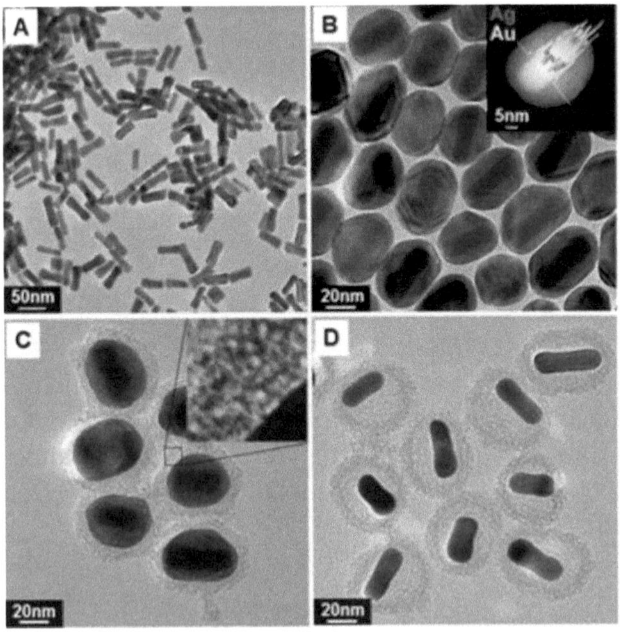

FIGURE 5.1 TEM images corresponding to all the steps for the fabrication of rattle-structured AuNR@mSiO_2: (A) Au NRs, (B) AuNR@Ag core–shell (inset: STEM-EDS line scan analysis), (C) AuNR@Ag@mSiO_2 core–shell–shell (inset: expanded image of the silica shell), and (D) the rattle-structured AuNR@mSiO_2 core–void–shell (Reprinted from Choi, E.; K.M.; Janga, B.; Piao, Y. Highly monodisperse rattle-structured nanomaterials with gold nanorod core–mesoporous silica shell as drug delivery vehicles and nanoreactors. *Nanoscale*. **2013**,5: 151–154.)

DOX is an effective anticancer drug that is used in many studies as a model drug for NP delivery. The hydrochloride form of DOX is water soluble, it is fluorescent, and can be monitored before and after loading into the NP. Elimination of excess DOX or unloading it can be done through centrifugation, dialysis, or magnetization.

In a study by Adeli et al. (2011), a nanostructured-based drug delivery system consisting of gold nanoparticle (Au NP) cores and polyrotaxane (PR) shells (Au NPs@PR) was synthesized. The core-shell structure facilitated controlled release of drugs within targeted tissues using the photothermal properties of the Au NPs. Further, it killed cancerous cells without damaging the surrounding healthy cells. Synthesis of polyrotaxanes was done through threading cyclodextrin rings onto PEG axes, and capping the resulting pseudopolyrotaxanes (Ps-PR) by Au NPs. Hydrophobic interactions between the end triazine groups of PEG and the cavity of the a-cyclodextrins (a-CDs) resulted in formation of Ps-PR complexes; and the noncovalent interactions between the citrate shell of the Au NPs and the end functional groups of PEG lead to formation of PRs. In order to investigate the effectiveness of this drug delivery system, cisplatin (CDDP) and DOX were conjugated to Au NPs PR hybrid nanomaterials. Infrared (IR) and nuclear magnetic resonance (NMR) spectroscopy analysis results indicated that the CDDP molecules were attached to the carboxylate groups of the citric acid shell; on the other hand, DOX molecules were conjugated to the hydroxyl functional groups of the Ps-PR. DLS results showed the size distribution of the Au NPs was very narrow (D_{mean} = 17.2 nm). The size and morphology of the synthesized hybrid nanomaterials were investigated using TEM and atomic force microscopy (AFM). It was observed that Au NPs/PR1 and Au NPsPR2 had a necklace-like (beads 150 and 200 nm in size) and a core–shell structure (300 nm in size), respectively. In vitro test results indicated that the toxicity of DOX loaded DDS was more than CDDP loaded DDS. This may be because CDDP molecules can attach to the shell again after cleavage from it, therefore the rate of their release is low; which is not the case for DOX molecules.

Attachment of drugs on the NM surface can be achieved through hydrophobic interaction (HI) by physical association of targeting ligands to the nanocarriers for hydrophobic NMs and hydrophobic drugs. An advantage of using HI is that it eliminates the need for application of rigorous and potentially destructive chemicals that weaken the efficacy of the drugs. On the other hand, there are some shortcomings such as low and weak binding, poor control of the NM–drug interactions and the ligands may not be in the desired orientation after binding.

Dagar et al. (2001) have used vasoactive intestinal peptide (VIP), a 28-amino acid mammalian neuropeptide, as a targeting moiety to cancer and inflamed tissues because receptors for vasoactive intestinal peptide are overexpressed in human breast cancer. They established that VIPs that are HI attached to liposomes were less able to target and attach to breast cancer cells. They used their sterically stabilized liposomes (SSLs) with covalently associated VIP on the surface. This was carried out by conjugating the VIP to DSPE–PEG 3400–NHS [1,2-dioleoylsn-glycero-3-phosphoethanolamine-n-[poly(ethyleneglycol)]N-hydroxy succinamide, PEG Mw 3400] under mild conditions to obtain a 1:1 conjugate of VIP and DSPE–PEG 3400 (DSPE–PEG 3400–VIP) which was confirmed by sodium dodecyl sulfate– polyacrylamide gel electrophoresis. This was followed by the insertion of DSPE–PEG 3400–VIP into preformed fluorescent cholesterol (BODIPY-Chol) labeled SSL by incubation at 37°C. The breast cancer targeting ability in vitro was carried out by incubating these, VIP–SSL with MNU-induced rat breast cancer tissue sections. Compared with fluorescent SSL without VIP or with non-covalently attached VIP, significantly more of the covalently formed VIP–SSL was attached to rat breast cancer tissues. This indicated that SSL with covalently attached VIP can be used for targeted attachment to rat breast cancer tissues. However, in another study, HI-attached VIP is preferred for the delivery of a therapeutic agent to inflammatory cells in animal joints with rheumatoid arthritis (Sethi et al., 2003).

5.5 DRUG RELEASE AND CELLULAR UPTAKE

The drug loaded in or on the NM is released in the body through diffusion or dissolution of the NMs matrix releasing the drug in solution. Drug release process and biodegradation of the NMs in the body are important factors that should be considered in developing a NMs drug delivery system. The Drug release process influences the efficacy of the drug, depending on its solubility and diffusion. When the particle size is small, the surface area-to-volume ratio is high, leading to faster drug release. In contrast, larger particles have large cores to encapsulate drugs, which results in slower release. As a result, the drug release rate can be controlled to an extent through controlling the particle size and size distribution.

In general, the drug release rate depends on drug solubility, desorption of the surface-bound or adsorbed drug, drug diffusion out of the NM matrix, and the NM matrix degradation. When drug is covalently attached to the NM, drug release is controlled by drug– NM diffusion (Fresta et al., 1995). When the drug is encapsulated inside a NM, the release is controlled by diffusion of the drug from the NM interior. If there is a polymer coating on the NM, it acts as a barrier for drug release; thus, the drug solubility and diffusion in or across the polymer membrane becomes a determining factor in drug release (Calvo et al., 1997).

The mechanisms of *in vitro* drug release were studied using model drugs tetracaine, etomidate, and prednisolone. Results indicated that lipid NMs exhibited burst release when incorporating tetracaine and etomidate. Drug release can be controlled as a function of the lipid matrix, surfactant concentration, and production parameters such as temperature achieving as long as 5–7 weeks (Muller et al., 2000).

The factors that affect the level of binding of blood components to the drug are the size of the NMs, surface hydrophobicity, and surface coating functionalities. Prolonging the half-life of the drug *in vivo* and preventing the osponization can enhance the efficacy of the DDS. Precoating of the NMs with hydrophilic polymers and surfactants or by using NMs with biodegradable hydrophilic copolymers such as PEG, can help prolonging the drug circulation. Studies have shown that attaching PEG on NM surfaces minimizes the chances of osponization (Bhadra et al., 2002).

NM particles >1 μm cause a phagocytic response in the body (Koval et al., 1998). Moghimi et al. (2003) reported that entrapment by hepatic and splenic endothelial fenestrations and subsequent clearance can be avoided by using nanocarriers that do not exceed 200 nm. Non-phagocytic cells such as tumor cells can uptake the drug if particles are <500 nm in size. Cellular uptake was not observed for particles >500 nm (Rejman et al., 2004).

5.6 DRUG TARGETING

Targeting nanocarriers for site-specific drug delivery offers some advantages over targeting ligand–drug conjugates (Emerich and Thanos, 2006). When using ligand-drug conjugates, by interactions between the ligand and its receptor, large payloads of drug is released at specific targeted site. In addition, a large number of ligand molecules can be attached to the nanocarrier to increase the probability of binding to target cells, which is beneficial especially for drugs with low binding affinities. Also, the return of drug back to the circulation caused by high intratumoral pressure can be reduced by active targeting, which enables efficient distribution of the carriers in the tumor (Sethi et al., 2003).

Targeting in nano drug delivery systems is achieved through labeling the NPs with receptors or biomolecules that specifically attach them to the target cells or tissues. Antibodies are the most common targeting molecules (Figure 5.2) against epithelial growth factor receptors, anti-epidermal growth factor receptor (EGFR). Because EGFR is expressed in all epithelial cells, this molecule is a special target when delivering drugs to epithelial cells. Human epithelial receptors (HERs) are commonly overexpressed in a number of different cancer cells and, therefore, commonly targeted in nano-enabled targeted drug delivery (Aguilar, 2012).

Antibodies have high target selectivity and binding affinities, but are potentially immunogenic. Due to these drawbacks, other alternative materials such as peptides, aptamers and other small molecules are used for

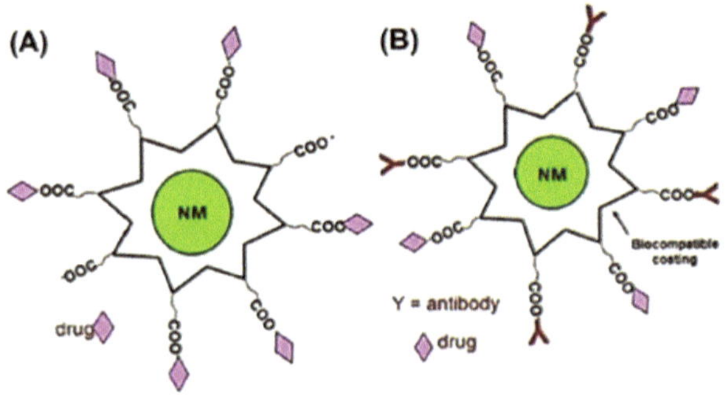

FIGURE 5.2 Schematic diagram of a NM for (A) drug delivery and (B) targeted drug delivery (Reprinted from Aguilar, Z. *Nanomaterials for Medical Applications*. Elsevier Science & Technology: Waltham, MA, 2012.)

drug targeting. Peptides are targeting moieties that do not posses immunogenicity and have lower cost to produce. However, they offer lower target affinities and an increased chance of nonspecific binding (MsCarthy et al., 2007). Aptamers are short oligonucleotides that have potential application as both therapeutic and targeting entities. Similar to antibodies, aptamers are highly selective and have high binding affinities that are largely due to the ability of the molecules complex three-dimensional structures. However, aptamers have easy synthesis procedure and minimal immunogenicity compared to antibodies. On the other hand, aptamers exhibit rapid blood clearance that is largely due to nuclease degradation (Farokhzad et al., 2006).

5.7 NMS USED FOR DRUG DELIVERY

The main application of NMs in drug delivery systems is for cancer treatment. Chemotherapy-based treatments are toxic to both cancer and healthy cells. Using NMs ad drug carriers can minimize the side effects of antitumor drugs on the healthy tissues, while maximizing the efficacy of the drug on the targeted tumor cells. In addition to that, NMs can also be used in delivering vaccines and other therapeutic agents for treatment of other diseases and disorders, and in sites that are hard to reach through conventional drug delivery systems. In this section some of the NMs that have been used in the studies as nanocarriers for drug delivery applications are briefly reviewed.

5.7.1 DENDRIMERS

Dendrimers are used in drug delivery systems for encapsulating the drug, but the release of drug from these NMs is difficult to control. The size of dendrimers is controllable; they exhibit monodispersity, and their functional groups can be modified, which makes them attractive candidates in biomedical applications. They provide concurrent delivery of water-soluble and insoluble drugs by adsorption to the surface through ionic interactions, encapsulation within their hydrophobic microcavities and inside branching clefts, and direct covalent conjugation to the surface functional groups (Figure 5.3). The bonding between dendrimer and the drug can be sensitive to environmental factors such as pH and radiation to help with release of therapeutic agents (Aguilar, 2012).

Due to the uncontrollable release of drug using dendrimers, new NMs have been developed to overcome this drawback; Dendronized polymers are linear polymers with dendrons at each repeat unit (Figure 5.4). They offer advantages for drug delivery applications such as their prolonged circulation time. Also, the release of the drug can be controlled through a degradable link that conjugates the drug to the dendrimers (Paez et al., 2012).

Studies have shown that drug conjugation to a biodegradable dendrimer with optimized blood circulation time can be achieved through the meticulous design of size and molecular architecture. The DOX–dendrimer drug loading was controlled through multiple attachment sites and its solubility was controlled through PEGylation; the drug release was controlled through the use of pH-sensitive hydrazonedendrimer linkages. *In vitro* studies using colon carcinoma cells showed that the DOX–dendrimers were more than 10 times less toxic than free DOX. *In vivo* study of this drug delivery system on mice showed that the tumor uptake of DOX–dendrimers was nine fold higher than free DOX and caused complete treatment of the tumor with 100% survival of the mice after 60 days (Lee et al., 2006).

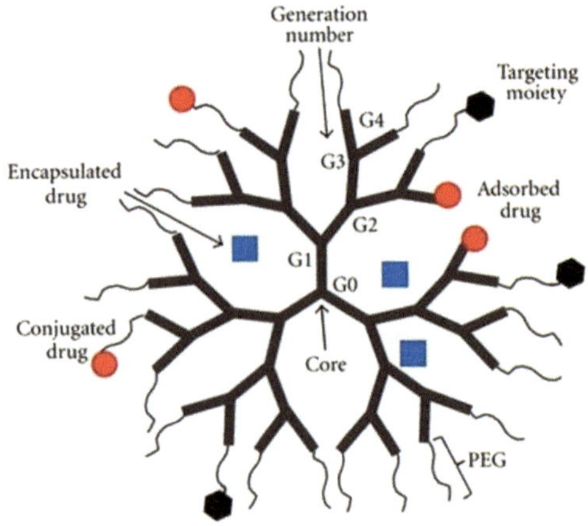

FIGURE 5.3 Various mechanisms of drug loading on a dendrimer (Reprinted from Jun H. Lee and Anjan Nan, "Combination Drug Delivery Approaches in Metastatic Breast Cancer," Journal of Drug Delivery, vol. 2012, Article ID 915375, 17 pages, 2012. doi:10.1155/2012/915375. https://creativecommons.org/licenses/by/3.0/)

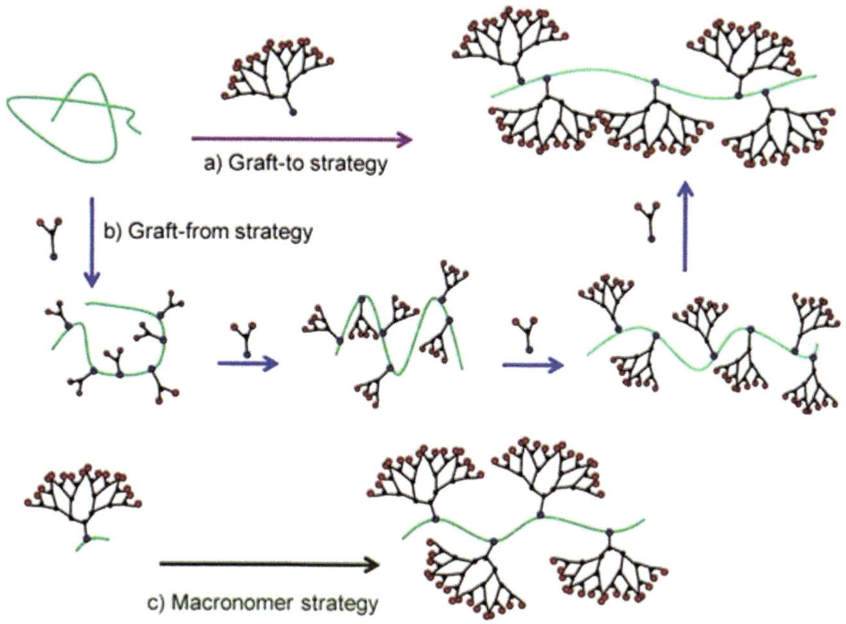

FIGURE 5.4 Methods for synthesis of Dendronized polymers (Reprinted from Paez, J.; M.M.; Brunetti, V.; Strumia, M. Dendronization: A useful synthetic strategy to prepare multifunctional materials. *Polymers*. **2012,** 4(1), 355-395. http://www.mdpi.com/2073-4360/4/1/355. https://creativecommons.org/licenses/by/3.0/)

5.7.2 CARBON NANOTUBES AND FULLERENE

Attaching many drug molecules to antibodies leads to limiting their targeting ability. Carbon nanotubes (CNTs) can be used to overcome this issue (McDevitt et al., 2007). CNTs can be loaded by multiple therapeutic agents both inside and on the surface. Spherical fullerenes (C60) are hollow spherical shells made of carbon atoms. These NMs can be loaded by drugs through encapsulation and also can be used as caps to close the open ends of CNTs for drug entrapment inside their hollow structure. Fullerenes can be loaded with several molecules of an anticancer drug such as CDDP (Mehdipoor et al., 2011). Moreover, distinct sites on the antibody are hydrophobic and attract the hydrophobic fullerenes in large numbers; therefore multiple drugs can be loaded into a single antibody, requiring no covalent bond. In this case the increased payload does not affect the targeting ability of the antibody significantly.

Mehdipoor et al. (2011) deposited γ-Fe_2O_3 nanoparticles (NP) onto the surface of CNTs to produce magnetic hybrid nanomaterials with potential applications in cancer therapy. TEM results showed that Fe_2O_3 nanoparticles with an average size of <5 nm were distributed on the surface of the CNTs with a 25 nm average diameter. *In vivo* studies revealed that 100 $mg.ml^{-1}$ concentration of the final drug delivery system killed more than 95% of cancer cells in mice.

5.7.3 MAGNETIC NANOPARTICLES

In magnetic NPs that have the size of magnetic domains, spontaneous magnetization and demagnetization occurs that results in a magnified response to an applied magnetic field called superparamagnetic behavior (Figure 5.6). Three elements exhibit ferromagnetic properties in physiological environment; nickel, cobalt and iron. Most studies have focused on Fe-based NPs due to its superior magnetic susceptibility (218 and 90 emu/g for elemental Fe and Fe_3O_4, respectively), as well as nontoxicity of this element inside the body (Klostergaard, 2012). Magnetic NPs are ideal nanocarriers of drugs and vaccines because they can also be useful in diagnosing the efficacy of treatment.

Dimercaptosuccinic acid (DMSA)-coated monodisperse magnetic nanoparticles (MNPs) were tested as a delivery system for the antitumorigenic cytokine interferon-gamma IFNγ in mouse models of cancer (Mejias et al., 2011). By applying an external magnetic field, the IFNγ -adsorbed DMSA-coated MNPs were targeted to the tumor site. This resulted in a notable reduction in tumor size indicating that IFNγ-adsorbed DMSA-coated MNPs are efficient *in vivo* drug delivery system for antitumor drugs.

In another study, the DOX-loaded magnetic nanoparticles (MNPs) were injected in mice and their biodistribution was assessed (Mykhaylyk et al., 2005). The results of electron spin resonance (ESR) showed that the DOX–MNPs decreased the DOX bioavailability in the heart and kidney compared with the free unconjugated DOX. By using a magnetic field of 210 mT, the DOX–MNPs bioavailability was effectively increased at the target cells.

Dramou et al. (2013) synthesized novel anticancer "epirubicin" (EPI) water-compatible magnetic molecularly imprinted polymers (MMIPs) and used it as a carrier for tumor targeted drug delivery. Magnetic properties of the synthesized nanomaterials were tested using a vibrating sample magnetometer (VSM). Results indicated that MMIP has slightly inferior magnetic properties compared to pure iron oxide because the polymeric coating had effectively shielded the magnetite. Yet the polymers could be rapidly isolated from the sample solution using a strong magnet in the separation step (Figure 5.5). Studies showed that *in vitro* release time is between 40 and 50 h and temperature and pH were found to affect the release of drug from MMIPs. As it has been reported that the local pH of the solid tumor is lower than normal tissue (pH 7.4), the pH-sensitive release behavior of EPI may be beneficial in tumor treatment (Zhang et al., 2009).

5.7.4 HYDROGELS

Hydrogel NPs encapsulate and deliver drugs, therapeutic proteins, or vaccine antigens by using hydrophobic polysaccharides (Singh and Lillard, 2009). A system using cholesterol to form a self-aggregating hydrophobic core resulting in cholesterol NPs stabilized entrapped proteins. These particles stimulate the immune system and are readily taken up by dendritic cells. Larger hydrogel NMs can also be used to encapsulate and release monoclonal antibodies (Aguilar, 2012).

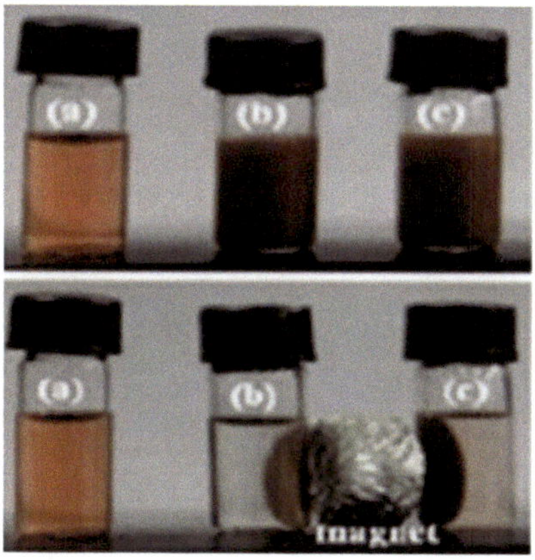

FIGURE 5.5 Dispersion photograph of the MMIP in EPI aqueous solution (top) and separation of the nanomaterials by a magnet after loading the drug (bottom) (Reprinted from Dramou, P.; Z.P.; He, H.; Pham, H.L.A.; Zou, W., Xiao, D.; Pham-Huyc, C.; Ndorbora, T. Anticancer loading and controlled release of novel water-compatible magnetic nanomaterials as drug delivery agents, coupled to a computational modeling approach. *J.Mater.Chem.* **2013,** 1: 4099-4109.)

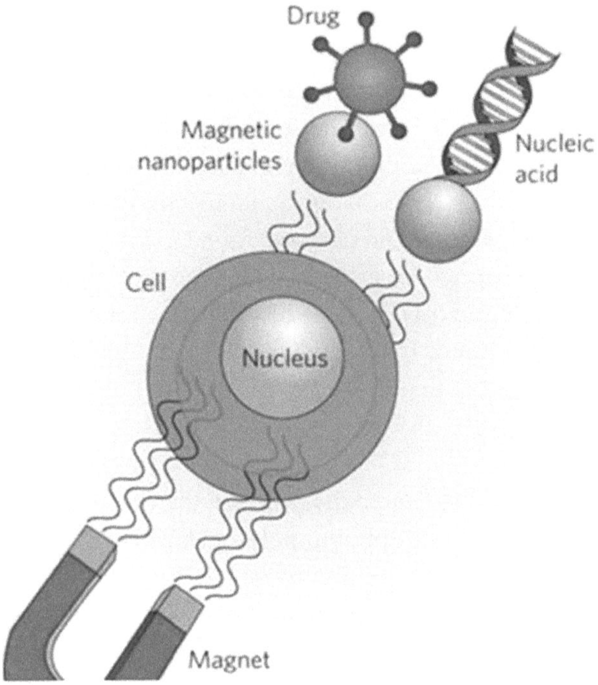

FIGURE 5.6 Schematic illustration of effect of applied external magnetic field on the drug carrying magnetic nanoparticles (Reprinted from Plank, C. Nanomedicine: silence the target. *Nature Nanotechnol.* **2009,** 4:, 544-545.).

Stimuli responsive or smart hydrogels are NMs that swell and deswell according to the change of environmental factors such as pH and temperature, which results in release of the entrapped drug (Figure 5.7).

5.7.5 MICELLES

Micelles are spherical supermolecules composed of an amphiphilic copolymer. They are composed of a core that can accommodate hydrophobic drugs and a shell that is a hydrophilic brush-like corona (Figure 5.8). The shell makes the micelle water-soluble and provides possibility of delivering poorly soluble contents (Singh and Lillard, 2009).

Polymer micelles offer some advantages over other NMs used in drug delivery systems, such as increased drug solubility, prolonged circulation, selective accumulation at target sites, and lower toxicity. However, controlling the release of entrapped drugs is an issue (Aguilar, 2012).

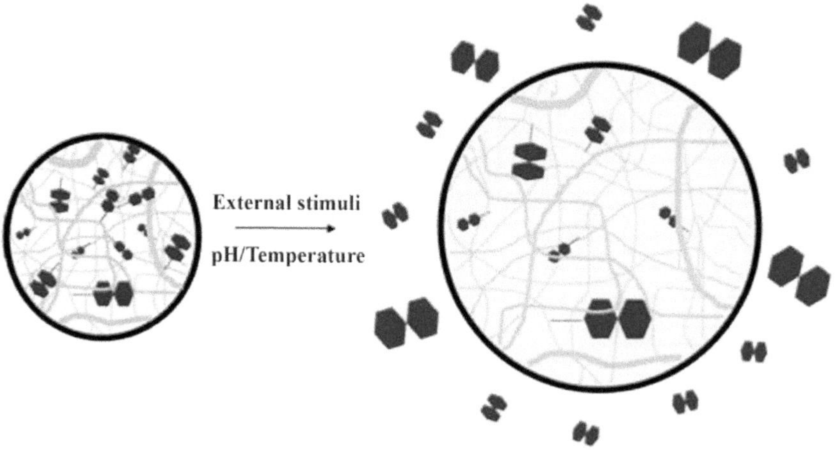

FIGURE 5.7 Release of drug from a hydrogel by an external stimuli (Reprinted from Ferreira, P.; C.J.; Almeida, J.; Gil, M. Pho*tocrosslinkable polymers for biomedical applications. In Biomedical Engineering - Frontiers and Challenges*; Rezai, F.R., Ed.; InTech: Rijeka, Croatia, 2011. With permission.).

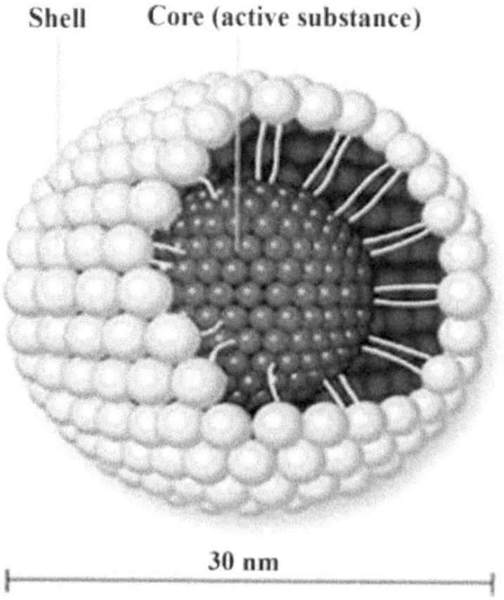

FIGURE 5.8 Schematic illustration of a micelle composed of a core and a shell (Reprinted from Center for Nanotechnology Research at Oakland University. http://www.nano-ou.net/Edu4ImagePages/micelle.aspx. [accessed 2014].)

Camptothecin (CPT) is a topoisomerase I inhibitor that is effective against cancer but has poor solubility, instability, and toxicity. To solve these issues, targeted sterically stabilized micelles (SSMs) have been used as nanocarriers for CPT (CPT–SSM). SSM solubilization of CPT is expensive yet reproducible and it prevents drug aggregate formation.

In addition, SSM composed of PEGylated phospholipids are attractive nanocarriers for CPT delivery because of their size (14 nm) and ability to extravasate through the leaky microvasculature of tumors and inflamed tissues. This results in high drug concentration in tumors and reduced drug toxicity to the normal tissues (Koo et al., 2006). PEG corona can stabilize stealth micelle formulations to minimize opsonization of the micelles and maximize serum half-life.

5.7.6 LIPOSOMES

Liposomes are synthesized spherical vesicles composed of a lamellar phase lipid bilayer (Figure 5.9). They offer prolonged release kinetics and long persistence at the target site.

Liposomes have been proven to be beneficial for drug delivery applications. These systems use "contact-facilitated drug delivery," which involves binding or interaction with the targeted cell membrane allowing for enhanced lipid– lipid exchange with the lipid monolayer of the NP, thereby accelerating the convective flux of lipophilic drugs (e.g., TAX) to dissolve through the outer lipid membrane of the NMs to targeted cells (Guzman et al., 1996).

Medina et al. (2011) developed positron emission tomography (PET) tracers, which irreversibly bound to EGFR. They used a liposomal NP delivery system to alter the pharmacokinetic profile and improve tumor targeting of highly lipophilic but otherwise promising cancer imaging

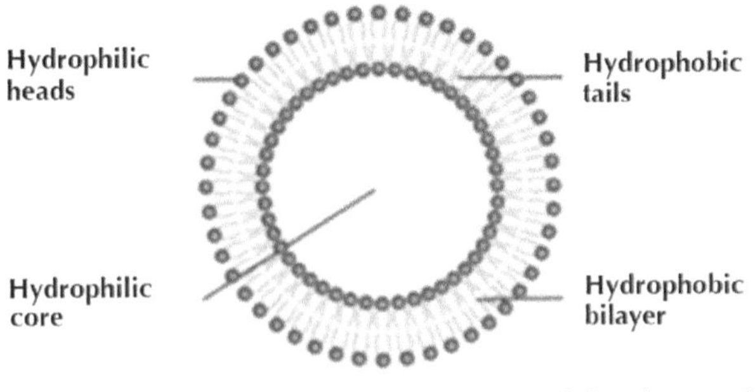

FIGURE 5.9 Schematic illustration of liposome used for drug and DNA delivery (Reprinted from Andros Pharmaceuticals Co., Ltd. http://www.andros.com.tw/en/technology_en.htm. [accessed 2014].)

tracers, such as the EGFR inhibitor SKI 243. In their study, they compared the pharmacokinetics and tumor targeting of the bare EGFR kinase targeting radiotracer SKI 212243 (SKI 243) with that of the same tracer embedded in liposomes. The results indicated that SKI 243 and liposomal SKI 243 were both taken up by tumor xenografts but liposomal SKI 243 remained in the blood longer and consequently exhibited a three to six fold increase in uptake in the tumor among several other organs.

Since liposomes have been proven to be biocompatible and biodegradable, the possibility of using liposomes as a drug delivery system for oral cavity application has been studied (Nguyen et al., 2011). Adsorption of charged liposomes to hydroxyapatite (HA) as a model for dental enamel was investigated in human parotid saliva to simulate oral-like conditions. The results indicated that precipitation occurred in the presence of lipids dipalmitoylphosphatidylcholine (DPPC)/dipalmitoyltrimethylammoniumpropane or DPPC/dipalmitoylphosphatidylglycerol– liposomes in parotid saliva with no HA, which indicated that constituents of parotid saliva reacted with the liposomes (Nguyen et al., 2011). According to the results of these studies, the constituents of saliva may interact with liposomes resulting in aggregation, which can be avoided by adding various ions to create ionic stability in the environment. Additional investigations should be pursued on the effect of liposomes in various physiological conditions in the oral cavity in vitro to evaluate the safety of these NMs for dental drug delivery.

5.8 CONCLUSION

NM-based drug delivery systems can optimize the effect of drugs and reduce toxic side effects in cancer therapy and treatment of other types of diseases. Nanocarriers used in biomedical applications must have specific features to be able to perform functions they are designed for. These NMs must be small (<100 nm), nontoxic, biodegradable, and biocompatible; not tend to aggregate or cause inflammatory response in the body, have prolonged circulation half-life and be cost effective. Physical and chemical characteristics of NM carrier systems that influence their efficacy include size, charge, shape, surface modifications, loading method, and chemical properties.

Many studies are available in literature, investigating the potential applications of various NMs such as liposomes, micelles and magnetic nanoparticles used for delivery of therapeutic agents to specifically targeted sites. However, more focused studies of drug distribution in the body; organ accumulation, toxicity, and genetic damage are yet to be pursued before entering NMs in the realm of human consumption.

ACKNOWLEDGEMENTS

Authors would like to acknowledge financial support from Agriculture and Agri-Food Canada, the Natural Sciences and Engineering Research Council of Canada (NSERC), and the Faculty of Engineering and Architectural Science at Ryerson University in Toronto, Canada.

KEYWORDS

- **Drug Delivery**
- **drug loading**
- **liposomes**
- **nanospheres**
- **nanocapsules**
- **nanoparticles**
- **cellular uptake**

REFERENCES

Adeli, M.; Sarabi, R. S.; Yadollahi, F. R.; Mahmoudi, M.; Kalantari, M. Polyrotaxane/gold nanoparticle hybrid nanomaterials as anticancer drug delivery systems. *J. Mater. Chem.* **2011**, *21*, 18686–18695.

Aguilar, Z. *Nanomaterials for Medical Applications.* Elsevier Science & Technology: Waltham, MA, 2012.

Aguilar, Z.; Aguilar, Y.; Xu, H.; Jones, B.; Dixon, J.; Xu, H. Nanomaterials in medicine. *Electrochem. Soc. Trans.* **2010**, *33*, 69–74.

Aguilar, Z.; Wang, Y. A.; Xu, H.; Hui, G.; Pusic, K. M. Nanoparticle-based immunological stimulation. U.S. Patent 20120189700 A1, July 26, **2012**.

Andros Pharmaceuticals Co., Ltd. http://www.andros.com.tw/en/technology_en.htm. (accessed 2014).

Au, J. L.; Jang, S. H.; Zheng, J.; Chen, C. T.; Song, S.; Hu, L.; Wientjes, M. G. Determinants of drug delivery and transport to solid tumors. *J. Controlled Release*. **2001**, *74*, 31–46.

Baranov, D.; Fiore, A.; Van, H. M.; Giannini, C.; Falqui, A.; Lafont, U. Assembly of colloidal semiconductor nanorods in solution by depletion attraction. *Nano Lett.* **2010**, *10*, 743–749.

Barratt, G. M. Therapeutic applications of colloidal drug carriers. *Pharm Sci. Tech. Today.* **2000**, *3*, 163–171.

Bhadra, D.; Bhadra, S.; Jain, P.; Jain, N. K. Pegnology: A review of PEG-ylated systems. *Pharmazie*. **2002**, *57*, 5–29.

Brigger, I.; Dubernet, C.; Couvreur, P. Nanoparticles in cancer therapy and diagnosis. *Adv. Drug Deliv. Rev.* **2002**, *54*, 631–651.

Calvo, P. R.; Remunan-Lopez, C.; Vila, J. J. L.; Alonso, M. J. Chitosan and Chitosan/Ethylene Oxide-Propylene Oxide Block Copolymer Nanoparticles as Novel Carriers for Proteins and Vaccines. *Pharm. Res.* **1997**, *14*, 1431–1436.

Carbone, L.; Kudera, S.; Carlino, E.; Parak, W. J.; Giannini, C.; Cingolani, R. Multiple wurtzite twinning in CdTe Nanocrystals induced by Methylphosphonic acid. *J. Am. Chem. Soc.* **2006**, 128, 748–755.

Center for Nanotechnology Research at Oakland University. [http://www.nano-ou.net/Edu4ImagePages/micelle.aspx. accessed 2014].

Cherukuri, P.; Bachilo, S.; Litovsky, S.; Wisman, R. Near-infrared fluorescence microscopy single-walled carbon nanotubes in phagocytic cells. *J. Am. Chem. Soc.* **2004**, 126, 15638–15639.

Choi, E.; Kwak, M.; Janga, B.; Piao, Y. Highly monodisperse rattle-structured nanomaterials with gold nanorod core–mesoporous silica shell as drug delivery vehicles and nanoreactors. *Nanoscale*. **2013**, *5*, 151–154.

Couvreur, P.; Barratt, G.; Fattal, E.; Legrand, P.; Vauthier, C. Nanocapsule technology: a review. *Crit. Rev. Ther. Drug. Carrier Syst.* **2002**, *19*, 99–134.

Cozzoli, P. D.; Manna, L.; Curri, M. L.; Kudera, S.; Giannini, C.; Striccoli, M. Shape and phase control of colloidal ZnSe nanocrystals. *Chem. Mater.* **2005**, *17*, 1296–1306.

Dagar, S.; Sekosan, M.; Lee, B. S.; Rubinstein, I.; Onyuksel, H. VIP receptors as molecular targets of breast cancer: implications for targeted imaging and drug delivery. *J. Controlled Release*. **2001**, *74*, 129–134.

Das, S.; Jagan, L.; Isiah, R.; Rajech, B.; Backianathan, S.; Subhashini, J. Nanotechnology in oncology: characterization and in vitro release kinetics of cisplatin-loaded albumin nanoparticles: Implications in anticancer drug delivery. *Indian J. Pharmacol.* **2011**, *43*, 409–413.

Deka, S.; Falqui, A.; Bertoni, G.; Sangregorio, C.; Poneti, G.; Morello, G. Fluorescent asymmetrically Cobalt-tipped CdSe@CdS core @shell nanorod heterostructures exhibiting room-temperature ferromagnetic behavior. *J. Am. Chem. Soc.* **2009**, *131*, 12817–12828.

Dekker, C. Carbon nanotubes as molecular quantum wires. *Phys. Today* **1999**, *52*, 22–28.

Demers, L.; Mirkin, C.; Mucic, R.; Reynolds, R.; Letsinger, R.; Elghanuan, R. A fluorescence-based method for determining the surface coverage and hybridization efficiency of thiol-capped oligonucleotides bound to gold thin films and nanoparticles. *Anal. Chem.* **2000**, *72*, 5535–5541.

Dramou, P.; Zuo, P.; He, H.; Pham, H. L. A.; Zou, W.; Xiao, D.; Pham-Huyc, C.; Ndorbora, T. Anticancer loading and controlled release of novel water-compatible magnetic nanomaterials as drug delivery agents, coupled to a computational modeling approach. *J. Mater. Chem.* **2013**, *1*, 4099–4109.

Drummond, D. C.; Mever, O.; Hong, K.; Kirpotin, D. B.; Papahadjopoulos, D. Optimizing liposomes for delivery of chemotherapeutic agents to solid tumors. *Pharmacol Rev.* **1999**, *51*, 691–743.

Emerich, D. F.; Thanos, C. G. The pinpoint promise if nanoparticle-based drug delivery and molecular diagnosis. *Biomol. Eng.* **2006**, *23*, 171–184.

Farokhzad, O. C.; Karp, J. M.; Langer, R. Nanoparticle-aptamer bioconjugates for cancer targeting. *Exp. Opin. Drug. Del.* **2006**, *3*, 311–324.

Ferreira, P.; Coelho, J.; Almeida, J.; Gil, M. *Photocrosslinkable polymers for biomedical applications.* In *Biomedical Engineering – Frontiers and Challenges.* Rezai, F. R., Ed.; InTech: Rijeka, Croatia, 2011.

Fresta, M.; Puglisi, G.; Giammona, G.; Cavallaro, G.; Micali, N.; Furneri, P. M. Pefloxacine Mesilateand Ofloxacin-Loaded Polyethylcyanoacrylate Nanoparticles: Characterization of the Colloidal Drug Carrier Formulation. *J. Pharm. Sci.* **1995**, *84*, 895–902.

Gong, J.; Liang, Y.; Huang, Y.; Chen, J.; Jiang, J.; Shen, G. Ag/SiO_2 core-shell nanoparticle-based surface-enhanced Raman probes for immunoassay of cancer marker using silica-coated magnetic nanoparticles as seperation tools. *Biosensors & Bioelectronics*. **2006**, *22*, 1501–1507.

Govender, T.; Stolnik, S.; Garnett, M. C.; Illum, L.; Davis, S. S. PLGA nanoparticles prepared by nanoprecipitation: drug loading and release studies of a water soluble drug. *J. Control Release*. **1999**, *57*, 171–185.

Guzman, L. A.; Labhasetwar, V.; Song, C.; Jang, Y.; Lincoff, A. M.; Levy, R.; Topol, E. J. Local Intraluminal Infusion of Biodegradable Polymeric Nanoparticles. A Novel Approach for Prolonged Drug Delivery after Balloon Angioplasty. *Circulation* **1996**, *94*, 1441–1448.

Huang, C.; Yang, Z.; Lee, K.; Chang, H. T. Synthesis of highly fluorescent gold nanoparticles for sensing mercury (II). *Angew. Chem. Int.* **2007**, *46*, 6824–6828.

Iscan, Y. Y.; Hekimoglu, S.; Kas, S.; Hinca, A. A. *Formulation and characterization of solid lipid nanoparticles for skin delivery in Conference lipid and surfactant dispersed systems*; Proceedings Book: Moscow, Russia, **1999**, 163–166.

Kamps, J. A.; Scherphof, G. L. Receptors versus non-receptor mediated clearance of liposomes. *Adv. Drug Deliv. Rev.* **1998**, *32*, 81–97.

Klostergaard, J. S. C. Magnetic nanovectors for drug delivery. *Nenomed. Nanotechnology. Biol. Med.* **2012**, *73*(1), 33–44.

Koo, O. M.; Rubinstein, I.; Onyuksel, H. Camptothecin in Sterically Stabilized Phospholipid Nano-micelles: A Novel Solvent pH Change Solubilization Method. *J. Nanosci. Nanotech.* **2006**, *6*, 2996–3000.

Koo, O. M.; Rubinstein, I.; Onyuksel, H. Role of nanotechnology in targeted drug delivery and imaging: A concise review. *Nanomedicine*. **2005**, *1*, 193–212.

Koval, M.; Preiter, K.; Adles, C.; Stahl, P. D.; Steinberg, T. H. Size of IgG opsonized particles determines macrophage response during internalization. *Exp. Cell. Res.* **1998**, *242*, 265–273.

Kreuter, J., Ed. *Nanoparticles, Encyclopedia of Nanotechnology.* Marcel Dekker Inc.: New York, 1994.

Kristl, J.; Volk, B.; Gasperlin, M.; Sentjure, M.; Jurkovic, P. Effect of colloidal carriers on a scorbyl palmitate stability. *Eur. J. Pharm. Sci.* **2003**, *19*, 181–189.

Kroll, R. A.; Pagel, M. A.; Muldoon, L. L.; Roman, G. S.; Fiamengo, S. A.; Neuwelt, E. A. Improving drug delivery to intracerebral tumor and surrounding brain in a rodent model: a comparison of osmotic versus bradykinin modification of the blood-brain and/or blood-tumor barriers. *Neurosurgery*. **1998**, *43*, 879–886.

Kubik, T.; Bogunia-Kubik, K.; Sugisaka, M. Nanotechnology on duty in medical applications. *Curr. Pharm. Biotechnol.* **2005**, *6*, 17–33.

Lee, C. C.; Gillies, E. R.; Fox, M. E.; Guillaudeu, S. J.; Fréchet, J. M.; Dy, E. E.; Szoka, F. C. A Single Dose of Doxorubicin-Functionalized Bow-Tie Dendrimer Cures Mice Bearing C-26 Colon Carcinomas. *Proc. Natl. Acad. Sci. U.S.A.* **2006**, *103*, 16649–16654.

Lee, J. H. and Nan, A. Combination Drug Delivery Approaches in Metastatic Breast Cancer, Journal of Drug Delivery, vol. 2012, Article ID 915375.

Linhardt, R. J. Biodegradable polymers for controlled release of drugs. In *Controlled Release of Drugs: Polymers and Aggregate Systems;* Rosoff, M., Ed.; VCH Publishers: New York, **1989**, 53–95.

Liversidge, G. Controlled release and nanotechnologies: recent advances and future opportunities. *Drug Dev. Deliv.* **2011**, *11*, 1.

Mahmoudi, M.; Sant, S.; Wang B.; Laurent, S.; Sen, T. Superparamagnetic iron oxide nanoparticles (SPIONs): Development, surface modification and applications in chemotherapy. *Adv. Drug Deliv. Rev.* **2011**, *63*, 24–46.

McDevitt, M. R.; Chattopadhyay, D.; Kappel, B. J.; Jaggi, J. S.; Schiffman, S. R.; A. C.; Njardarson, J. T.; Brentjens, R.; Scheinberg, D. A. Tumor Targeting with Antibody-Functionalized, Radiolabeled Carbon Nanotubes. *J. Nucl. Med.* **2007**, *48*, 1180–1189.

Medina, O. P.; Pillarsetty, N.; Glekasa, A.; Punzalan, B.; Longo, V.; Gönen, M.; Zanzonico, P.; Smith, J. P.; Larson, S. M. Optimizing Tumor Targeting of the Lipophilic EGFR-Binding Radiotracer SKI 243 Using a Liposomal Nanoparticle Delivery System. *J. Control Release* **2011**, *149*, 292–298.

Mehdipoor, E.; Adeli, M.; Bavadi, M.; Sasanpour, P.; Rashidian, B. A possible anticancer drug delivery system-based on carbon nanotube-dendrimer hybrid nanomaterials. *J. Chem.* **2011**, *21*, 15456–15463.

Mejias, R. P.; Pérez-Yagüe, S.; Gutierez, L.; Cabrera, L. I.; Spada, R.; Acedo, P.; Serna, C. J.; Lazaro, F. J.; Villanueva, A.; Morales, M. P.; Barber, D. F. Dimercaptosuccinic Acid-Coated Magnetite Nanoparticles for Magnetically Guided In Vivo Delivery of Interferon Gamma for Cancer Immunotherapy. *Biomaterials.* **2011**, *32*, 2938–2952.

Milhelm, C.; Gazeau, F. Universal Labeling with Anionic Magnetic Nanoparticles. *Biomaterials.* **2008**, 3161–3174.

Moghimi, S. M.; Szebeni, J. Stealth liposomes and long circulating nanoparticles: Critical issues in pharmacokinetics, osponization and protein-binding properties. *Prog. Lipid Res.* **2003**, *42*, 463–478.

Moghimi, S. M.; Hunter, A. C.; Murray, J. C. Long-circulating and target-specific nanoparticles: Theory to practice. *Pharmacol Rev.* **2001**, *53*, 283–318.

MsCarthy, J. R.; Kelly, K. A.; Sun, E. Y.; Weiddleder, R. Targeted delivery of multifunctional magnetic nanoparticles. *Nanomedicine* **2007**, *2*, 153–167.

Muller, R. H.; Maassen, S.; Weyhers, H.; Mehnert, W. Phagocytic uptake and cytotoxicity of solid lipid nanoparticles (Sln) sterically stabilized with poloxamine 908 and poloxamer 407. *J. Drug Target.* **1996**, *4*, 161–170.

Muller, R. H.; Schwarz, C.; Mehnert, W; Lucks, J. S. Production of solid lipid nanoparticles (SLN) for controlled release drug delivery. *Proc. Int. Symp. Control Release Bioact. Mater.* **1993**, *20*, 480–481.

Muller, R.; Mader, K.; Gohla, S. Solid lipid nanoparticles (SLN) for controlled drug delivery: a review of the state-of-the-art. *Eur. J. Pharm. Biopharm.* **2000**, *50*, 161–177.

Mykhaylyk, O. M.; Dudchenko, N. O.; Dudchenko, A. K. Pharmacokinetics of the Doxorubicin Magnetic Nanoconjugate in Mice Effects of the Nonuniform Stationary Magnetic Field. *Ukr. Biokhim. Zh.* **2005**, *77*, 80–92.

Nguyen, S.; Hiorth, M.; Rykke, M.; Smistad, G. The potential of liposomes as dental drug delivery systems. *Eur. J. Pharm. Biopharm.* **2011**, *77*, 75–83.

Nobs, L.; Buchegger, F.; Gurny, R.; Allemann, E. Current methods for attaching targetting ligands to liposomes and nanoparticles. *J. Pharm. Sci.* **2004**, *93*, 1980–1992.

Olivier, J. C. Drug transport to brain with targeted nanoparticles. *NeuroRX.* **2005**, *2*, 108–119.

Osterfield, S. J.; Yu, H.; Gaster, R. S.; Caramatu, S.; Xu L.; Han, S. Multiplex protein assays-based on real-time magnetic nanotag sensing. *PNAS.* **2008**, *105*, 20637–20640.

Paez, J.; Martinelli, M.; Brunetti, V.; Strumia, M. Dendronization: A useful synthetic strategy to prepare multifunctional materials. *Polymers.* **2012**, *4*(1), 355–395.

Panyam, J.; Labhasetwar, V. Biodegradable nanoparticles for drug and gene delivery to cells and tissue. *Adv. Drug Deliv. Rev.* **2003**, *55*, 329–347.

Panyam, J.; Sahoo, S.; Prabha, S.; Bargar, T.; Labhasetwar, V. Fluorescence and electron microscopy probes for cellura and tissue uptake of poly(D-, L-lactide-coglycolide) nanoparticles. *Int. J. Pharm.* **2003**, 262: 1–11.

Panyam, J.; Williams, D.; Dash, A.; Lesli, P. D.; Labhasetwar, V. Solid-state solubility influences encapsulation and release of hydrophobic drugs from PLGA/PLA nanoparticles. *J. Pharm. Sci.* **2004**, *93*, 1804–1814.

Park, J. H.; Maltzahn, V. G.; Zhang, L.; Schwartz, M. P.; Ruoslahti, E.; Bhatia, S. Magnetic iron oxide nanoworms for tumor targeting and imaging. *Adv. Mater.* **2008**, *20*, 1630–1635.

Peracchia, M. T.; Gref, R.; Minamitae, Y.; Domb, A.; Lotan, N.; Langer, R. PEG-coated nanospheres from amphiphilic diblock and multiblock copolymers: investigation of their drug encapsulation and release characteristics. *J. Control Release.* **1997**, *46*, 223–231.

Pitt, C. G.; Gratzl, M. M.; Kimmel, G. L.; Surles, J.; Schindler, A.. Aliphatic polyesters II. The degradation of poly(DL-lactide), poly(epsilon-caprolactone), and their copolymers in vivo. *Biomaterials.* **1981**, *2*, 215–220.

Plank, C. Nanomedicine: silence the target. *Nat Nanotechnol.* **2009**, *4*, 544–545.

Pusic, K.; Xu, H.; Stridiron, A.; Aguilar, Z.; Wang, A.; Hui, G. Blood stage merozite surface protein conjugated to nanoparticles induce potent parasite inhibitory antibodies. *Vaccine.* **2011**, *29*, 8898–8908.

Rejman, J.; Oberle, V.; Zuhorn, I. S.; Hoekstra, D. Size dependent internalization of particles via the pathways of clathrin- and caveolae-mediated endocytosis. *Biochem. J.* **2004**, *377*, 159–169.

Sethi, V.; Rubinstein, I. A novel therapy for rheumatoid arthiritis using a-helix VIP. *FASEB Conference Proceedings*. **2003**, 660.

Singh, R.; Lillard, J. W. Jr. Nanoparticle-based targeted delivery. *Exp. Mol. Pathol.* **2009**, *86*, 215–223.

Slowing, I. I.; Trewyn, B. G.; Giri, S; Lin, V. Mesoporous Silica Nanoparticles for Intracellular Delivery of Membrane-Impermeable Proteins. *Advanced Functional Materials*. **2007**, *17*, 1225–1236.

Su, H.; Xu, H.; Gao, S.; Dixon, J.; Aguilar, Z.; Wang, A. Microwave synthesis of nearly monodisperse core/multishell quantum dots with cell imaging applications. *Nanoscale Res. Lett.* **2010**, *5*, 625–630.

Ueno, Y.; Futagawa, H.; Takagi, Y.; Ueno, A.; Mizuzshima, Y. Drug-incorporating calcium carbonate nanoparticles for a new delivery system. *J. Control Release*. **2005**, *103*, 93–98.

Vo, D. T.; Kasili, P.; Wabuyele, M. Nanoprobes and nanobiosensors for monitoring and imaging individual living cells. *Nanomedicine*. **2006**, *2*, 22–30.

Wagner, W.; Dullaart, A.; Bock, A.; Zweck, A. The emerging nanomedicine landscape. *Nat. Biotechnol.* **2006**, *24*, 1211–1217.

Yang, L.; Cao, Z.; Sajja, H.; Mao, H.; Wang, L.; Geng, H. Development of receptor targeted iron oxide nanoparticles for efficient drug delivery and tumor imaging. *J. Biomed. Nanotech.* **2008**, *4*, 1–11.

Yang, Y.; Wang, S.; Wang, Y.; Wang, X.; Wang, Q.; Chen, M. Advances in self-assembled chitosan nanomaterials for drug delivery. *Biotechnology Advances*. **2014**, *32*(7), 1301–1316.

Zauner, W.; Farrow, N. A.; Haines, A. M. In vitro uptake of polystyrene microspheres: effect of particle size, cell line and cell density. *J. Control Release*. **2001**, *71*, 39–51.

Zhang, X.; Meng, L.; Lu, Q.; Fei, Z.; Dyson, P. J. Targeted delivery and controlled release of doxorubicin to cancer cells using modified single wall carbon nanotubes. *Biomaterials*. **2009**, *30*, 6041–6047.

CHAPTER 6

CARBON NANOTUBES USED AS NANOCARRIERS IN DRUG AND BIOMOLECULE DELIVERY

HUA HE,[1,2] DELI XIAO,[1] LIEN AI PHAM-HUY,[3] PIERRE DRAMOU,[1] and CHUONG PHAM-HUY[4,*]

[1]*China Pharmaceutical University, Nanjing 210009, China*

[2]*Key Laboratory of Drug Quality Control and Pharmacovigilance, Ministry of Education, China Pharmaceutical University, Nanjing 210009, China*

[3]*Department of Pharmacy, Lucile Packard Children's Hospital Stanford, Palo Alto, CA, USA*

[4]*Faculty of Pharmacy, University of Paris V, 4 Avenue de l'Observatoire, 75006 Paris, France, *E-mail: phamhuychuong@yahoo.com*

CONTENTS

ABSTRACT

Carbon nanotubes (CNTs), an artificial allotrope of carbon recently discovered in 1991, are made of graphite and constructed in cylindrical tubes having nanometer in diameter and several millimeters in length. The structures of CNTs are classified in two types: single-walled carbon nanotubes (SWCNTs) and multiwalled carbon nanotubes (MWCNTs). Due to their small size and mass, strong mechanical potency, and high electrical and thermal conductivity, CNTs have been successfully applied in different areas of pharmacy and medicine for drug preparation and therapy. They have been first proven to be an excellent nano-vehicle for the delivery of different therapeutic agents (drugs, biomolecules, etc.) directly into cells without metabolism by the body. Thanks to their tiny structure, CNTs can effectively cross the cell membrane and directly deliver the transported drug into the cell. CNTs are able to maintain the drug intact during this transport and protect the drug against the metabolism by the body. The important characteristics of this nanotechnology in pharmaceutics are to revolutionize the methods of drug delivery since traditional drug administration cannot resolve the problems of toxicity and/or bioavailability of numerous effective drugs, thereby limiting their use in therapeutics. As CNTs are not water soluble, surface functionalization is required before their linkage with drugs or biomolecules. CNTs have also been proposed in biomolecule delivery such as DNA, proteins, antibodies, etc., for gene therapy, tissue regeneration, artificial implants and diagnosis of human diseases.

This review focuses the applications of CNTs used as nanocarriers in drug and biomolecule delivery for chemotherapeutic use and also studies the pharmacokinetics, metabolism and toxicity of different forms of CNTs. Finally, it discusses the prospect of this promising bio-nanotechnology in the future clinical exploitation.

6.1 INTRODUCTION

At the beginning of the twenty-first century, the development of nanomaterials in general and of carbon-based nanostructures in particular, such as carbon nanotubes (CNTs) has attracted a number of scientists worldwide for their applications in different areas of medicine and pharmacy. One important application of this nanotechnology in pharmaceutics is to revolutionize the ways of drug delivery because traditional drug administration cannot resolve in many cases the problems of toxicity and/or bioavailability for numerous effective drugs, thereby limiting their use in therapeutics. A nanocarrier is nanomaterial being used as a vehicle to transport a drug or another substance to the target cell or organ. Main nanocarriers with sizes of diameter from 1–100 nm include carbon-based nanomaterials (carbon nanoparticles, carbon nanofibers, and CNTs), polymeric micelles, polymers, superparamagnetic Fe_3O_4 nanoparticles, liposomes, dendrimers, etc. (Babu et al., 2014; Bamrungsap et al., 2012; Cho et al., 2008; Rawat et al., 2006; Safari and Zarnegar, 2014; Surendiran et al., 2009).

Since the discovery of CNTs in 1991 by Japanese scientist Sumio Iijima (1991), awarded the Kavli Prize for nanoscience in 2008, an explosion of their diverse applications has been sparked worldwide from academic research to industrial development. Artificial CNTs are built of graphite in the form of cylindrical tubes with nanometer-scale in diameter and several millimeters in length and are classified as allotropes of carbon (He et al., 2013; Kumar et al., 2012; Usui et al., 2012). Structures of CNTs are divided in two types: single-walled carbon nanotubes (SWCNTs) and multiwalled carbon nanotubes (MWCNTs). Due to their unique combinations of chemical and physical properties (i.e., thermal and electrical conductivity, high mechanical strength, and optical properties) and also to their small size and mass, CNTs were initially used as additives to various structural materials for electronics, optics, plastics and other materials in the nanotechnology fields (Baughman et al., 2002). They were introduced to the field of medicine only within the last decade, and have proven to be an excellent nano-vehicle for drug delivery because they can transport drug directly to the target cells without metabolism by the body. Thanks to its tiny structure and also to its special physicochemical properties such as rich electronic polyaromatic structure, excellent chemical stability, high surface area, a CNT can adsorb or conjugate with different therapeutic agents (drugs, biochemicals, biomacromolecules, etc.) and cross the cell membrane without being destroyed during this penetration (Liu et al., 2007; Mehra et al., 2014; Singh et al., 2012; Zhang et al., 2010). The CNT then directly delivers the transported molecule into the cell without being recognized as an undesirable intruder. Briefly, a CNT can keep the drug intact during the transport and can also be used effectively as a vehicle to deliver drug to the ill target cell with high accuracy and efficiency. Moreover, the amount of drug used for this preparation would be very low since it is not metabolized during transport in the body and is only released in the target cell, thereby avoiding drug toxicity to healthy tissues or organs (He et al., 2013; Liu et al., 2007; Mehra et al., 2014; Singh et al., 2012; Zhang et al., 2010). As CNTs are not soluble in water, and cannot be consequently circulated in the body, surface functionalization is required to solubilize CNTs before their linkage with drugs or biomolecules for the delivery process (Kateb et al., 2010; Liu et al., 2007; Mehra et al., 2014; Pastorin et al., 2005; Rosen and Elman, 2009; Singh et al., 2012; Zhang et al., 2010, 2011). Many comparison studies in drug delivery have proven that drugs are delivered more effectively and safely into cells by this nanotechnology using CNTs as carriers than by traditional methods (He et al., 2013; Kateb et al., 2010; Liu et al., 2007; Mehra et al., 2014; Singh et al., 2012; Zhang et al., 2010, 2011). A new procedure for drug preparation that is completely different to the traditional techniques has opened and can radically change previous concepts of drug pharmacology and toxicology. This original nanotechnology using CNTs as carriers has been applied not only to toxic drugs for cancer or infection therapy, but also to other biomolecules (genes, proteins, DNA, antibodies, vaccines, biosensors, cells, etc.) for gene therapy, immunotherapy, tissue regeneration, diagnosis, etc. (Digge et al., 2012; Kumar et al., 2014; Liao et al., 2011; Liu et al., 2009; Pastorin et al., 2005; Rosen and Elman, 2009; Tan et al., 2014; Zhang et al., 2011). Nevertheless, these medicinal findings are being in experimental stages and still not applied in clinical study. Besides the capacity of CNTs to act as carriers for the delivery of a wide range of therapeutic agents, they have also been used for the analysis and extraction of numerous drugs and biomolecules in pharmaceutical industry as well as in laboratory for their galenic or therapeutic controls (He et al., 2013; El-Sheikh and Sweileh et al., 2011). Our group has recently contributed to some applications of CNTs in the pharmaceutical field such as the development of different novel techniques for the functionalization of CNTs used in drug delivery as well as for the study on the interactions between CNTs and albumin or drugs and also on the drug analysis (Chen et al., 2011; Dai et al., 2015; He et al., 2013; Jiang et al., 2012; Li et al., 2008, 2010, 2013; Xiao et al., 2012, 2013, 2014; Zha et al., 2011).

In summary, this review is focused on the applications of CNTs as nanocarriers in the delivery process of drugs and other biomolecules used in therapeutics. It describes the structure of CNTs, their surface functionalization techniques and their linkage process with drugs or biomolecules. It examines some main methodologies using CNTs as vehicle for drug and biomolecule delivery in the treatment of different diseases as well as in the tissue regeneration therapy. Other aspects of CNTs such as pharmacokinetics, metabolism and toxicity are also described. The perspectives and obstacles of this promising bio-nanotechnology using CNTs in the future medicine are also commented in the conclusion.

6.2 STRUCTURE, TYPE, PHYSICOCHEMICAL PROPERTIES OF CNTs

In the periodic table, carbon is the only element that occurs in allotropic forms from 0 dimensions to 3 dimensions, due to its different hybridization capabilities. Carbon nanotubes (CNTs) are tiny cylindrical tubes formed exclusively of carbon atoms arranged in a series of condensed benzene rings or graphite sheets rolled-up into a tubular structure with 1–3 nanometers in diameter, and hundreds to thousands of nanometers long. The artificial CNTs belong to the family of fullerenes, the third allotropic form of carbon along with their two natural carbons: graphite and diamond which are both sp2 (planar) and sp3 (cubic) forms, respectively (Baughman et al., 2002; He et al., 2013; Singh et al., 2012; Usui et al., 2012). The chemical bonding of CNTs is composed entirely of *sp*2 bonds, similar to those of graphite, but stronger than the *sp*3 bonds found in alkanes and diamond (Zhang et al., 2011).

Purified CNT can exhibit metallic conductivity, chemical and thermal stability and extremely high tensile strength and elasticity (Mehra et al., 2014; Pastorin et al., 2005). They are among the stiffest and strongest fibers known, and have remarkable electronic properties and many other unique characteristics like high flexibility, specific surface area, low density, and superior optoelectronic characteristics (Kumar et al., 2014). CNTs are considered as the strongest and stiffest materials yet discovered (Zhang et al., 2011).

Depending on the number of layers in their wall, CNTs are classified in two types: single-walled carbon nanotubes (SWCNTs) and multiwalled carbon nanotubes (MWCNTs) (Figure 6.1) (He et al., 2013). When their wall is formed by rolling up of single sheet, CNTs obtained are called SWCNTs, and when their wall is constructed by more than one sheet,

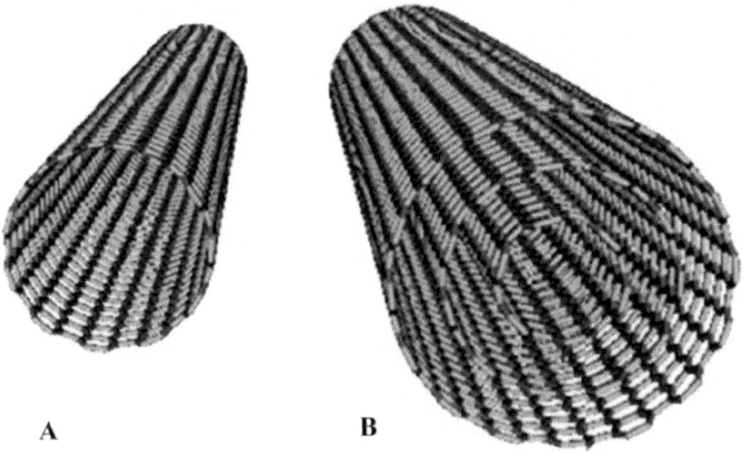

FIGURE 6.1 Schematic illustration of Single Walled Carbon Nanotube (SWCNT) (A) and Multiple Walled Carbon Nanotube (MWCNT) (B). (From He, H.; Pham-Huy, A. L.; Dramou, P.; Xiao, D-L.; Zuo, P.; Pham-Huy, C. Carbon Nanotubes: Applications in Pharmacy and Medicine, *BioMed. Res Int.* **2013,** *ID 578290*, 1–12. Creative Commons Attribution License.)

they are called MWCNTs. Both SWCNTs and MWCNTs are capped at both ends of the tubes called tips in a hemispherical arrangement of carbon networks (Figure 6.2) (Singh et al., 2012). From a chemical reactivity point of view, CNT can be differentiated in two zones: the tips and the sidewalls. An important factor that controls these unique properties comes from a variation of tubule structures that are caused by the rolling up of the graphene sheet into a tube. Depending on the rolling direction formed by three distinct ways, CNTs can be further classified as zigzag, chiral and armchair types (Baviskar et al., 2012; He et al., 2013). The first of the three structural types is called "zigzag" because the pattern of hexagon moves circumferentially around the body of the tube. The second form is named "chiral" because the tube may twist in either direction and that means handedness. The third form is designed as "armchair" because it describes one of the two conformers of cyclo-hexane, hexagon of carbon atoms and indicates the shape of the hexagons as one moves around the body of the tube (Baviskar et al., 2012; He et al., 2013; Kumar et al., 2012; Usui et al., 2012). Armchair nanotubes are really metallic, whereas zigzag and chiral forms are semiconductors (Baviskar et al., 2012) (Figure 6.3).

SWCNTs consist of a single graphene cylinder with diameter ranging from 0.4 to 2 nm. Their sidewalls are usually formed as hexagonal close-packed bundles, but their tips as pentagonal ones (Figure 6.2). MWCNTs consist of two to several coaxial cylinders, each made of a single graphene sheet surrounding a hollow core. The interlayer separation of the graphene layers of MWCNTs is 0.34 nm in average, their outer diameter

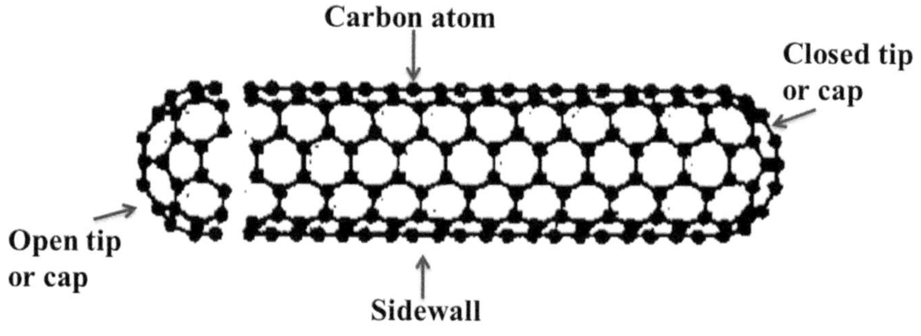

FIGURE 6.2 General representation of an open single walled carbon nanotube.

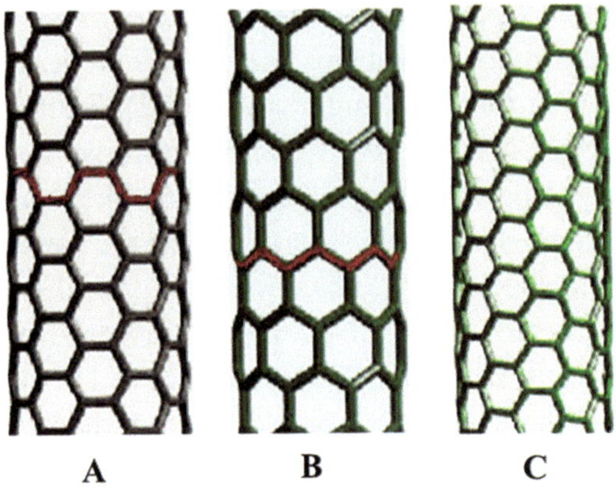

FIGURE 6.3 Carbon nanotube structures of armchair, zigzag and chiral configurations (A = Armchair; B = Zigzag; C = Chiral). They differ in chiral angle and diameter: Armchair carbon nanotubes share electrical properties similar to metals. The zigzag and chiral carbon nanotubes possess electrical properties similar to semiconductors.

varies between 2–100 nm, while their inner diameter ranges from 1 to 3 nm, and their length is 0.2 to several μm (Singh et al., 2012; Zhang et al., 2011). SWCNTs have a better-defined wall, whereas MWCNTs are more likely to have structural defects, resulting in a less stable nanostructure (Zhang et al., 2011). However, for their use as drug carriers, no conclusive advantages of SWCNTs relative to MWCNTs are notified (Zhang et al., 2011). CNTs vary significantly in length and diameter depending on the chosen synthetic procedure. Individual SWCNTs and MWCNTs have natural tendency to bundle together into ropes by attractive Van der Waals forces (Mishra and Mishra, 2013; Zhang et al., 2011). Bundles contain many nanotubes and can be considerably longer and wider than the original ones from which they are formed. This phenomenon could be of important toxicological significance (Mishra and Mishra, 2013; Zhang et al., 2011). Other differences between the structures and characterization of SWCNTs and MWCNTs are summarized in Table 6.1 according to the references following (He et al., 2013; Kumar et al., 2012, 2014; Singh et al., 2012).

For the production of CNTs, three main techniques generally used are: arc-discharge method, laser ablation method and chemical vapor deposition. Arc-discharge is the most common and easiest way to produce

TABLE 6.1 Differences between Pristine SWCNT and MWCNT

SWCNT	MWCNT
1. Single cylindrical layer of graphene	Multiple cylindrical layer of graphene
2. Diameter ranging from 0.4 to 2 nm	Diameter ranging from 2 to 100 nm
3. Catalyst is required for synthesis	Catalyst is not required for synthesis
4. Bulk synthesis is difficult	Bulk synthesis is easy
5. More defection during functionalization	Less defection, but difficult to improve
6. Purity is poor	Purity is high
7. Less accumulation in body	More accumulation in body
8. Easy characterization and evaluation	Difficult characterization and evaluation
9. Highly toxic	Relatively less toxic

CNTs. By this method, CNTs are produced through arc-vaporization of two carbon rods placed end to end, distanced about 1 mm, in an enclosure filled with inert gas at low pressure (Basu and Mehta, 2014; He et al., 2013). In laser ablation, CNTs are obtained from graphite under laser vaporization pulses (Basu and Mehta, 2014; He et al., 2013). Chemical vapor deposition can produce large amounts of CNTs by reduction of hydrocarbon sources (CO, methane, ethylene, acetylene) under the action of a metal catalyst (cobalt, nickel, iron on magnesium) at high temperature (800–1000°C). This technique can give a very high proportion of SWCNTs and few MWCNTs (Basu and Mehta, 2014; He et al., 2013). After obtaining, CNTs are then purified by acid refluxing, or surfactant aided sonication or air oxidation procedure in order to eliminate impurities such as amorphous carbon, fullerenes, transition metals used as catalysts during the synthesis (Digge et al., 2012; Kumar et al., 2012; Usui et al., 2012). The lengths of CNTs can vary from several hundreds of nanometers to several micrometers and can be shortened chemically or physically for their uses as drug carriers by opening their two tips (caps) for intratube drug loading and chemical functionalization (Zhang et al., 2011). The properties of CNTs may vary depending upon the types of methods employed in the synthesis of CNTs (Mehra et al., 2014). Pristine CNTs (raw material) are now synthetized by many chemical firms and marketed worldwide.

6.3 FUNCTIONALIZATION OF CNTS FOR DRUG AND BIOMOLECULE DELIVERY

Due to their highly hydrophobic surfaces, the solubility of pristine CNTs in water and organic solvents is extremely low therefore the use of this intact nanomaterial in biomedicine is limited even impossible. Moreover, their sediment in tissues or organs could become toxic. Several trials have been tested to render CNTs water soluble, mainly focusing on different surface functionalization (Arora et al., 2014; Bianco et al., 2005; He et al., 2013; Liu et al., 2009). Sidewall functionalization is required to solubilize CNTs, and to render biocompatibility and low toxicity for their medical applications (Liu et al., 2009). Used as drug carriers, the solubility of CNTs in water or biological liquids is an essential condition for gastrointestinal absorption, blood transportation, biocompatibility; and secretion. So, before the linkage with therapeutic agent, pristine CNTs must be rendered hydrophilic by different techniques (He et al., 2013; Zhang et al., 2011). There are at least two main approaches to increase their solubility, depending on the nature of drug or biomolecule linked to carbon nanotube. The first approach corresponds to the covalent attachment by chemical bond formation and the second one to the noncovalent attachment by physio-adsorption (Pastorin et al., 2005; Yang et al., 2007; Zhang et al., 2010).

6.3.1 COVALENT FUNCTIONALIZATION OF CNTs

The covalent functionalization of CNTs is generally obtained by oxidation with strong acids. Oxidation of CNTs is carried out by reflexing pristine CNTs in strong acidic media, e.g., HNO_3/H_2SO_4 (He et al., 2013; Kateb et al., 2010; Mehra et al., 2014; Liu et al., 2007; Zhang et al., 2010). During the process, the end caps of the CNTs are opened, and carboxylic groups are formed at these end caps (tips) and at some defect sites on the sidewalls of SWCNTs or MWCNTs. The carboxylic groups obtained can provide further derivatization of CNTs by esterification or amidation reactions. For example, drugs or biomolecules containing amine groups can be directly linked to the carboxylic groups present on the surface of the CNTs (Tasis et al., 2006). Otherwise, the carboxyl moieties can be activated with thionyl chloride, then react with amine groups to form a covalent amide bond (Jiang et al., 2012; Naficy et al., 2009). These reactions are widely applied for conjugation of water-soluble drugs or biomolecules, hydrophilic polymers, nucleic acid (DNA or RNA), or peptides to the oxidized CNTs, which result in multifunctional CNTs (Liu et al., 2009; Tasis et al., 2006; Yang et al., 2007). It is notified that this oxidation technique can only generate carboxyl groups at the open sides (tips) and at the defect sites on the sidewalls of SWCNTs or MWCNTs (He et al., 2013).

For the creation of –COOH on the side walls of CNTs, nitrene cycloaddition, arylation using diazonium salts or 1,3-dipolar cycloadditions are usually employed (Kateb et al., 2010; Yang et al., 2007; Zhang et al., 2010). Cycloaddition reaction is a very powerful methodology, in which the 1,3-dipolar cycloaddition of azomethineylides can easily attach a large amount of pyrrolidine rings on sidewalls of nanotubes. Thus, the resulting functionalized CNTs are highly soluble in water. In addition, pyrrolidine ring can be substituted with many functional groups for different applications (He et al., 2013; Jiang et al., 2012). Because covalent functionalization of CNTs is robust and easy to control, this procedure is often used in drug delivery system such as drugs for cancer therapy (epirubicin, doxorubicin, cisplatin, methotrexate, paclitaxel, quercetin), infection therapy (antibiotics, vaccines), neurodegenerative diseases (acetylcholine), etc. Moreover, covalent functionalization can also diversify CNTs' surface properties, from which favorable CNTs can be selected (13).

Schematic illustration of common methods for chemical functionalization of carbon nanotubes is summarized in Figure 6.4.

6.3.2 NONCOVALENT FUNCTIONALIZATION OF CNTS

In contrast to covalent functionalization, non-covalent functionalization of CNTs can be carried out by coating CNTs with amphiphilic surfactant molecules or polymers (polyethyleneglycol). The large aromatic (π-electrons) hydrophobic surface of carbon nanotubes makes them ideal partners for non-covalent interactions with suitable complementary

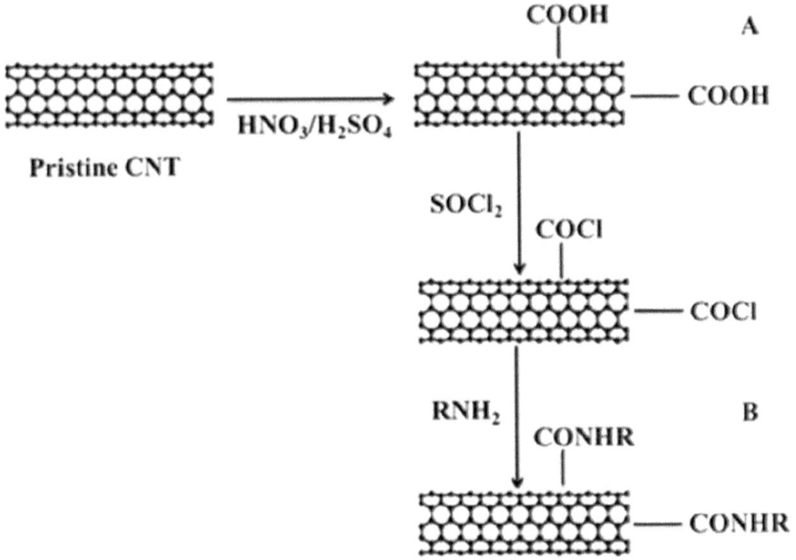

FIGURE 6.4 Schema of covalent functionalization of CNTs by: (a) Oxidation reaction by strong acid (HNO_3/H_2SO_4); (b) Further attaching hydrophilic molecules by amidation reactions.

molecules and macrobiomolecules (DNA) (He et al., 2013; Zhang et al., 2010). These interactions can take place both on the inside and outside of CNTs. However, macromolecules cannot be linked on their inside (Digge et al., 2012; Liu et al., 2009; Yang et al., 2007). As the chemical structure of the π-network of CNTs is not disrupted, the physical properties of CNTs are essentially preserved by the non-covalent approach (Liu et al., 2009; Zhang et al., 2011). Non-covalent functionalization is realized via enthalpy-driven interactions, such as π-π, CH-π, NH-π, etc., between the CNT surface and the dispersants and/or entropy-driven interaction, i.e., hydrophobic interaction using surfactants (Fujigaya and Nakashima, 2015). Forces that govern such adsorption are the hydrophobic and π-π stacking interactions between the chains of the adsorbed molecules and the surface of CNTs (Zhang et al., 2011). As many anticancer drugs are hydrophobic in nature or have hydrophobic moieties, the loading of such drugs into or onto CNTs is due to these hydrophobic forces. The presence of charge on the nanotube surface due to chemical treatment can enable the adsorption of the charged molecules through ionic interactions (Fujigaya and Nakashima, 2015; Zhang et al., 2011). Non-covalent functionalization of CNT is particularly attractive because it offers the possibility of attaching chemical handles without affecting the electronic network of CNTs, thereby favorable for diverse biomedical applications. The polyaromatic graphitic surface of a carbon nanotube is available to the binding of aromatic molecules via π-π stacking (Bianco et al., 2005; He et al., 2013). The advantage of the noncovalent methodology permits to preserve the aromatic structure of the nanotubes and thus their electronic characteristics. Different reagents and compounds often used in this technique are: organic solvents, organic polymers, amphiphilic peptides, detergents, nucleic acids and sugars. Generally, this functionalization is often used to associate CNTs with several molecules and to produce nanotube biosensors in particular for potential medical diagnostic and biological applications. However, the major problem of aggregation and precipitation of CNTs after the release of the bioactive molecule from its CNT complex, and the related undesirable effects, remain to be solved (Bianco et al., 2005).

Many biomolecules (proteins, metallothionein proteins, DNA, amylose, polysaccharides) can interact with CNTs without producing of covalent conjugates because they possess high affinity with the graphitic network and are capable to be adsorbed strongly on the external sides of CNTs; thus, the products obtained can be visualized clearly by microscopy techniques (Zhang et al., 2011). Schemes of different noncovalent functionalization of CNTs are illustrated in Figure 6.5.

Generally speaking, the covalent modifications are superior to the noncovalent modification in terms of the stability of the functionalization (Fujigaya and Nakashima, 2015). However, the covalent modification changes the intrinsic properties of the CNTs, such as conductivity and mechanical toughness, and often cuts the CNTs into shorter tubes; thus, the noncovalent modification is superior in most cases in order to use the inherent properties of the CNTs (Fujigaya and Nakashima, 2015). In addition, noncovalent modifications are characterized as their simple procedure, typically just by the mixing of CNTs with biomolecules under a shear force treatment such as sonication (Fujigaya and Nakashima, 2015).

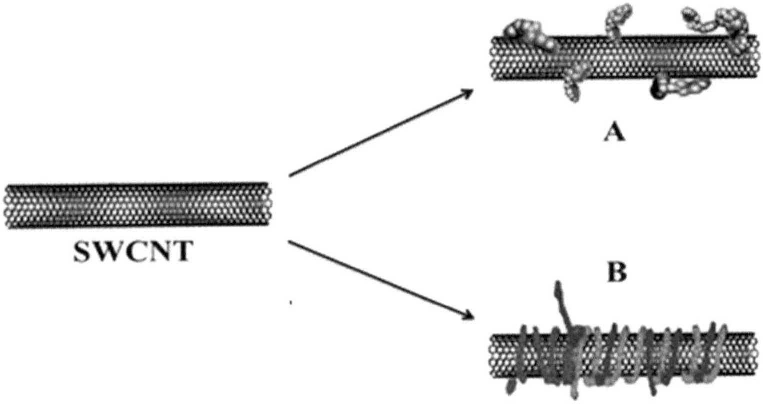

FIGURE 6.5 Schema of noncovalent functionalization of CNTs by: (a) Surfactants such as protein adsorption; (b) Polymers such as DNA wrapping.

Briefly, the choice between covalent or noncovalent functionalization for CNTs depends on the physicochemical nature of the therapeutic agent used, its quantity loaded as well as its quantity released. Detailed techniques for different functionalization of CNTs can be found in the literature and in our recent publications cited herein (Chen et al., 2011; Jiang et al., 2012; Kateb et al., 2010; Liu et al., 2007, 2009; Mehra et al., 2014; Zhang et al., 2010).

After suitable modifications, CNTs become hydrophilic and less toxic and are ready to be linked with drugs or biomolecules (genes, DNA, proteins, enzymes, biosensors, etc.) for their delivery into the target cells or organs.

6.3.3 MULTIFUNCTIONAL CNTs COMPOSITES

Another emerging issue of CNTs for use in biomedical areas is to provide multifunctionality. Carbon nanotubes present remarkable opportunities to meet the future drug delivery challenges because they are stable chemically and mechanically. The material incorporation into their inner hollow space is convenient, and their outside wall can be chemically modified according to the desired purpose (Singh et al., 2014). These strategies allow the generation of novel versatile systems that can be employed in many biotechnological or biomedical fields. One example of this is to endow magnetic properties, which have been implemented through the combination of magnetic nanoparticles (MNPs) with CNTs (Singh et al., 2014). The preparation of magnetic CNTs (Mag-CNTs) opens new avenues in nanobiotechnology and biomedical applications as a consequence of their multiple properties embedded within the same moiety (Masotti and Caporali, 2013). Several preparation techniques have been developed during the last few years to obtain magnetic CNTs: grafting or filling nanotubes with magnetic ferrofluids or attachment of magnetic nanoparticles to CNTs or their polymeric coating (Masotti and Caporali, 2013). Therefore, multifunctional CNTs composites, especially Mag-CNTs, have attracted considerable researcher attention over the past decades for different biomedical applications (Eatemadi et al., 2014; Gao et al., 2006; Masotti and Caporali, 2013; Singh et al., 2014; Xiao et al., 2012, 2013, 2014). Indeed, magnetic nanomaterials including Fe, Co, Ni and their alloys have been used as magnetic carriers for drug targeting. The magnetic delivery of CNTs through an external magnetic field is considered a promising approach to achieve specific delivery of drugs when directing these nanosystems to diseased organs. However, these nano-sized materials are easy to gather together due to their small size and their unstable active surface. In order to improve the stability and to expand the application of magnetic nanostructures, effective combining or surface coating is efficient strategy. Several preparation techniques for magnetic CNTs have been developed by our group or by the others during the last few years (Gao et al., 2006; Lee et al., 2010; Masotti and Caporali, 2013; Singh et al., 2014; Xiao et al., 2013, 2014) and their applications in biomedical fields will be described in different paragraphs below.

6.4 MECHANISMS OF DRUG OR BIOMOLECULE DELIVERY BY FUNCTIONALIZED CNTs

Functionalized CNTs have been assayed by different in vitro and/or in vivo experiments for studying the transport of various therapeutic agents, ranging from small drug molecules to bio-macromolecules, such as protein and DNA/RNA, into different types of cells. Many reports of these investigations could be resumed as the following.

Generally, the conjugate of functionalized CNTs with targeted therapeutic agent is able to cross the cytoplasmic membrane and nuclear membrane without destruction during this penetration and also without obvious toxic effect (Liu et al., 2009; Tripathi et al., 2015). After reaching the targeted cell, two possibilities to deliver drug: either the drug enters the cells without internalization of the CNT carrier or both the drug and the CNT carrier enters the cells. The latter internalization method is more efficacy than the first one, because after entering the cells, the intracellular environment degrades the drug-carrier conjugate releasing drug molecules *in situ*, i.e., inside the cells. While, in the non-internalization method, the extracellular environment helps degrade drug-carrier conjugates and the drug then crosses itself the lipid membrane to enter the cells, thereby, possibility of drug degradation during this penetration by itself (Basu and Mehta, 2014; He et al., 2013; Liu et al., 2009). To date, two major mechanisms of CNT internalization have been widely considered: (a) the endocytosis/phagocytosis pathway and (b) the insertion and diffusion pathway, i.e., via the endocytosis-independent pathway. However, the mechanism of the last pathway is still not well-known (Basu and Mehta, 2014; He et al., 2013; Liu et al., 2007; Tripathi et al., 2015). There are different interpretations about the drug or biomolecule delivery by functionalized CNTs into cells found in the literature, almost are not too divergent, here are some resumes of them.

According to Rastogi et al. (2014), the exact cellular uptake pathway of CNTs is complex and to resume there are two possible pathways to cross the cellular membrane (Figures 6.6A–6.6C). The first is endocytosis dependent pathway which may be either receptor mediated or nonreceptor mediated and the second is-based on endocytosis independent pathway which includes diffusion, membrane fusion, or direct pore transport of the extracellular material into the cell (Rastogi et al., 2014). The process of internalization of CNTs depends on several parameters such as the size, length, nature of functional groups, hydrophobicity, and surface chemistry of CNTs (Lacerda et al., 2012; Lee and Geckeler et al., 2010). Endocytosis dependent pathway is an energy and temperature dependent transport process (Rastogi et al., 2014). Once taken up by cells via endocytosis, CNTs are able to exit cells through exocytosis (Liu et al., 2009).

The understanding of the mechanisms involved in the interaction of biological systems with CNTs is of interest to both fundamental and

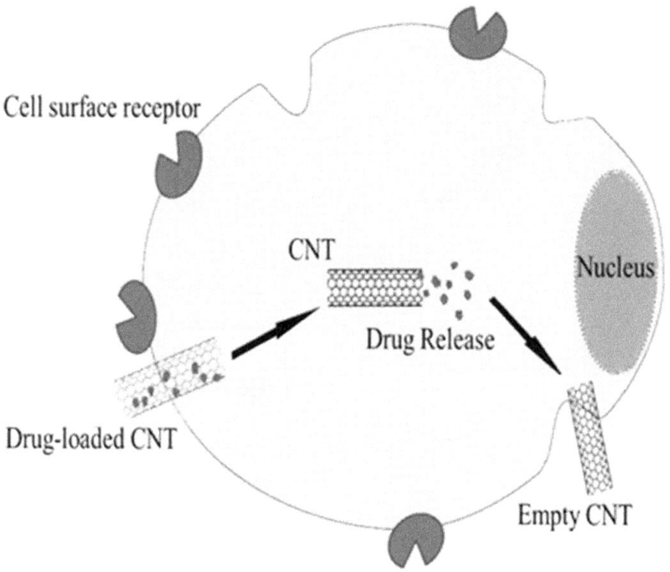

FIGURE 6.6-A By endocytosis independent pathway or direct penetration of drug-CNTs into the cell. In this passive process, drug-CNTs diffuse across the cellular membrane, then, release drugs *in situ*. This penetration looks like a tiny needle crossing the cellular membrane (Tripathi et al., 2015).

applied researches. Data demonstrates that shorter nanotubes have a stronger propensity to passively penetrate the cell membrane and reach the cytoplasm (Tripathi et al., 2015). Although much progress has been made in this domain, the details of different mechanisms about the penetration of functionalized CNTs through cell membrane are still debated. Cellular internalization has always been a preferred mechanism of drug delivery. Hydrophobicity and membrane asymmetry are the key factors responsible for the insertion and complete translocation (Tripathi et al., 2015). Different studies illustrated the significant contribution of plasma membrane translocation in the cellular uptake of CNTs. Multiple internalization pathways may simultaneously operate and determine CNT cellular uptake and trafficking (Tripathi et al., 2015). In addition, the last one will be strongly dependent on the type of CNT functionalization, the physicochemical nature of CNT dispersions and the type of target cells. In general, it could be concluded that there was no single, unique mechanism

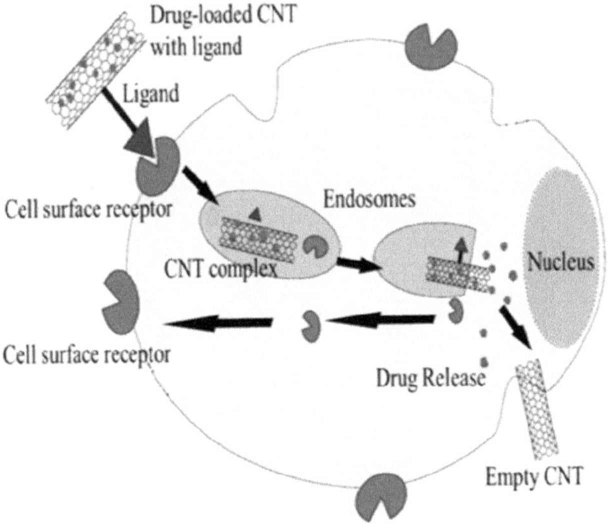

FIGURE 6.6-B By receptor mediated endocytosis, i.e., by ligand-receptor interactions. CNT surface is linked with drug and a ligand which is attracted by its surface cell receptor such as folate and cell folate receptor. After penetration into the cell by forming endosomes followed by internalization, drug is released *in situ* and cell receptor is regenerated (Rastogi et al., 2014). (Adapted from Rastogi, V.; Yadav, P.; Bhattacharya, S. S.; Mishra, A. K.; Verma, N.; Verma, A.; Pandit, J. K. Carbon Nanotubes: An Emerging Drug Carrier for Targeting Cancer Cells. *J. Drug Deliv.*, **2014,** *ID 670815*, 23 pp. https://creativecommons.org/licenses/by/3.0/)

responsible for CNT cellular uptake, and that chemical functionalization could represent a way to tailor the ulterior fate of CNTs such as internalization, cellular processing and elimination/degradation depending upon the desired application (Lacerda et al., 2012).

According to some other authors (He et al., 2013; Madani et al., 2011), the exact mechanism of CNT uptake is determined by various factors, such as size, shape, degree of dispersion, and the formation of supramolecular CNT complexes. They have reported that small CNTs with a length of upto 400 nm are internalized by a diffusion mechanism and that CNTs larger than 400 nm in length are internalized by endocytosis. It has also been suggested that CNTs attached to large proteins, such as streptavidin, staphylococcal protein A, or bovine serum albumin, are taken up via endocytosis, whereas CNTs attached to small molecules, such as ammonium, methotrexate, or amphotericin B, enter cells by a diffusion process (Madani et al., 2011).

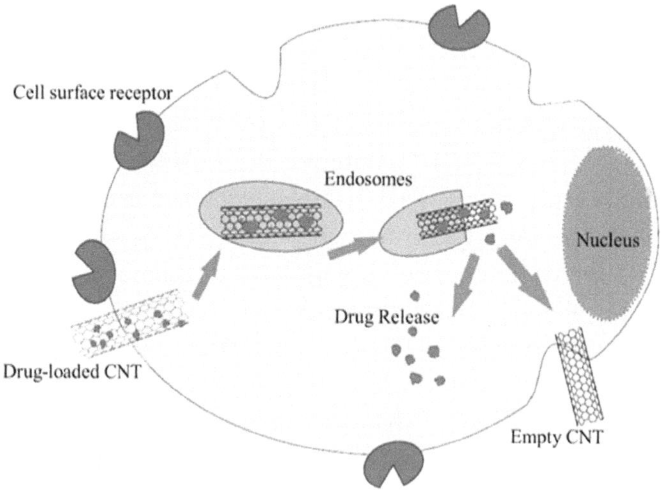

FIGURE 6.6-C By non receptor mediated endocytosis, i.e., without ligand-receptor interactions. In this case, the penetration of drug-CNTs is mediated by endocytosis, with formation of endosomes, but without the intervention of ligand-receptor process (Rastogi et al., 2014). (Adapted from Rastogi, V.; Yadav, P.; Bhattacharya, S. S.; Mishra, A. K.; Verma, N.; Verma, A.; Pandit, J. K. Carbon Nanotubes: An Emerging Drug Carrier for Targeting Cancer Cells. *J. Drug Deliv.*, **2014,** *ID 670815*, 23 pp. https://creativecommons.org/licenses/by/3.0/)

Briefly, the conjugate of drug or biomolecule attached to functionalized CNTs by covalent or noncovalent bonding, including hydrophobic interaction, π–π stacking interaction, and electrostatic adsorption is ready to cross cell membranes to deliver its active compound into the cytoplasm (He et al., 2013; Jiang et al., 2012). Besides, CNTs have capacity to keep drug or biomolecule intact during transportation and cellular penetration. The last property permits to decrease the dosages of drugs used, and consequently their toxicity, especially for anticancer drugs. Therefore, the drug-CNT conjugate proves to be safer and more effective than drug used alone by traditional preparation.

For the delivery of a drug to a specific target cell, such as cancer cell, chemotherapeutic agent can be bound to a complex formed by CNT and a marker that is able to recognize the cancer cell or to be directly driven to this tumor. These techniques will be developed in the following paragraph.

6.5 APPLICATIONS OF FUNCTIONALIZED CNTs IN DRUG DELIVERY

The functionalized CNTs are mainly applied in drug delivery. Drug delivery system is designed to improve the pharmacological, bioavailable and therapeutic profiles of a drug molecule. Most drugs chosen in these applications are potent drugs with high efficacy, but also with high toxicity and/or low bioavailability because traditional techniques cannot resolve these problems, thereby limiting their therapeutic activities. Furthermore, the large inner volume of CNTs allows encapsulation of drugs (Fujigaya and Nakashima, 2015). CNTs can also be used for multidrug therapy by loading the tubes with more than one drug. Moreover, CNTs can also control the release system by delivering drugs over a long period of time (Kantamneni and Gollakota et al., 2013). Several drugs tested in vitro and/or in vivo experiments using CNTs as drug carriers include antineoplastic agents for cancer treatment, antibiotics or vaccines for infection therapy or prevention, neurotransmitters for chronic neurodegenerative diseases, etc.

6.5.1 FOR CANCER THERAPY

The most common types of cancer treatment are chemotherapy, radiation therapy, surgery, hyperthermia, immunotherapy, targeted therapy and many others. However, the effectiveness of chemotherapy drugs is often limited by the toxicity to other tissues in the body such as immune system, a vital organ for fighting cancer and infection. This is because most chemotherapy drugs do not specifically kill cancer cells, they act to kill all cells undergoing fast division.

6.5.1.1 By Drug Delivery or Chemotherapy

The efficacy of chemotherapy drugs used alone is often limited not only by their toxicity to other tissues in the body and narrow therapeutic window but also as a result of drug resistance and limited cellular penetration. The exploration of CNTs used as drug carrier to treat tumor has been recently investigated by different in vitro and in vivo studies (6, 8, 17, 55, 58, 60).

Because CNTs can easily across the cytoplasmic membrane and nuclear membrane, anticancer drug transported by this vehicle will be liberated *in situ* with intact concentration and consequently, its action in the tumor cell will be higher than that administered alone by traditional therapy. Thus, the development of efficient delivery systems with the ability to enhance cellular uptake of existing potent drugs is needed. The high aspect ratio of CNTs offers great advantages over existing delivery vectors, because the high surface area provides multiple attachment sites for drugs (8, 30). Many anticancer drugs have been conjugated with functionalized CNTs and successfully tested in vitro and in vivo such as epirubicin, doxorubicin, cisplatin, methotrexate, quercetin, paclitaxel, etc. (Chen et al., 2011; Dhar et al., 2008; Elhissi et al., 2012; Lay et al., 2011; Liu et al., 2008; Madani et al., 2011; Pineda et al., 2014; Rastogi et al., 2014; Sahoo et al., 2011; Zhang et al., 2011).

Drugs can either attach to the outer surface of the CNTs via functional groups or be loaded inside the CNTs. Attachment of the anticancer drug to the outer surface of the CNTs can be through either covalent or noncovalent bonding, including hydrophobic, π–π stacking, and electrostatic interactions (Elhissi et al., 2012; He et al., 2013; Pineda et al., 2014). Indeed, the poorly water-soluble anticancer camptothecin has been loaded into polyvinyl alcohol-functionalized MWCNTs and reported to be potentially effective in the destruction of human breast and skin cancer cells in comparison of the same drug used alone (Sahoo et al., 2011). Other authors have incorporated carboplatin, a platinum anticancer drug, into CNTs and obtained better results with this conjugate than drug used alone in the inhibition of the urinary bladder cancer cells proliferation in vitro (Dhar et al., 2008; Zhang et al., 2011). Liu et al. (Liu et al., 2008) demonstrated that the in vivo administration of SWCNT–paclitaxel (Taxol) conjugate in a murine breast cancer model has been observed with higher efficacy in suppressing tumor growth and less toxic effects to normal organs. The higher therapeutic efficacy and lower side effects could be attributed to prolonged blood circulation, higher tumor uptake and slower release of drug from SWCNTs (Liu et al., 2008). In another study, anticancer effects have been shown to be dependent on the method used to entrap the drug in the CNTs, which highlighted the possible effects of preparation conditions on the therapeutic activity of antineoplastic agents associated with CNTs (Elhissi et al., 2012).

For specific delivery of a drug to targeted cancer cells, chemo agent can be linked to a CNT complex containing a marker or link that is either directly guided to the tumor or attractively recognized by the cancer cell surface of an organ. The first marker is Magnetic-CNTs used to drive linked drug to the target cell by an external magnet, while the second type is an antibody against antigen on cancerous cell surface by the phenomenon of antigen-antibody attraction.

Thus, various investigators have prepared different Magnetic-CNTs linked to chemotherapeutic agents for targeting lymph node cancers (Elhissi et al., 2012; Gao et al., 2006; He et al., 2013). Lymphatic cancer metastasis occurs frequently even after extended lymph node dissection. No efficient therapeutic methods have been so far developed to target lymphatic metastasis (Gao et al., 2006). Therefore, the delivery of anticancer drugs to the lymph nodes may be of great interest. Nanoparticles can be effectively taken up into lymphatics, but only few nanosystems can be retained in the draining lymph node. Ni et al. (Yang et al., 2008) have developed a technology using a magnetic carbon nanotube (MCNT) delivery system in which chemo agents were incorporated into the pores of functionalized MCNTs synthesized with a layer of magnetite nanoparticles on the inner surface of the nanotubes. To improve drug delivery to cancer cells in the lymph nodes, individualized MCNTs were noncovalently functionalized by folic acid (FA). By using an externally placed magnet to guide the drug matrix to the regional targeted lymph nodes, the MCNTs can be retained in the draining targeted lymph nodes for several days and continuously release anticancer drugs. In another study, Yang et al. (2009) have successfully prepared a hydrophilic MWCNTs decorated with magnetic nanoparticles as lymphatic targeted chemo drug delivery vehicles under the guidance of a magnetic field. Xiao et al. (2012) have presented a new synthesis technique of magnetic MWCNTs by a simple solvothermal process for tumor-specific targeting of epirubicin delivery. Panczyk et al. (2009) have studied a computer simulation of the behavior of a carbon nanotube in which both ends are connected to magnetic nanoparticles by a short alkane chain. Their survey results showed that the access to the nanotube interior can be easily controlled by switching the external magnetic field on or off. Such nanodevices might be very promising candidates for drug delivery systems or storage materials with controllable release of encapsulated molecules. Recently, Yang et al. (2011) have subcutaneously injected in mice the anticancer molecule gemcitabine loaded into magnetic MWCNTs; they reported that high activity against lymph node metastasis has been obtained, in contrast with drug injected alone.

Another procedure for the specific delivery of a drug to a target cell in vivo, such as cancer cell, involves using an antibody against antigen generally overexpressed on the cancerous cell surface. Such method consists in linking a chemotherapeutic agent to a conjugate of functionalized CNT containing this antibody. By the attraction of antigen-antibody, this complex can be taken up by the tumor cell only before the anticancer drug is cleaved off CNTs; thus, specific targeting delivery is performed (Elhissi et al., 2012; He et al., 2013; Kantamneni and Gollakota et al., 2013; Lay et al., 2011; Madani et al., 2011). Some cancer cells express large amounts of p-glycoprotein (P-gp), an important protein of the cell membrane, which renders these cancers multidrug resistance and is a major obstacle to effective anticancer drug therapy (He et al., 2013). To resolve this problem, Li and co-workers (2010) have shown that SWCNTs can be functionalized with p-glycoprotein antibodies and loaded with the anticancer agent doxorubicin. Using these antibodies, the designed conjugates first neutralize the obstacle, p-glycoprotein on the cell membrane, then, deliver the drug into the cancer cells. Compared with free doxorubicin, this formulation demonstrated higher cytotoxicity by 2.4-fold against K562R leukemia cells.

6.5.1.2 By Antitumor Immunotherapy

CNTs play an important role in modulating immunological functions. Prior study has revealed that viral peptides conjugated to CNTs can elicit strong antipeptide antibody responses in mice with no detectable cross reactivity to the CNTs (He et al., 2013; Kantamneni and Gollakota et al., 2013; Lay et al., 2011; Madani et al., 2011; Meng et al., 2008). It is also reported that functionalized CNTs are noncytotoxic to immune cells. The significance of immunotherapy as an adjuvant anticancer treatment is well recognized (Meng et al., 2008). Recently, some in vitro and in vivo studies have found that CNTs used as carrier can effectively improve antitumor immunotherapy (Digge et al., 2012; Elhissi et al., 2012; Kantamneni and Gollakota et al., 2013; Lay et al., 2011; Madani et al., 2011). This therapeutic consists of stimulating the patient's immune system to attack the malignant tumor cells. This stimulation can be achieved by the administration of a cancer vaccine or a therapeutic antibody as drug. Some authors have validated the use of CNTs as vaccine delivery tools (Meng et al., 2008). Yang's group reported that the conjugate of MWCNTs and tumor lysate protein [tumor cell vaccine (TCV)] can considerably and specifically enhance the efficacy of an antitumor immunotherapy employing TCV in a mouse model bearing the H22 liver cancer (Meng et al., 2008). In vitro, the conjugate of CNTs and tumor immunogens can act as natural antigen presenting cells (such as mature dendritic cells) by bringing tumor antigens to immune effector T cells; this action is due to the high avidity of antigen on the surface and the negative charge. The complement system activation effects of CNTs and also their adjuvant effects may play a role in the stimulation of antitumor immunotherapy; however, the mechanism remains unknown (Meng et al., 2008).

Given the excellent features of CNTs as transporters for bioactive molecules, it will be of great interest to evaluate the role of CNTs in antitumor immunotherapy.

6.5.1.3 By Local Antitumor Hyperthermia Therapy

Thermal ablation represents a potential form of cancer therapy that is noninvasive and harmless to normal cells, with high efficacy. Exposure to modalities, such as near infrared, leads to cell death by irreversible protein denaturation or plasma membrane damage as a result of temperatures reaching over 40°C. This form of therapy has been shown to be efficacious for the treatment of numerous malignancies, including those of the lung, liver, and prostate. It is proposed to use the novel paradigm of treating cancer with hyperthermia therapy using either MWCNTs or SWCNTs stimulated with near infrared (NIR) irradiation (NIR from 700–1100 nm) (Madani et al., 2011). These nano-materials are considered as potent candidate for hyperthermia therapy since they generate significant amounts of heat upon excitation with NIR light. It was demonstrated that addition of MWCNTs to Erlich ascitic carcinoma (EAC) cell suspension by in vitro experiment results in the photo-ablative destruction of cells exposed to short time NIR irradiation (Burlaka et al., 2010). In an in vivo experiment (Madani et al., 2011), a group of mice was injected with human epidermoid cancer cells. Once the size of the tumor reached 70 mm^3, functionalized

CNTs were injected into the tumors which were then exposed to near infrared. The results showed that the cancerous cells treated with CNTs and near infrared light disappeared within 20 days (Madani et al., 2011). It was also shown that a local rise in temperature increased the permeability of the tumor vasculature which may be advantageous for selective delivery of drugs to the tumor site from the systemic circulation. Other authors reported that the photo-thermal effect can induce the local thermal ablation of in vivo tumor cells by excessive heating of SWCNTs shackled in tumor cells such as pancreatic cancer (Elhissi et al., 2012; Kantamneni and Gollakota et al., 2013; Lay et al., 2011). In addition, magnetic-CNTs can also be used for thermal treatment of cancer. These nanomaterials guided by an external magnetic field can reach the cancerous tissue of an organ, then, destroy this tumor by thermal ablation (Madani et al., 2011). It has also been recently reported that breast cancer stem cells, highly resistant to conventional thermal treatments, can be successfully treated with CNT-based photothermal therapies by promoting necrotic cell death (Shao et al., 2013). Further studies in this direction show that DNA-encased MWCNTs are more efficient at converting NIR irradiation into heat compared to non-encased MWCNTs and that this method can be effectively used in vivo for the selective thermal ablation of cancer cells (Arora et al., 2014).

Some progress in hyperthermia technique using CNTs has been achieved in recent years, and it has shown feasibility in clinical application.

6.5.2 FOR INFECTION THERAPY

Due to the uncontrolled use of antibiotics as well as of other antiviral and antibacterial drugs, a number of infectious agents become resistant to these treatments. It is also observed for certain vaccine inefficacy in the prevention of epidemic. To resolve the problem of drug resistance as well as to avoid the toxicity or/and to increase the narrow therapy of certain antiinfectious agents, CNTs have been assayed by many researchers recently.

Functionalized CNTs have been demonstrated to be able to act as carriers for antimicrobial agents such as the antifungal amphotericin B (AmB) (He et al., 2013; Rosen et al., 2011). Due to its lower aqueous solubility, the systemic toxicity of AmB restricts its application for chronic fungal infections. CNTs can attach covalently to amphotericin B and transport it into mammalian cells. This conjugate has reduced the antifungal toxicity about 40% as compared to the free drug (Rosen et al., 2011). Our group has successfully combined an antimicrobial agent Pazufloxacin mesilate with amino-MWCNT with high adsorption and will be applied to experimental assays for infection treatment (Jiang et al., 2012).

The antimycobacterium activity of dapsone is limited by its high lipophilic nature that exposes dapsone to extensive metabolism by the liver with the formation of diverse toxic metabolites responsible for adverse reactions like methemoglobinemia, hemolytic anemia, agranulocytosis and dapsone hypersensitivity syndrome (Kumar et al., 2014). Conjugated with MWCNTs, the antimycobacterium activity of dapsone is evidenced by delaying apoptosis of peritoneal macrophages indicating selective accumulation of drug in the endosomal localization. Thanks to the local concentration of the drug, such an approach using CNTs may reduce the systemic toxicity of dapsone, and thus make it more efficient in the treatment of tuberculosis (Kumar et al., 2014). Isoniazid, a powerful antitubercular drug shows limited action due to its lower permeability into the mycobacterial cell wall resulting in drug accumulation and hence aggravated toxicity. The delivery of isoniazid under conjugated form with functionalized CNTs has allowed the successful permeation of drug into the mycobacterial cell wall (Gallo et al., 2007; Kumar et al., 2014).

Functionalized CNTs can also act as vaccine delivery procedures (Liao et al., 2011; Rosen et al., 2011). The linkage of a bacterial or viral antigen with CNTs permits of keeping intact antigen conformation, thereby, inducing antibody response with the right specificity (Digge et al., 2012). The fixation of functionalized CNTs with B and T cell peptide epitopes can generate a multivalent system able to induce a strong immune response, thereby becoming a good candidate for vaccine delivery (Usui et al., 2012; Yang et al., 2007). Thus, functionalized CNTs can be used as a good nanocarrier for the delivery of vaccine antigens. Besides, CNTs themselves might have antimicrobial activity since bacteria may be adsorbed onto the surfaces of CNTs, such as the case of *E. coli*. The antibacterial effect was attributed to carbon nanotube-induced oxidation of the intracellular antioxidant glutathione, resulting in increased oxidative stress on the bacterial cells and eventual cell death (Digge et al., 2012). The applications of functionalized CNTs used either as drug carriers or as antimicrobials by themselves for infection therapy may play an important role in the successful combat against a number of dreadful ailments due to different microbes.

6.5.3 FOR ANTIINFLAMMATORY THERAPY

Dexamethasone (DEX) is a synthetic glucocorticoid hormone commonly used as immunosuppressant and antiinflammatory drug to treat many autoimmune and inflammatory diseases. SWCNTs have been used as host-carrier film for the electrically stimulated delivery of DEX. An accelerated cellular uptake and a complete drug release of DEX were observed due to electrostatic repulsions between SWCNTs and DEX when −0.8 V potential was applied. The passive release of DEX was decreased by the addition of SWCNTs, due to the possible attractive interactions between the drug and SWCNTs. This new technique of drug delivery may improve the antiinflammatory therapy of dexamethasone in the future (Luo et al., 2011; Naficy et al., 2009; Tripathi et al., 2015). Ketoprofen, a nonsteroidal antiinflammatory drug (NSAID), is used for the treatment of inflammatory diseases (arthritis, headache, etc.) by its inhibition of the prostaglandin production in the body. An electro-sensitive transdermal DDS, composed of a semiinterpenetrating polymer network (polyethylene oxide-pentaerythritol triacrylate) as the matrix and MWCNTs was demonstrated to increase the electrical sensitivity of (S)-(+)-ketoprofen. The amount of released drug increases with enhanced applied potentials, which can be attributed to higher electrical conductivity of CNTs (Tripathi et al., 2015).

6.5.4 FOR NEURODEGENERATIVE DISEASES

Because of their tiny dimensions and accessible external modifications, CNTs are able to cross the blood–brain barrier by various targeting mechanisms for acting as effective delivery carriers for target brain. Thus, CNTs are promising biomedical materials for drug delivery system (DDS) in many dreadful neurodegenerative ailments (He et al., 2013).

Yang et al. (2010) have observed that SWCNTs were successfully used to deliver acetylcholine in mouse brains affected by Alzheimer's disease with high safety range. Acetylcholine is natural neurotransmitter of the cholinergic nervous system and related with high-level nervous activities, such as learning, memory, and thinking. Because of the synthesis impairment, acetylcholine is decreased in the neurons in the Alzheimer disease brain, leading to the incapability of intellectual working. However, providing acetylcholine into brain is a problem because this neurotransmitter with strong hydrophilic property cannot cross the blood-brain barrier (BBC). To overcome this obstacle, Yang's group (2010) has prepared a conjugate by adsorption of acetylcholine on SWCNT. This system successfully delivered acetylcholine into the neurons in the brain through the axoplasma transformation of neurites and significantly improved the learning and memory capabilities of the model animals with Alzheimer diseases (Yang et al., 2010). Many other functionalized SWCNTs or MWSCNTs have successfully been used as suitable delivery systems for treating neurodegenerative diseases or brain tumors (Digge et al., 2012; He et al., 2013; Liao et al., 2011; Zhang et al., 2010). Overall the results of these studies have indicated that conjugates of CNTs with therapeutic molecules have better effects on neuronal growth than drugs used alone. Another possible therapeutic application of CNT in the CNS is in the treatment of glioblastoma tumors. The treatment of brain tumors remains a challenge despite advances in tumor therapy and the increasing understanding of carcinogenesis. Some antitumor drugs linked with new CNT modalities have been administered by systemic way to cross the blood-brain barrier (BBB) (Tripathi et al., 2015). For example, Zhao et al. (2011) have recently demonstrated in vitro and in mice that the CNT delivery system significantly enhanced CpG oligodeoxynucleotides immunotherapy; eradicating the glioma and protecting against tumor rechallenge (GL261 and GL261.egfp, efgp = encoding green fluorescent protein, models). These findings suggest that CNTs can potentiate CpG immunopotency by enhancing its delivery into tumor-associated inflammatory cells.

CNT-mediated therapy is a valuable option for the treatment of neurodegenerative diseases, including the treatment of stroke (Tripathi et al., 2015). In this context, the potential use of amine-functionalized SWCNT was evaluated by Lee et al. (2014), using a middle cerebral artery occlusion (MCAO) stroke model, to enhance the survival of neurons following ischemic injury. It was observed that the intracerebroventricular injections of SWCNT without any therapeutic molecule enhanced the motor function recovery of the animals; however, the mechanism for such activity remains elusive. No considerable toxicity was noticed in mice and only minor changes in tumor cytokine expression were observed, in spite of a transient increase in inflammatory cell infiltration into both normal and tumor-bearing brains following MWCNT injection. This study suggested that MWCNTs could be used as nontoxic vehicles to target tumor-associated macrophages (MP) in brain (Tripathi et al., 2015). Many other functionalized CNTs have been used successfully as suitable delivery systems for treating neurodegenerative diseases or brain tumors (He et al., 2013). Overall, the results of these in vitro and in vivo studies in drug delivery have indicated that drugs are delivered more effectively and safely into cells by CNTs as nanocarriers than by traditional methods (Bamrungsap et al., 2012; Safari and Zarnegar, 2014; Surendiran et al., 2009).

6.6 APPLICATIONS OF FUNCTIONALIZED CNTs IN BIOMOLECULE DELIVERY

Biomolecules including macromolecules such as proteins, lipids, nucleic acids, DNA/RNA, genes, polysaccharides, as well as small molecules such as hormones, alkaloids, vitamins, antioxidants, etc. can be used in therapy and diagnosis of many human ailments. Applications of functionalized CNTs in biomolecule delivery are recent and have been proposed for gene therapy, tissue regeneration, artificial implants and diagnosis of human diseases.

6.6.1 CARBON NANOTUBES FOR GENE THERAPY BY DNA DELIVERY

Gene therapy is an experimental techniquc that uses genes to treat or prevent disease. This novel discovery could be a way to fix a genetic problem at its source by correcting a defective gene which is the cause of some chronic or hereditary diseases by introducing DNA molecule into the cell nucleus. However, classical delivery of nucleic acids in vivo is challenging due to the lack of stability, unwanted toxicity and their inability to cross cell membranes. Some new nanocarrier systems for DNA transfer including liposomes, cationic lipids and CNTs recently discovered have been assayed in vitro and in vivo in order to resolve these previous obstacles (He et al., 2013; Liao et al., 2011; Pantarotto et al., 2004; Usui et al., 2012). When bound to SWCNTs, DNA probes are protected from enzymatic cleavage and interference from nucleic acid binding proteins, consequently, DNA-SWCNT complex exhibits superior bio-stability and increases self-delivery capability of DNA in comparison to naked DNA (Usui et al., 2012). CNTs conjugated with DNA were found to release DNA before it was destroyed by cell defense system, boosting transfection significantly (Liu et al., 2009). Gene therapy by CNTs as nanocarrier has demonstrated that these nanostructures can effectively transport the genes inside mammalian cells and keep them intact because the CNT-gene complex has conserved the ability to express proteins (Li et al., 2008). CNTs can be modified with positive charges to bind DNA plasmids for gene transfection (Liu et al., 2009). Pantarotto et al. and Singh et al. (2004, 2005) used amino-SWCNTs and amino-MWCNTs functionalized by 1,3-dipolar cycloaddition to bind DNA plasmids, and have achieved reasonable transfection efficiency. In the work of Gao et al. (2006), amine groups were introduced to oxidized-MWCNTs for DNA binding and transfection, successfully expressing green fluorescence protein (GFP) in mammalian cells. Although the MWCNT-based method was less efficient than commercial gene transfection agents, such as lipofectamine 2000, the MWCNTs exhibited much lower toxicity. In another study carried out by Liu et al. (2005) polyethylenimine (PEI) grafted MWCNTs were used for DNA attachment and delivery, which afforded comparable efficacy to the standard PEI transfection method with the benefit of reduced cytotoxicity.

Unlike various small molecules (drugs, hormones, etc.) which are able to diffuse into cells, macrobiomolecules (proteins, DNA, and RNA) rarely cross cell membranes by themselves. Intracellular delivery is thus required in order to use these molecules for therapeutic applications. For example, proteins can be either covalently conjugated or noncovalently adsorbed on CNTs for intracellular delivery (Li et al., 2008; Pantarotto et al., 2004). The hydrophobic surface of partially oxidized SWCNTs allows nonspecific binding of proteins. After being translocated into cells by nanotubes, proteins can become bioactive once they are released from endosomes

(Li et al., 2008; Pantarotto et al., 2004). The use of functionalized CNTs as nanocarriers in macrobiomolecule delivery is a promising tool for the treatment of different dreadful ailments such as cancer in the future.

6.6.2 CARBON NANOTUBES FOR TISSUE REGENERATION AND ARTIFICIAL IMPLANTS

Tissue engineering recently developed is the study of the growth of new connective tissues, or organs, from cells and a collagenous scaffold to produce a fully functional organ for implantation back into the donor host. The aim of regenerative medicine is repair and regeneration of human body tissues and organs affected or lost because of disease, trauma, and the like (He et al., 2013). Whichever means are used, no tissue can be regenerated without a scaffold. Thus, the scaffold is of paramount importance in therapy, and research aimed at developing CNTs as scaffold material has been increasing (Saito et al., 2014). Carbon nanotubes may be an important tissue engineering material for delivering of transfection agents, tracking of cells, sensing of microenvironments, and scaffolding for incorporating with the host's body (Kumar et al., 2012; Haniu et al., 2012). The knowledge advances of cell and organ transplantation and of CNT chemistry in recent years have contributed to the sustained development of CNT-based tissue engineering and regenerative medicine. Carbon nanotubes may be the best tissue-engineering candidate among numerous other materials for tissue scaffolds since this nanomaterial is biocompatible, resistant to biodegradation and can be functionalized with biomolecules for enhancing the organ regeneration. In this field, CNTs can be used as additives to reinforce the mechanical strength of tissue scaffolding and conductivity by incorporating with the host's body (Haniu et al., 2012; He et al., 2013; Liao et al., 2011; Saito et al., 2014).

The use of CNT composites as scaffold in regenerative medicine has been vigorously investigated in vitro (Saito et al., 2014). Indeed, MacDonald et al. (2005) have successfully combined in vitro a carboxylated-SWCNTs with a polymer or type I collagen to form a composite nanomaterial used as scaffold for myocyte culture in tissue regeneration. Other in vitro studies showed that a CNT/polyurethane composite could be used as a scaffold for fibroblasts growth and biosynthesis and also for culturing vascular endothelial cells in order to promote their proliferation and to suppress thrombus formation (Saito et al., 2014). A composite of MWCNTs with regenerated silk fibroin films were shown to support the adhesion and growth of human bone marrow stem cells. SWCNTs nonwoven films enhanced long-term proliferation of many cell types. Besides numerous in vitro studies examining the reactions between cells and CNT composites used as scaffolds, some in vivo animal experiments have recently started (Haniu et al., 2012; Paratala et al., 2011; Saito et al., 2014). It has been reported that CNTs can effectively enhance bone tissue regenerations in mice and neurogenic cell differentiation by embryonic stem cells in vitro (He et al., 2013). Other scientists reported for the first time that CNTs promote bone tissue formation in vivo by inducing ectopic osteogenesis in mouse back muscle (Saito et al., 2014). For artificial implants, Saito et al. (2014) conjugated CNTs to polyethylene for use in sliding parts and rotating parts of artificial joints. Other tissue engineering applications of CNTs concerning cell tracking and labeling, sensing cellular behavior, and enhancing tissue matrices are also studied recently (He et al., 2013; Kumar et al., 2012; Li et al., 2008; Paratala et al., 2011). For nerve regeneration, some scientists enhanced and stimulated the regeneration of injured nerve cells and fibers by using CNTs composites. Another study found that CNTs were useful in the differentiation of embryonic stem cells to nerve cells. Other applications of CNTs are expected to develop new methods using Schwann cells (neurolemmocytes of the peripheral nervous system) for nerve regeneration. Regenerative medicine for nerve is an interesting research field that aims to apply in combination the electrical and mechanical properties of CNTs to biomaterials.

The contribution of CNTs to bone and nerve tissue regenerations is of paramount interest. Other applications of CNTs in regenerative medicine have been found in the literature such as differentiation of mesenchymal stem cells to cardio-myocyte lineage cells, regeneration of cartilage, skeletal muscle and heart muscle (Haniu et al., 2012; Kumar et al., 2012; Li et al., 2008; MacDonald et al., 2005; Paratala et al., 2011; Saito et al., 2014). In the future, CNTs will be used to stimulate the regeneration of many other tissues and organs, so regenerative medicine is quite an interesting field of applied research.

6.7 OTHER APPLICATIONS OF CNTs IN MEDICINE AND PHARMACY

6.7.1 FUNCTIONALIZED CARBON NANOTUBES AS ANTIOXIDANTS

Recently, Galano et al. (2010) observed that carboxylated-SWCNTs possess antioxidant property in nature and may have useful biomedical applications for prevention of chronic ailments and aging and food preservation (Galano et al., 2010). Nevertheless, the potential role of CNTs as free-radical scavengers is still an emerging area of research. Their antioxidant property has been used in anti-aging cosmetics and sunscreen creams to protect skin against free radicals formed by the body or by UV sunlight (He et al., 2013; Digge et al., 2012). More investigations of different functionalized CNT forms in the future are needed, since free radicals are well known to be very damaging species (Galano et al., 2010; Pham-Huy et al., 2008).

6.7.2 CARBON NANOTUBES AS BIOSENSOR VEHICLES FOR DIAGNOSTIC AND DETECTION

Nanotechnology-based biosensors by CNTs are an analytical procedure designed to the detection of an analyte that combines a biological component with a physicochemical detector. This recent application represents a most exciting area for therapeutic monitoring and in vitro and in vivo diagnostics because the accuracy and sensitivity of biosensors coupled with CNTs are higher than biosensors used alone (He et al., 2013; Wang, 2005; Zhu et al., 2011).

Many CNT-enzyme biosensors such as CNT-based dehydrogenase, peroxidase, catalase or glucose-oxidase biosensors have been developed for different therapeutic monitoring and diagnostics (Digge et al., 2012; Usui et al., 2012; Wang, 2005). For electrical detection of DNA, the sensitivity of the assay using SWCNT-DNA sensor obtained by integration of SWCNTs with single-strand DNAs (ssDNA) was considerably higher than traditional fluorescent and hybridization assays (He et al., 2013; Wang, 2005; Zhu et al., 2011).

6.7.3 CARBON NANOTUBES FOR ENANTIOSEPARATION OF CHIRAL DRUGS AND BIOCHEMICALS

In pharmaceutical industries, 56% of the drugs currently in use are chiral products and 88% of the last ones are marketed as racemates consisting of an equimolar mixture of two enantiomers (Nguyen et al., 2006). Many analysts have developed new chiral selectors using either SWCNTs or MWCNTs alone or combined with other chiral stationary phases such as hydroxypropyl-β-cyclodextrin for the separation of numerous drug racemic drugs such as β-blockers, clenbuterol, etc. (He et al., 2013; Silva et al., 2012; Tan et al., 2014).

6.7.4 CARBON NANOTUBES FOR SOLID-PHASE EXTRACTION OF DRUGS AND BIOCHEMICALS

Due to their strong interaction with other molecules, particularly with those containing benzene rings, CNTs surfaces possess excellent adsorption ability. Non-functionalized or functionalized CNTs have been investigated as Solid-Phase Extraction (SPE) adsorbents used alone or in conjugation with classical SPE sorbents for the analytical extraction of drugs, biological compounds or pesticides in different media such as biological fluids, drug preparations, environment, etc. (El-Sheikh and Sweileh et al., 2011; He et al., 2013). Sorbents containing CNTs exhibit similar or higher adsorption capacity than silica-based sorbents or macroporous resins used alone. Many applications of CNTs in SPE can be found in our articles recently published or in the literature (Dai et al., 2015; El-Sheikh and Sweileh et al., 2011; Xiao et al., 2013, 2014).

6.8 PHARMACOKINETICS, METABOLISM AND TOXICITY OF CARBON NANOTUBES

6.8.1 PHARMACOKINETICS OF CARBON NANOTUBES

Studies in pharmacokinetics, metabolism and toxicity of diverse forms of CNTs play important role to evaluate the utilities of CNTs as a drug delivery vehicle. The blood circulation time and accumulation of CNTs and their eventual metabolites in the body affect their biomedical applications and toxicity (Arora et al., 2014; Yang et al., 2012). The bio-distribution and pharmacokinetics of CNTs mainly depend on their physicochemical characteristics such as functionalization mode, solubility, shape, aggregation, chemical composition and also on their in vivo administration routes (He et al., 2013). To study the pharmacokinetics and bio-distribution of CNTs, various analytical techniques have been developed for measuring CNTs in bio-samples (Yang et al., 2012). Many reports about the bio-distribution and pharmacokinetics of different forms of CNTs are found in the literature (Arora et al., 2014; He et al., 2013; Singh et al., 2006; Wang et al., 2004; Yang et al., 2012). As pristine CNTs are hydrophobic and easy to aggregate, they are not used as drug carriers and their pharmacokinetics are not discussed herein. They are only studied in toxicology.

For water-soluble CNTs, i.e., functionalized CNTs, the results of their pharmacokinetic and bio-distribution studies are abundant and sometimes divergent. In a recent review article (Yang et al., 2012), it was reported that the blood circulation half-life of Carboxyl-SWCNTs covalently linked to polyethylene-glycol (PEG1500) increased to 15.3 h. This good pharmacokinetics profile of PEG-SWCNTs makes them promising in biomedical applications. In another study (Liu et al., 2009), it was observed that when CNTs were dispersed in polyethylene glycol-phospholipid (PEG-PL) for thermal therapy and drug delivery, the blood circulation half-life of this complex PEG-PL/SWCNTs was prolonged, while the reticuloendothelial system (RES) capture of SWCNTs was reduced in vivo. PEG chains were linked to PL molecules to form a hydrophilic-hydrophobic structure. PEG-PL suspended CNTs undergo hydrophobic interaction between PL and CNTs. The blood circulation half-life increased along with increasing the PEG density, while the RES uptake decreased accordingly (Singh et al., 2006; Wang et al., 2004; Yang et al., 2012). The good behaviors of PEG-PL/SWCNTs in vivo make them very useful in biomedical areas (Elhissi et al., 2012; Kantamneni and Gollakota et al., 2013; Lay et al., 2011; Madani et al., 2011). By carefully choosing suspending reagents, the pharmacokinetics and bio-distribution of CNTs can be modified significantly.

Two other studies were performed with water soluble CNTs (SWCNT or/and MWCNT) for their bio-distribution in mice (He et al., 2013). None of these studies report toxic side effects or mortality. Both experiments have used either ^{125}Iodine or ^{111}Indium as radio-tracers for observing their bio-distribution in mice (Singh et al., 2006; Wang et al., 2004). In the first study, the results showed that the CNT bio-distribution was not significantly influenced by the administration route and that the ^{125}I-SWCNT- OH distributed quickly throughout the whole body with 94% of the unchanged nanotubes excreted into the urine and 6% in the feces (Wang et al., 2004). The preferred organs for accumulation were the stomach, kidneys and bone. No tissue damage or distress was reported. Second study used two forms of ^{111}Indium-functionalized SWCNT or MWCNT by IV administration only in mice. The bio-distribution profiles obtained were found very similar for both types of functionalized CNTs which showed an affinity for kidneys, muscle, skin, bone and blood 30 min after administration (Singh et al., 2006; Yang et al., 2012). However, all types of CNTs were rapidly cleared from all tissues and a maximum blood circulation half-life of 3.5 h was determined. Both SWCNTs and MWCNTs were found to be excreted through the renal route and observed intact in the excreted urine by transmission electron microscopy (Yang et al., 2012).

6.8.2 METABOLISM AND TRANSFORMATION OF CNTs

CNTs employed for biomedical purposes are in forms of either dispersed or functionalized ones. After introducing to biosystems, it is possible that these CNTs are metabolized into other substances. The metabolism of CNTs definitely affects their biomedical applications and also might lead to unwanted toxicity. Therefore, the in vivo metabolism of CNTs should be highly concerned (Yang et al., 2012). According to Zhang et al. (2011), functionalized SWCNTs seem to be metabolizable in animal body. Yang et al. (2012) found that the skeleton of CNTs is relatively more stable than their functional groups. The stability of CNTs is also reflected by the long-term accumulation in vivo without being metabolized, where CNTs were visualized in mouse organs upon transmission electron microscopy (TEM) at several months postexposure. The stability of carbon skeleton also implies that CNTs are hard to be metabolized to small molecules, which may easily be excreted via urine and feces. In contrast to the stability of carbon skeleton, the functional groups fixed on CNTs by both kinds of covalent and noncovalent functionalization are much easier to fall off from CNTs. Current results suggest that both kinds of functional groups could be detached from CNTs in vivo and this makes functionalized CNTs transformed into less functionalized or pristine CNTs. The metabolism of noncovalently functionalized CNTs is generally regarded as desorption of the suspending reagents. Yang et al. (2012) suggested that the metabolism of CNTs in vivo is organ-dependent, which requires individual evaluation for each accumulation organ. Because of the high stability of carbon skeleton, the metabolism of CNTs mainly means defunctionalization of the surface functional groups. Therefore, the liver is the most possible metabolic organ in body for CNTs. Less functionalized CNTs are believed to be more toxic. Therefore, the hepatic toxicity of CNTs needs more attentions. Comparing the stability of noncovalently and covalently functionalized CNTs in vivo, the covalently functionalized CNTs seem more stable than noncovalently suspended ones. This phenomenon indicates that covalently modified CNTs are more suitable for biomedical use in vivo from the stability view. For the noncovalently suspended CNTs, PEG-PL is preferred among other stable suspending reagents (Yang et al., 2012; Zhang et al., 2011). These findings imply that the biodegradation of CNTs may be a key determinant of the degree and severity of the inflammatory responses in individuals exposed to them. Nevertheless, some scientists have recently shown that CNTs can be broken down by myeloperoxidase (MPO), an enzyme found in neutrophils of mice (Kagan et al., 2010). Their discoveries contradict what was previously believed, that CNTs are not degraded in the body. This action of how MPO converts CNTs into water and carbon dioxide can be significant to toxicology and thus represents a major advance in nanomedicine, since it clearly shows that CNTs can be metabolized by endogenous MPO (He et al., 2013; Singh et al., 2012). However, further studies are still required in order to draw an appropriate conclusion.

6.8.3 TOXICITY OF CARBON NANOTUBES

It is important to distinguish the three forms of CNT toxicity-based on their structures including pristine CNTs, i.e., insoluble raw SWCNTs and MWSCNTs, water soluble CNTs, i.e., covalent or noncovalent functionalized CNTs, and CNTs conjugated with drug or biomolecule.

The safety of raw and functionalized CNTs as well as of their conjugates with drugs or biomolecules is still debatable due to the lack of systematic and complete toxicity evaluation (Liu et al., 2009). A large number of studies have been performed in the past several years to explore the potential toxic effects of carbon nanotubes. The conclusions of these reports varied drastically, showing a large dependence on the type of nanotube materials as well as functionalization approaches. So, toxicological studies of different CNT types found in the literature seem still controversial.

6.8.3.1 Toxicity of Pristine CNTs

Pristine CNTs are hydrophobic and easy to aggregate, so it is difficult to disperse them in biosystems. Toxicological study of pristine CNTs is mainly focused in the professional risk assessments as well as the pure toxicological evaluations (Yang et al., 2012). Pristine CNTs are now produced worldwide for their use as additives in various structural materials of nanotechnology fields; their presence is only found in the workplace of CNT production or of different industries using CNT-based materials. Most publications found in the literature suggested that pristine CNTs could be the source of occupational lung diseases in workers of CNT industries like asbestos pathology previously observed in man (Ali-Boucetta et al., 2013; Donaldson et al., 2006; Fisher et al., 2012; He et al., 2013; Lacerda et al., 2006; Poland et al., 2008; Yang et al., 2012). For in vitro experiments, pristine, water insoluble CNTs have been found to be highly toxic to many different types of cells, including human keratinocytes, rat brain neuronal cells, human embryonic kidney cells, and human lung cancer cells (Madani et al., 2013; Yang et al., 2012). For in vivo experiments-based on several rodent studies in which test dusts were administered intratracheally or intrapharyngeally to assess the pulmonary toxicity of manufactured CNTs, a number of authors observed that CNTs were capable of producing inflammation, epithelioid granulomas, fibrosis, and biochemical changes in the lungs (Madani et al., 2013; Yang et al., 2012). Folkmann et al. (2009) reported that oxidative damages to DNA was observed in the colon mucosa, liver, and lung of rats after intragastric administration of pristine SWCNTs or C60 fullerenes by measuring the level of oxidative damage to DNA as the premutagenic 8-oxo-7,8-dihydro-2'-deoxyguanosine. Recently, Kostarelos et al. (2013) reported that the asbestos-like reactivity and pathogenicity reported for long pristine nanotubes can be completely alleviated if their surface is modified and their effective length is reduced as a result of chemical treatment, such as with triethylene glycol (TEG). However, opinions about the potential hazards of exposures to pristine CNTs and their residual metal impurities are still discussed (He et al., 2013). Moreover, the apparent similarity between multiwalled carbon nanotubes (MW-CNTs) and asbestos fibers has raised many questions about their safety profile. The toxicity of pristine SWCNTs is closely related to the oxidative stress in despite of the administration routes (Madani et al., 2013). Other scientists found that CNTs introduced into the abdominal cavity of mice showed asbestos-like pathogenicity (Poland et al., 2008). Indeed, there are several parameters affecting the toxicity of pristine CNTs in vivo. Metal impurities might contribute partially to the oxidative stress, thus, careful purification of CNTs is necessary (He et al., 2013; Yang et al., 2012). However, Kagan et al. (2010) reported that the degradation by neutrophil myeloperoxidase reduced the pulmonary toxicity of CNTs significantly. Chemically functionalized CNTs show higher biocompatibility than pristine CNTs. The reports of the National Institute of Advanced Industrial Science and Technology (AIST) in Japan, showed that, as compared to asbestos, CNTs have much lower inhalation toxicity (Saito et al., 2014). The currently projected goal of toxicity assessment is to determine the threshold level of exposure triggering inflammation in the lung. In the near future, international criteria of exposure to inhaled CNTs in industry will be established (Saito et al., 2014).

6.8.3.2 Toxicity of Functionalized CNTs

Functionalization of hydrophobic pristine CNTs is the process to create hydrophilic functional groups on their surface for their ulterior applications in nanomedicine. It has been established that surface functionalization is critical to the behaviors of CNTs in biological systems and can significantly improve the dispersibility and biocompatibility, thereby reducing the toxicity of CNTs (Madani et al., 2013). In vitro and in vivo toxicity studies reveal that well water soluble and serum stable nanotubes are biocompatible, nontoxic and potentially useful for biomedical applications (Liu et al., 2009). In vivo bio-distributions vary with the functionalization and possibly also sizes of nanotubes, with a tendency of accumulation in the reticuloendothelial systems including the liver and spleen, after intravenous administration. If well functionalized, nanotubes may be excreted mainly through the biliary pathway in feces (He et al., 2013; Liu et al., 2009).

According to Yang et al. (2012), many in vivo toxicological assessments have been performed by IV or SC injections and gastrointestinal exposure with functionalized or dispersed SWCNTs or/and MWCNTs in different animals (rats, mice). The available safety data collectively indicate that CNTs are of low toxicity via various exposure pathways for biomedical applications (He et al., 2013; Yang et al., 2012). Results from other studies have suggested that functionalization will result in the complete disappearance of the CNTs' toxicity (Liu et al., 2009; Madani et al., 2013). Dumortier et al. (2006) observed that water soluble SWCNTs marked with fluorescein were nontoxic to cultures of mouse B- and T-lymphocytes and macrophages and preserved the function of these immune cells. The toxicity of MWCNTs with end defects critically depends on their density of functionalization. Jain and co-workers found that the pristine and acid functionalized MWCNTs were devoid of any obvious nephrotoxicity (Mehra et al., 2014). The CNTs with larger dimension and lower degrees of functionalization cleared out from the body through the renal excretion pathway, and also excreted via biliary pathway through feces (Mehra et al., 2014). It was also clear that the oxidized CNTs with surface carboxyl density >3 mmol/mg were not retained in any RES organs (Mehra et al., 2014). Other study aimed to evaluate the potential toxicity and the general mechanism involved in two diameter kinds of water soluble multiwalled carbon nanotubes-induced cytotoxicity in C6 rat glioma cell line (Han et al., 2012). Results demonstrated that smaller size of f-MWCNT seemed to be more toxic than larger one. F-MWCNT-induced cytotoxicity in C6 rat glioma cells was probably due to the increased oxidative stress. Notably, many published articles have suggested that the toxicity of CNT biomaterials is extremely low (Haniu et al., 2012; Saito et al., 2014).

6.8.3.3 Toxicity of CNTs Conjugated With Drugs or Biomolecules

As applications of functionalized CNTs linked with therapeutic molecules are still not assayed in man for clinical trials, most publications found in the literature demonstrated that conjugated CNTs were safe in many in vitro and in vivo experiments. This finding may play an important role in the therapy of various dreadful ailments in the future. For example, when anticancer drugs are transported by functionalized CNTs, the toxicity of the conjugates is lower than that of the same drugs used alone as cited above in the paragraph of drug delivery. The mechanism of this phenomenon is explained by the direct delivery of drug into the targeted cells and also by the exocytosis of free functionalized CNTs after drug release (He et al., 2013; Madani et al., 2013). So, no accumulation of CNTs is found in tissues. Almost of them are eliminated either by urine or feces. In other hand, when CNTs have been used as tissue engineering materials for cell growth by implanting subcutaneously, CNTs exhibited very good biocompatibility and did not arise any serious toxicity, except very limited inflammation (Saito et al., 2014; Yang et al., 2012). Other reports on the toxicity of CNTs as tissue engineering to skin suggested that CNTs were biocompatible to skin with good results after subcutaneously planting (Haniu et al., 2012). For the carcinogenicity evaluations, Takanashi et al. (2012) also reported that no neoplasms were developed after implanting multiwalled CNTs (MWCNTs) into the subcutaneous tissue of rasH2 mice. More toxicity evaluations are encouraged to give the safety threshold value of different CNTs and clarify the toxicological mechanism.

Recently, a group of Japanese researchers (Saito et al., 2014) has envisaged a clinical study to evaluate the safety and application of CNTs in the near future and planned to proceed in four stages according to the magnitude of risk involved. The first stage is characterized by the use of a CNTs composite material for implantation in humans. In Stage 2, CNTs particles are used within the human body. In Stage 3, the assay will be expanded to include the treatment of diseases requiring higher level of safety than in Stage 2. Finally, Stage 4 will evaluate the treatment of diseases involving the injection of CNTs and their systemic circulation via the bloodstream for the purpose of drug delivery and whole-body imaging (Saito et al., 2014). It is very hopeful that this clinical project would be successful in saving the lives of many patients affected with deadly ailments.

6.9 CONCLUSION AND PERSPECTIVES

Thanks to the recent discovery of carbon nanotubes, the door of new and more effective alternatives than compared to traditional drug formulation has opened in different areas of pharmacy and medicine. The fantastic characteristics of CNTs over other existing nanomaterials are their strong penetration through cell membranes as well as their great capacity for carrying drugs, genes, biomolecules, vaccines, etc. and transporting them deep into the target cells or organs previously unreachable. Another novel approach is the fabrication of collagen-CNTs materials as scaffolds in tissue generation and artificial implants because CNTs resist to biodegradation and are a powerful engineering material to repair defective organs. Thus, this nanotechnology has opened new therapeutic concepts in the future and gives a glimmer of hope for the treatment of many deadly ailments such cancer, infection and so on.

Although there are numerous encouraging results of CNTs obtained during the beginning of this research field, much more work is still needed before CNTs can be applied in humans. Indeed, there are tremendous questions of clinical pharmacology and toxicology effects concerning long-term treatment by drug or biomaterial linked with CNTs in the future. For example, the use of CNT- anticancer drug conjugate linked to an antibody may raise the question about the reaction of the host immune system that could neutralize this complex in long-term therapy. The incorporation of CNTs with biological components of allogeneic or xenogeneic origin as well as stem cells into tissue-engineered or regenerative approaches may also open up a myriad of other challenges. Many different in vitro and in vivo assays with CNT conjugates found in the literature have only been performed in the short-term, more efforts are required to also confirm these findings in more long-term assays. Therefore, more innovations are needed to elaborate new multifunctional CNTs and their drug conjugates with high efficacy and safety for the treatment of different incurable diseases. For example, future preparation of drug-CNT conjugates should be attached with new sensitive markers or ligands so that they could directly reach the target cells without causing intolerance or serious high side-effects in the body. Moreover, the choice between SWCNTs and MWCNTs with their functionalization mode and their size and/or length also is a problem to resolve because literature reports about each option are still inconsistent. A standardized conjugation rule of CNTs with drugs or biomolecules is necessary in the future. Moreover, carefully optimizing the physicochemical parameters of CNTs to minimize their toxicity is highly recommended before they could be applied in men.

In summary, due to their enormous potential for therapeutic application, conjugates of CNTs with drugs or biomolecules could be effectively used in clinical study and then be marketed worldwide as long as they have shown to be devoid of serious adverse effects.

KEYWORDS

- **carbon nanotubes**
- **drug and biomolecule delivery**
- **metabolism**
- **nanocarriers**
- **pharmacokinetics**
- **toxicity**

REFERENCES

Ali-Boucetta, H.; Nunes, A.; Sainz, R.; Herrero, M. A.; Tian, B.; Prato, M.; Bianco, A.; Kostarelos, K. Asbestos-Like Pathogenicity of Long Carbon Nanotubes Alleviated by Chemical Functionalization. *Angew Chem. Inter. Ed.*, **2013**, *52*, 2274–2278.

Arora, S.; Kumar, V.; Yadav, S.; Singh, S.; Bhatnagar, D.; Kaur, I. Carbon Nanotubes as Drug Delivery Vehicles. *Solid State Phenom.*, **2014**, *222*, 145–158.

Babu, A.; Templeton, A. K.; Munshi, A.; Ramesh, R. Nanodrug Delivery Systems: A Promising Technology for Detection, Diagnosis, and Treatment of Cancer. *AAPS PharmSciTech.* **2014**, *15*, 709–721.

Bamrungsap, S.; Zhao, Z.; Chen, T.; Wang, L.; Li, C.; Fu, T.; Tan, W. Nanotechnology in Therapeutics: a Focus on Nanoparticles as a Drug Delivery System. *Nanomedicine*, **2012**, *7*, 1253–1271.

Basu, B.; Mehta, G. K. Carbon Nanotubes: A Promising Tool in Drug Delivery. *Int. J. Pharm. Bio. Sci.*, **2014**, *5*, 533–555.

Baughman, R. H.; Zakhidov, A. A.; de Heer W. A. Carbon Nanotubes – the Route Toward Applications. *Science*, **2002**, *297*, 787–792.

Baviskar, D. T.; Tamkhane, C. M.; Maniyar, A. H.; Jain, D. K. Carbon Nanotubes: An Emerging Drug Delivery Tool in Nanotechnology. *Int. J. Pharm. Pharm. Sci.* **2012**, *4*, 11–15.

Bianco, A.; Kostarelos, K.; Prato, M. Applications of Carbon Nanotubes in Drug Delivery. *Curr. Opin. Chem. Biol.*, **2005**, *9*, 674–679.

Burlaka, A.; Lukin, S.; Prylutska, S.; Remeniak, O.; Prylutskyy, Yu.; Shuba, M.; Maksimenko, S.; Ritter, U.; Scharff, P. Hyperthermic Effect of Multi-Walled Carbon Nanotubes Stimulated with Near Infrared Irradiation for Anticancer Therapy: In Vitro Studies. *Exp. Oncol.*, **2010**, *32*, 48–50.

Chen, Z.; Dramou, P.; He, H.; Tan, S.; Pham-Huy, C.; Hong, H.; Huang, J. Adsorption Behavior of Epirubicin Hydrochloride on Carboxylated Carbon Nanotubes. *Int. J. Pharm.*, **2011**, *405*, 153–161.

Chen, Z.; He, H.; Tan, S-H.; Zha, J.; Huang, J-L. Comparative Study on Contents of Oxygen-containing Groups on Multi-walled Carbon Nanotubes Functionalized by Three Kinds of Acid Oxidative Methods. *Chin. J. Anal. Chem.*, **2011**, *39*, 718–722.

Cho, K.; Wang, X.; Nie, S.; Chen, Z. G.; Shin, D-M. Therapeutic Nanoparticles for Drug Delivery in Cancer. *Clin. Cancer Res.* **2008**, *14*, 1310–1316.

Dai, H.; Xiao, D-L.; He, H.; Li, H.; Yuan, D. H.; Zhang, C. Synthesis and Analytical Applications of Molecularly Imprinted Polymers on the Surface of Carbon Nanotubes: a Review. *Microchim. Acta*, **2015**, *182*, 893–908.

Dhar, S.; Liu, Z.; Thomale, J.; Dai, H.; Lippard, S. J. Targeted Single-Wall Carbon Nanotube-Mediated Pt(IV) Prodrug Delivery Using Folate as a Homing Device. *J. Am. Chem. Soc.*, **2008**, *130*, 11467–11476.

Digge, M. S.; Moon, R. S.; Gattani, S. G. Applications of Carbon Nanotubes in Drug Delivery: A Review. *Int. J. Pharm. Tech. Res.*, **2012**, *4*, 839–847.

Donaldson, K.; Aiken, R.; Tran, L.; Stone, V.; Duffin, R.; Forrest, G.; Alexander, A. Carbon Nanotubes: A Review of Their Properties in Relation to Pulmonary Toxicology and Workplace Safety. *Toxicol. Sci.* **2006**, *92*, 5–22.

Dumortier, H.; Lacotte, S.; Pastorin, G.; Marega, R.; Wu, W.; Bonifazi, D.; Briand, J. P.; Prato, M.; Muller, S.; Bianco, A. Functionalized Carbon Nanotubes Are Non-Cytotoxic and Preserve the Functionality of Primary Immune Cells. *Nano Lett.* **2006**, *6*, 1522–1528.

Eatemadi, A.; Daraee, H.; Karimkhanloo, H.; Kouhi, M.; Zarghami, N.; Akbarzadeh, A.; Abasi, M.; Hanifehpour, Y.; Joo, S. W. Carbon Nanotubes: Properties, Synthesis, Purification and Medical Applications. *Nanoscale Res. Lett.*; **2014**, *9*:393–405.

El-Sheikh, A. H.; Sweileh, J. A. Recent Applications of Carbon Nanotubes in Solid Phase Extraction and Preconcentration: A Review. *Jord. J. Chem.*, **2011**, *6*, 1–16.

Elhissi, A. M. A.; Ahmed, W.; Hassan, I. U.; Dhanak, V. R.; D'Emanuele, A. Carbon Nanotubes in Cancer Therapy and Drug Delivery. *J. Drug Deliv.* **2012**, *ID 837327*, 1–10.

Fisher, C.; Rider, A. E.; Han, Z. J.; Kumar, S.; Levchenko, I.; Ostrikov, K. K. Applications and Nanotoxicity of Carbon Nanotubes and Graphene in Biomedicine. Review. *J. Nanomater.*, **2012**, *ID 315185*, 19 pp.

Folkmann, J. K.; Risom, L.; Jacobsen, N. R.; Wallin, H.; Loft, S.; Moller, P. Oxidatively Damaged DNA in Rats Exposed by Oral Gavage to C_{60} Fullerenes and Single-Walled Carbon Nanotubes. *Environ. Health Perspect.*, 2009, *117*, 703–708.

Fujigaya, T.; Nakashima, N. Non-covalent Polymer Wrapping of Carbon Nanotubes and the Role of Wrapped Polymers as Functional Dispersants. *Sci. Technol. Adv. Mater.*, **2015**, *16*, 1–21.

Galano, A. Carbon Nanotubes: Promising Agents against Free Radicals. *Nanoscale*, **2010**, *2*, 373–380.

Gallo, M.; Favila, A.; Mitnik, D. G. DFT Studies of Functionalized Carbon Nanotubes and Fullerenes as Nanovectors for Drug Delivery of Antitubercular Compounds. *Chem. Phys. Lett.*, **2007**, *447*, 105–109

Gao, C.; Li, W.; Morimoto, H.; Nagaoka, Y.; Maekawa, T. Magnetic Carbon Nanotubes: Synthesis by Electrostatic Self-Assembly Approach and Application in Biomanipulations. *J. Phys. Chem. B*, **2006**, *110*, 7213–7220.

Gao, L. Z.; Nie, L.; Wang, T. H.; Qin, Y. J.; Guo, Z. X.; Yang, D. L; Yan, X. Y. Carbon Nanotube Delivery of the GFP Gene into Mammalian Cells. *ChemBioChem.* **2006**, *7*, 239–242.

Han, Y-G.; Xu, J.; Li, Z-G.; Ren, G-G.; Yang, Z. In Vitro Toxicity of Multi-Walled Carbon Nanotubes in C6 Rat Glioma Cells. *NeuroToxicol.* **2012**, *33*, 1128–1134.

Haniu, H.; Saito, N.; Matsuda, Y.; Tsukahara, T.; Usui, Y.; Narita, N.; Hara, K.; Aoki, K.; Shimizu, M.; Ogihara, N.; Takanashi, S.; Okamoto, M.; Kobayashi, S.; Ishigaki, N.; Nakamura, K.; Kato, H. Basic Potential of Carbon Nanotubes in Tissue Engineering Applications. *J. Nanomaterials*, **2012**, *ID 343747*, 1–10.

He, H.; Pham-Huy, A. L.; Dramou, P.; Xiao, D-L.; Zuo, P.; Pham-Huy, C. Carbon Nanotubes: Applications in Pharmacy and Medicine, *BioMed. Res Int.* **2013**, *ID 578290*, 1–12.

Iijima, S. Helical microtubules of graphitic carbon. *Nature* **1991**, *354*, 56–58.

Jiang, L.; Li, L-L.; He, H.; Xiao, D-L.; Liu, T-B. Preparation Methods of Amino-functionalized Carbon Nanotubes and Their Application in Pharmaceutical Field. *Prog. Pharmac. Sci.*, **2012**, *36*, 400–405.

Jiang, L.; Liu, T-B.; He, H.; Pham-Huy, A. L.; Li, L-L.; Pham-Huy, C.; Xiao, D-L. Adsorption Behavior of Pazufloxacin Mesilate on Amino-Functionalized Carbon Nanotubes. *J. Nanosci. Nanotechnol.*, **2012**, *12*, 7271–7279.

Kagan, V. E.; Konduru, N. V.; Feng, W.; Allen, B. L.; Conroy, J.; Volkov, Y.; Vlasova, I. I.; Belikova, N. A.; Yanamala, N.; Kapralov, A.; Tyurina, Y. Y.; Shi, J.; Kisin, E. R.; Murray, A. R.; Franks, J.; Stolz, D.; Gou, P.; Klein-Seetharaman, J.; Fadeel, B.; Star, A.; Shvedova, A. A. Carbon Nanotubes Degraded by Neutrophil Myeloperoxidase Induce Less Pulmonary Inflammation. *Nature Nanotech.*, **2010**, *5*, 354–359.

Kantamneni, H.; Gollakota, A. Carbon Nanotubes Based Systems for Targeted Drug Delivery: A Review *Int. J. Eng. Res. Technol.*, **2013**, *2*, 1–8.

Kateb, B.; Yamamoto, V.; Alizadeh, D.; Zhang, L.; Manohara, H. M.; Bronikowski, M. J.; Badie, B. Multi-walled Carbon Nanotube (MWCNT) Synthesis, Preparation, Labeling, and Functionalization. *Methods Mol. Biol.*, **2010**, *651*, 307–317.

Kumar, R.; Dhanawat, M.; Kumar, S.; Singh, B. N.; Pandit, J. K.; Sinha, V. R. Carbon Nanotubes: A Potential Concept for Drug Delivery Applications. *Recent Pat Drug Deliv Formul.* **2014**, *8*, 1–15.

Kumar, S. P.; Prathibha, D.; Gowri Shankar, N. L.; Parthibarajan, R.; Mastyagiri, L.; Shankar M. Pharmaceutical Applications of Carbon Nanotube-Mediated Drug Delivery Systems. *Int. J. Pharm. Sci. Nanotech.* **2012**, *5*, 1685–1696.

Lacerda, L.; Bianco, A.; Prato, M.; Kostarelos, K. Carbon Nanotubes as Nanomedicines: From Toxicology to Pharmacology. *Adv. Drug Deliv. Rev.*, **2006**, *58*, 1460–1470.

Lacerda, L.; Russier, J.; Pastorin, G.; Herrero, M. A.; Venturelli, E.; Dumortier, H.; Al-Jamal, K. T.; Prato, M.; Kostarelos, K.; Bianco, A. Translocation Mechanisms of Chemically Functionalized Carbon Nanotubes across Plasma Membranes. *Biomaterials*, **2012**, *33*, 3334–3343.

Lay, C. L.; Liu, J.; Liu, Y. Functionalized Carbon Nanotubes for Anticancer Drug Delivery. *Expert Rev. Med. Devices*, **2011**, *8*, 561–566.

Lee, P. L.; Chiu, Y. K.; Sun, Y. C.; Ling, Y. C. Synthesis of a Hybrid Material Consisting of Magnetic Iron-Oxide Nanoparticles and Carbon Nanotubes As a Gas Adsorbent. *Carbon*, **2010**, *48*, 1397–1404.

Lee, S.; Lee, M.; Hong, Y.; Won, J.; Lee, Y.; Kan, S-G.; Chan, K-T.; Hong, Y. Middle Cerebral Artery Occlusion Methods in Rat *versus* Mouse Models of Transient Focal Cerebral Ischemic Stroke. *Neural Regen. Res.*, **2014**, *9*, 757–758.

Lee, Y.; Geckeler, K. E. Carbon Nanotubes in the Biological Interphase: The Relevance of Noncovalence. *Adv. Mater.*, **2010**, *22*, 4076–4083.

Li, L-L.; Lin, R.; He, H.; Jiang, L.; Gao, M-M. Interaction of Carboxylated Single-Walled Carbon Nanotubes with Bovine Serum Albumin. *Spectrochim. Acta A*. **2013**, *105*, 45–51.

Li, R.; Wu, R.; Zhao, L.; Wu, M.; Yang, L.; Zou, H. P-glycoprotein Antibody Functionalized Carbon Nanotube Overcomes the Multidrug Resistance of Human Leukemia Cells. *ACS Nano*, **2010**, *4*, 1399–1408.

Li, S.-S.; He, H.; Chen, Z.; Zha, J.; Pham-Huy, C. Fluorescence Study on the Interactions between Carbon Nanotubes and Bovine Serum Albumin. *Spectrosc. Spect. Anal.*, **2010**, *30*, 2689–2692.

Li, S.-S.; He, H.; Jiao, Q.; Pham-Huy, C. Applications of Carbon Nanotubes in Drug and Gene Delivery. *Prog. Chem.*, **2008**, *20*, 1798–1803.

Liao, H.; Paratala, R.; Sitharaman, B.; Wang Y. Applications of Carbon Nanotubes in Biomedical Studies. *Methods Mol. Biol.*, **2011**, *726*, 223–241.

Liu, Y.; Wu, D. C.; Zhang, W. D.; Jiang, X.; He, C. B.; Chung, T. S.; Goh, S. H.; Leong, K. W. Polyethylenimine Grafted Multiwalled Carbon Nanotubes for Secure Noncovalent Immobilization and Efficient Delivery of DNA. *Angew. Chem. Int. Ed.* **2005**, *44*, 4782–4785.

Liu, Z. Sun, X.; Nakayama-Ratchford, N.; Dai, H. Supramolecular Chemistry on Water-Soluble Carbon Nanotubes for Drug Loading and Delivery. *ACS Nano.*, **2007**, *1*, 50–56.

Liu, Z.; Chen, K.; Davis, C.; Sherlock, S.; Cao, Q.; Chen, X.; Dai, H. Drug Delivery with Carbon Nanotubes for in vivo Cancer Treatment. *Cancer Res.*, **2008**, *16*, 6652–6660.

Liu, Z.; Tabakman, S. M.; Chen, Z.; Dai, H. Preparation of Carbon Nanotubes Bioconjugates for Biomedical Applications. *Nat. Protoc.*, **2009**, *4*, 1372–1382.

Liu, Z.; Tabakman, S.; Welsher, K.; Dai, H. Carbon Nanotubes in Biology and Medicine: In vitro and in vivo Detection, Imaging and Drug Delivery. *Nano Res.*, **2009**, *2*, 85–120.

Luo, X.; Matrangar, C.; Tan, S.; Alba, N.; Cui, X. T. Carbon Nanotube Nano-reservoir for Controlled Release of Anti-inflammatory Dexamethasone. *Biomaterials*, **2011**, *32*, 6316–6323.

MacDonald, R. A.; Laurenzi, B. F.; Viswanathan, G.; Ajayan, P.; Stegemann, J. P. Collagen-Carbon Nanotube Composite Materials as Scaffolds in Tissue Engineering. *J. Biomed. Mater. Res.*, Part A, **2005**, *74*, 489–496.

Madani S. Y.; Naderi, N.; Dissanayake, O.; Tan, A.; Seifalian, A. M. A New Era of Cancer Treatment: Carbon Nanotubes as Drug Delivery Tools. *Int. J. Nanomed.*, **2011**, *6*, 2963–2979.

Madani, S. Y.; Mandel, A.; Seifalian, A. M. A Concise Review of Carbon Nanotube's Toxicology. *Nano Rev.*, **2013**, *4*, 21521–21534.

Masotti, A.; Caporali, A. Preparation of Magnetic Carbon Nanotubes (Mag-CNTs) for Biomedical and Biotechnological Applications. *Int. J. Mol. Sci.* **2013**, *14*, 24619–24642.

Mehra, N. K.; Mishra, V.; Jain, N. K. A Review of Ligand Tethered Surface Engineered Carbon Nanotubes. *Biomaterials*, **2014**, *35*, 1267–1283.

Meng, J.; Meng, J.; Duan, J.; Kong, H.; Li, L.; Wang, C.; Xie, S.; Chen, S.; Gu, N.; Xu, H.; Yang, X-D. Carbon Nanotubes Conjugated to Tumor Lysate Protein Enhance the Efficacy of an Antitumor Immunotherapy. *Small*, **2008**, *4*, 1364–1370.

Mishra, R.; Mishra, A. Review on Potential Applications of Carbon Nanotubes and Nanofibers. *Int. J. Pharm. Rev. Res.*, **2013**, *3*, 12–17.

Naficy, S.; Razal, J. M.; Spinks, G. M.; Wallace, G. G. Modulated Release of Dexamethasone from Chitosan-Carbon Nanotube Films. *Sens. Actuators A*, **2009**, *155*, 120–124.

Nguyen, A. L.; He, H.; Pham-Huy, C. Chiral Drugs. An Overview. *Int. J. Biomed. Sci.*, **2006**, *2*, 85–100.

Panczyk, T.; Warzocha, T. P. Monte Carlo Study of the Properties of a Carbon Nanotube Functionalized by Magnetic Nanoparticles. *J. Phys. Chem. C.* **2009**, *113*, 19155–19160.

Pantarotto, D.; Singh, R.; McCarthy, D.; Erhardt, M.; Briand, J-P.; Prato, M.; Kostarelos, K.; Bianco, A. Functionalized Carbon Nanotubes for Plasmid DNA Gene Delivery. *Angew Chem. Int. Ed.*, **2004**, *43*, 5242–5246.

Paratala, B. S.; Sitharaman, B. Carbon Nanotubes in Regenerative Medicine. In: *Carbon Nanotubes for Biomedical Applications*, Klingeler, R.; Sim, R. B.; Eds.; Springer-Verlag: Berlin, **2011**, pp. 27–40.

Pastorin, G.; Kostarelos, K.; Prato, M.; Bianco, A. Functionalized Carbon Nanotubes: Towards the Delivery of Therapeutic Molecules. *J. Biomed. Nanotech.* **2005**, *1*, 1–10.

Pham-Huy, A. L.; He, H.; Pham-Huy, C. Free Radicals, Antioxidants in Disease and Health. *Int. J. Biomed. Sci.*, **2008**, *4*, 89–96.

Pineda, B.; Hernández-Pedro, N. Y.; Maldonado, R. M.; Pérez-De la Cruz, V.; Sotelo, J. Carbon Nanotubes: A New Biotechnological Tool on the Diagnosis and Treatment of Cancer. In: *Nanobiotechnology*, Phoenix, D. A.; Ahmed, W.; Eds.; OCP: Manchester, **2014**, pp. 113–131.

Poland, C. A.; Duffin, R.; Kinloch, I.; Maynard, A.; Wallace, W. A. H.; Seaton, A.; Stone, V.; Brown, S.; MacNee, W.; Donaldson, K. Carbon Nanotubes Introduced into the Abdominal Cavity of Mice Show Asbestos-Like Pathogenicity in a Pilot Study. *Nature Nanotechnol.*, **2008**, *3*, 423–428.

Rastogi, V.; Yadav, P.; Bhattacharya, S. S.; Mishra, A. K.; Verma, N.; Verma, A.; Pandit, J. K. Carbon Nanotubes: An Emerging Drug Carrier for Targeting Cancer Cells. *J. Drug Deliv.*, **2014**, *ID 670815*, 23 pp.

Rawat, M.; Singh, D.; Saraf, S. Nanocarriers: Promising Vehicle for Bioactive Drugs. *Biol. Pharm. Bull.* **2006**, *29*, 1790–1798.

Rosen, Y.; Elman, N. M. Carbon Nanotubes in Drug Delivery: Focus on Infectious Diseases. *Expert Opin. Drug Deliv.*, **2009**, *6*, 517–530.

Rosen, Y.; Mattix, B.; Rao, A.; Alexis, F. Carbon Nanotubes and Infectious Diseases. In: *Nanomedicine in Health and Disease*. Hunter, R. J., Ed.; Science Publishers: London, **2011**, pp. 249–267.

Safari, J.; Zarnegar, Z. Advanced Drug Delivery Systems: Nanotechnology of Health Design. A Review. *J. Saudi Chem. Soc.* **2014**, *18*, 85–99.

Sahoo, N. G.; Bao, H.; Pan, Y.; Pal, M.; Kakran, M.; Cheng, H. K. F.; Li, L.; Tan, L. P. Functionalized Carbon Nanomaterials as Nanocarriers for Loading and Delivery of a Poorly Water-soluble Anticancer Drug: A Comparative Study, ***Chem. Commun.***, **2011**, ***47***, 5235–5237.

Saito, N.; Haniu, H.; Usui, Y.; Aoki, K.; Hara, K.; Takanashi, S.; Shimizu, M.; Narita, N.; Okamoto, M.; Kobayashi, S.; Nomura, H.; Kato, H.; Nishimura, N.; Taruta, S.; Endo, M. Safe Clinical Use of Carbon Nanotubes as Innovative Biomaterials. *Chem. Rev.* **2014**, *114*, 6040–6079.

Shao, W.; Arghya, P.; Yiyong, M.; Rodes, L.; Prakash, S. Carbon Nanotubes for Use in Medicine: Potentials and Limitations. In: *Syntheses and Applications of Carbon Nanotubes and Their Composites*. Suzuki, S., Ed.; InTech: Rijeka, **2013**, pp. 285–311.

Silva, R. A.; Talio, M. C.; Luconi, M. O.; Fernandez, L. P. Evaluation of Carbon Nanotubes as Chiral Selectors for Continuous-Flow Enantiomeric Separation of Cardevilol with Fluorescent Detection. *J. Pharm. Biomed. Anal.*, **2012**, *70*, 631–635.

Singh, B. G. P.; Baburao, C.; Pispati, V.; Pathipati, H.; Muthy, N.; Prassana, S. R. V.; Rathode B. G. Carbon Nanotubes. A Novel Drug Delivery System. *Int. J. Res. Pharm. Chem.*, **2012**, *2*, 523–532.

Singh, R. K.; Patel, K. D.; Kim, J-J.; Kim, T-H.; Kim, J-H.; Shin, U. S.; Lee, E-J.; Knowles, J. C.; Kim; H-W. Multifunctional Hybrid Nanocarrier: Magnetic CNTs Unsheathed with Mesoporous Silica for Drug Delivery and Imaging System. *ACS Appl. Mater. Interfaces*, **2014**, *6*, 2201–2208.

Singh, R.; Pantarotto, D.; Lacerda, L.; Pastorin, G.; Klumpp, C.; Prato, M.; Bianco, A.; Kostarelos, K. Tissue Biodistribution and Blood Clearance Rates of Intravenously Administered Carbon Nanotube Radiotracers. *Proc. Natl. Acad. Sci. USA*. **2006**, *103*, 3357–3362.

Singh, R.; Pantarotto, D.; McCarthy, D.; Chaloin, O.; Hoebeke, J.; Partidos, C. D.; Briand, J. P.; Prato, M.; Bianco, A.; Kostarelos, K. Binding and Condensation of Plasmid DNA onto Functionalized Carbon Nanotubes: Toward the Construction of Nanotube-Based Gene Delivery Vectors. *J. Am. Chem. Soc.* **2005**, *127*, 4388–4396.

Surendiran, A.; Sandhiya, S.; Pradhan, S. C.; Adithan, C. Novel Applications of Nanotechnology in Medicine. *Indian J. Med. Res.* **2009**, *130*, 689–701.

Takanashi, S.; Hara, K.; Aoki, K.; Usui, Y.; Shimizu, M.; Haniu, H.; Ogihara, N.; Ishigaki, N.; Nakamura, K.; Okamoto, M.; Kobayashi, S.; Kato, H.; Sano, K.; Nishimura, N.; Tsutsumi, H.; Machida, K.; Saito, N. Carcinogenicity Evaluation for the Application of Carbon Nanotubes as Biomaterials in RasH2 Mice. *Sci. Rep.* **2012**, *2*, 1–7.

Tan, J. M.; Arulselvan, P.; Fakurazi, S.; Ithnin, H.; Hussein, M. Z. A Review on Characterizations and Biocompatibility of Functionalized Carbon Nanotubes in Drug Delivery Design. *J. Nanomater*. **2014**, *ID 917024*, 20 pp.

Tasis, D.; Tagmatarchis, N.; Bianco, A.; Prato, M. Chemistry of Carbon Nanotubes. *Chem. Rev.*, **2006**, *106*, 1105–1136.

Tripathi, A. C.; Saraf, S. A.; Saraf, S. K. Carbon Nanotropes: A Contemporary Paradigm in Drug Delivery. *Materials* **2015**, *8*, 3068–3100.

Usui, Y.; Haniu, H.; Tsuruoka, S.; Saito, N. Carbon Nanotubes Innovate on Medical Technology. *Med. Chem.*, **2012**, *2*, 1–6.

Wang, H. F.; Wang, J.; Deng, X. Y.; Sun, H. F.; Shi, Z. J. Biodistribution of Carbon Single–Walled Carbon Nanotubes in Mice. *J. Nanosci. Nanotechnol.*, **2004**, *4*, 1019–1024.

Wang, J. Carbon-Nanotube Based Electrochemical Biosensors: A Review. *Electroanalysis*, **2005**, *17*, 7–14.

Xiao, D-L.; Dramou, P.; He, H.; Pham-Huy, A. L.; Li, H.; Pham-Huy, C. Magnetic Carbon Nanotubes: Synthesis by a Simple Solvothermal Process and Application in Magnetic Targeted Drug Delivery System. *J. Nanopart. Res.*, **2012**, *14*, 984–995.

Xiao, D-L.; Dramou, P.; Xiong, N.; He, H.; Li, H.; Yuan, D.; Dai, H. Development of Novel Molecularly Imprinted Magnetic Solid-Phase Extraction Materials Based on Magnetic Carbon Nanotubes and Their Application for The Determination of Gatifloxacin in Serum Samples Coupled with High Performance Liquid Chromatography. *J. Chromatogr. A*. **2013**, *1274*, 44–53.

Xiao, D-L.; Dramou, P.; Xiong, N.; He, H.; Yuan, D.; Dai, H.; Li, H.; He, X-M.; Peng, J.; Li, N. Preparation of Molecularly Imprinted Polymers on the Surface of Magnetic Carbon Nanotubes with a Pseudo Template for Rapid Simultaneous Extraction of Four Fluoroquinolones in Egg Samples, *Analyst*, **2013**, *138*, 3287–3296.

Xiao, D-L.; Li, H.; He, H.; Lin, R.; Zuo, P. Adsorption Performance of Carboxylated Multi-Wall Carbon Nanotube-Fe3O4 Magnetic Hybrids for Cu(II) in Water, *New Carbon Mater.*, **2014**, *29*, 15–25.

Xiao, D-L.; Yuan, D.; He, H.; Pham-Huy, C.; Dai, H.; Wang, C-X.; Zhang, C. Mixed Hemimicelles Solid-Phase Extraction Based on Magnetic Carbon Nanotubes and Ionic Liquid for the Determination of Flavonoids, *Carbon*, **2014**, *72*, 274–286.

Yang, D.. Yang, F.; Hu, J.; Long, J.; Wang, C.; Fu, D.; Ni, Q. Hydrophilic Multi-Walled Carbon Nanotubes Decorated with Magnetite Nanoparticles as Lymphatic Targeted Drug Delivery Vehicles. *Chem. Commun.*, **2009**. *7*, 4447–4449.

Yang, F.; Fu D.; Long, J.; Ni, Q. Magnetic Lymphatic Targeting Drug Delivery System using Carbon Nanotubes. *Med. Hypotheses* **2008**, *70*, 765–767.

Yang, F.; Jin, C.; Yang, D.; Jiang, Y.; Li, J.; Di, Y.; Hu, J.; Wang, C.; Ni, Q.; Fu, D. Magnetic Functionalized Carbon Nanotubes as Drug Vehicles for Cancer Lymph Node Metastasis Treatment. *Eur. J. Cancer*, **2011**, *47*, 1873–1882.

Yang, S.-T.; Luo, J.; Zhou, Q.; Wang, H. Pharmacokinetics, Metabolism and Toxicity of Carbon Nanotubes for Bio-medical Purposes. *Theranostics*, **2012**, *2*, 271–282.

Yang, W.; Thordarson, P.; Gooding, J. J.; Ringer, S. P.; Braet, F. Carbon Nanotubes for Biological and Biomedical Applications. *Nanotechnology*, **2007**, *18*, 1–12.

Yang, Z.; Zhang, Y.; Yang, Y.; Sun, L.; Han, D.; Li, H.; Wang, C. Pharmacological and Toxicological Target Organelles and Safe Use of Single-Walled Carbon Nanotubes as Drug Carriers in Treating Alzheimer Disease. *Nanomedicine*, **2010**, *6*, 427–441.

Zha, J.; He, H.; Liu, T.-B.; Li, S-S.; Jiao, Q-C. Studies on the interaction of Gatifloxacin with Bovine Serum Albumin in the presence of Carbon Nanotubes by Fluorescence Spectroscopy. *Spectrosc. Spect. Anal.*, **2011**, *31*, 149–153.

Zhang, W.; Zhang, Z.; Zhang, Y. The Application of Carbon Nanotubes in Target Drug Delivery Systems for Cancer Therapies. *Nanoscale Res. Lett.*, **2011**, *6*, 555–577.

Zhang, Y.; Bai, Y.; Yan, B. Functionalized Carbon Nanotubes for Potential Medicinal Applications. *Drug Disc. Today*, **2010**, *15*, 428–435.

Zhao, D.; Alizadeh, D.; Zhang, L.; Liu, W.; Farrukh, O.; Manuel, E.; Diamond, D. J.; Badie, B. Carbon Nanotubes Enhance CpG Uptake and Potentiate Anti-Glioma Immunity. *Clin. Cancer Res.*, **2011**, *17*, 771–782.

Zhu, Y.; Wang, L.; Xu, C. Carbon Nanotubes in Biomedicine and Biosensing. In: *Carbon Nanotubes – Growth and Applications*. Naragh, M., Ed.; InTech, Shanghai, **2011**, pp. 135–162.

CHAPTER 7

DENDRIMERS: A GLIMPSE OF HISTORY, CURRENT PROGRESS, AND APPLICATIONS

SURYA PRAKASH GAUTAM,[1,*] ARUN KUMAR GUPTA,[2] REVATI GUPTA,[2] TAPSYA GAUTAM,[1] and MANINDER PAL SINGH[1]

[1]*CT Institute of Pharmaceutical Sciences, Shahpur Campus, Jalandhar, India, Tel.: +91-8894447581, +91-7696813994;*
**E-mail: suryagautam@ymail.com, gautamsuryaprakash@gmail.com*

[2]*RKDF Institute of Pharmaceutical Sciences, Indore, India*

CONTENTS

ABSTRACT

Dendrimers acclaimed its fascinating position in the nanoworld. Its unique polymeric architecture, it exhibits precise compositional and constitutional properties. The combination of a discrete number of functionalities and their high local densities make dendrimers as multifunctional platforms for amplified substrate binding. As a result of their unique architecture and construction, dendrimers possess inherently valuable physical, chemical and biological properties. Versatility of dendrimers, showing promise in the therapeutics arena, and there is a great deal of commercial activity, with a few products hitting the market. Together with recent progress in the design of biodegradable chemistries, has enabled the application of these branched polymers as antiviral drugs, tissue repair scaffolds, targeted carriers of chemotherapeutics and optical oxygen sensors. Pattering other functional groups, will facilitate further development of this system for novel applications. It is beginning to make significant inroads into the commercial world. Several products using dendrimers as platform have been developed and commercialized upon the approval of the FDA. There are also some others applications like: for cellular transport, as artificial cells, for diagnostics and analysis, as protein/enzyme mimics or modeling, for manufacture of artificial bones, for development of topical microbicide creams; antimicrobial, antiviral (e.g., for use against HIV) and antiparasitic agents, for biomedical coatings (e.g., for artificial joints), as artificial antibodies and biomolecular binding agents, for carbon fiber coatings and ultra thin films, as polymer and plastics additives (e.g., for lowering viscosity, increasing stiffness, incorporating dyes, compatibilizers, etc.) for creation of foams (i.e., synthetic zeolites or insulating material), as building blocks for nanostructured materials, as dyes and paints, as industrial adhesives, for manufacture of nanoscale batteries and lubricants, as decontamination agents (trapping metal ions), for ultrafiltration, molecular electronics for data storage, 3D optical materials, for light-harvesting systems, quantum dots, liquid crystals, printed wire boards, etc.

7.1 INTRODUCTION

Dendrimers acclaimed its most fascinating position in the nano world. By virtue of its unique polymeric architecture, it exhibits precise compositional and constitutional properties. Multifunctional platform provides ligand conjugation and surface modification possibilities. Dendrimers manifests various clinical applications (Tomalia et al., 1985). Versatility of dendrimers, showing promise in the therapeutics arena, and there is a great deal of commercial activity, with a few products hit the market. The ability to engineer and control these parameters provides an endless list of possibilities for using dendrimers as modules for nanodevice design (Figure 7.1) (Tomalia et al., 1990).

Dendrimers are hyper branched, highly ordered 3-D structure, having definite molecular weight, size, shape, in which all the bonds are converging to a focal point. Dendrimers synthesis involves iterative fashion of reaction steps, in which each additional iteration leads to a higher generation dendrimer. Building on a central core, dendrimers are formed by the step-wise sequential addition of concentric shells consisting of branched molecules and connector groups (Figure 7.1). The basic structure of dendrimers enables for a lot of flexibility, which is leverage in the case of the application of dendrimers. Past several years there has been huge progress in the utilization of polymeric carriers for a wide variety of drugs (Smith et al., 1987). The surface properties of dendrimers may be manipulated by the use of appropriate 'capping' reagents on the outermost generation (Esfand et al., 2001; Greenwald et al., 2003). The combination of a discrete number of functionalities and their high local densities make dendrimers as multifunctional platforms for amplified

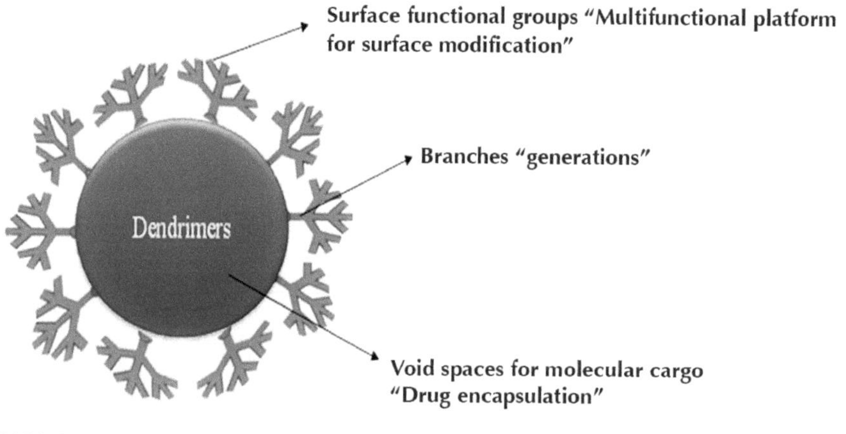

FIGURE 7.1 Core structure of dendrimers.

substrate binding. As a result of their unique architecture and construction, dendrimers possess inherently valuable physical, chemical and biological properties (Namazi et al., 2005; Liu et al., 2000). Synthesizing mono-disperse polymers demands a high level of synthetic control. Dendrimers branch out in a highly predictable fashion to form amplified three-dimensional structures with highly ordered architectures which assist in achieving monodispersity (Santo et al., 1999). Internal cavities of dendrimers structures can be used to carry and store a wide range of metals, organic, or inorganic molecules (Tomalia et al., 1986). The outer shell of each dendrimers can be manipulated to contain a large number of reactive groups (Kallos et al., 1991). Each of these reactive sites has the potential to interact with a target entity, often resulting in poly valets interactions (Hummeln et al., 1997; Maiti et al., 2004). Dendrimers literature suggests that ionic interaction, hydrogen bonding and hydrophobic interactions are possible mechanism by which dendrimers exerts its solubilizing property (Duncan et al., 1996; Ihre et al., 2002; Kono et al., 2005; Newkombe et al., 2001; Stiriba et al., 2002). Dendrimers have well defined globular structures with controlled surface functionality. By chemical linkage of ligand and surface modification, it may add to their potential as a new scaffold for drug delivery. High surface density and empty inner spaces provide a flexible option for diverse application of dendrimers in drug/gene delivery (Chauhan et al., 2004; Devarakonda et al., 2004; Jevprasesphant et al., 2003; Kojima et al., 2000; Konda et al., 2005; Malik et al., 2002; Yang et al., 2004; Yiyun et al., 2005;). These bioactive dendrimers open a new avenue in dendrimer research. By virtue of unique properties, dendrimers may address common problems for treatment for cancer patient's, i.e., limited accessibility of drugs to the tumor tissue, unfavorable biodistributation, their intolerable toxicity, development of multidrug resistance, and the dynamic heterogeneous biology of the growing tumors.

7.2 UNIQUENESS OF DENDRIMERS: INTRINSIC QUALITIES OF DENDRIMERS

The system serves to fulfill following objectives:

1. **Enhanced permeability and retentation effect:** Size of dendrimers, i.e., (generation 4–4.4 nm) is in nano range. Cancer cells have leaky membranes and having higher biopermeability. Lymphatic system is

one way and drug loaded dendrimers get retained inside (Bae et al., 2005; Patri et al., 2005; Purohit et al., 2001).

2. **High permeability**: Dendrimers cross bio- barriers like blood brain barrier, cell membrane. Nanometer range and uniformity in size enhance their ability to cross cell membranes and reduce the risk of undesired clearance from the body through the liver or spleen (Jansen et al., 1994; Newkome et al., 1996; Zhuo et al., 1999).
3. **Sustained/extended effect:** Dendrimers having 3-D network structure release drug in a sustained manner. PAMAM dendrimers exhibited slower release, higher accumulation in solid tumors, and lower toxicity. Conjugation with Polyethylene glycol (PEG) improves the bio distribution by suppressing any unfavorable nonspecific interaction with biomolecules. Extension in circulation time is essential to produce desired clinical effect (Newkome et al., 1991).
4. **Higher Solublization Potential:** Ionic interaction, hydrogen bonding and hydrophobic interactions are possible mechanism by which dendrimers exerts its solubilizing property (Chauhan et al., 2003; Galia et al., 1998; Jani et al., 2009; Kolhe et al., 2006; Milhem et al., 2000; Twyman et al., 1999).
5. **High uniformity and purity:** The synthetic process used produces dendrimers with uniform sizes, precisely defined surface functionality, and very low impurity levels. Monodispersed molecules like dendrimers would enable us to achieve more precise drug delivery (Sayed et al., 2001, 2002).
6. **Multifunctional platform:** Free surface groups can form complex or conjugates with drug molecules or ligands by using cross linkers. The surface of dendrimers provides an excellent platform for the attachment of cell-specific ligands, solubility modifiers, and stealth molecules reducing the interaction with macromolecules from the body defense system (Devarakonda et al., 2005; Erk et al., 2003; Stevemans et al., 1996).
7. **High loading capacity**: Dendrimers structures can be used to carry and store a wide range of metals, organic or inorganic molecules by encapsulation and absorption. Drug can get entrapped inside the internal cavities as well as electro statically in the surface of dendrimers (Nekkanti et al., 2009).

8. **High stability**: Dendrimers drug complex or conjugate exhibits better stability.
9. **Low toxicity**: Most dendrimers systems display very low cytotoxicity levels (Michaels et al., 1975; Sathesh et al., 2008).
10. **Low immunogenicity:** Dendrimers commonly manifest a very low or negligible immunogenic response when injected or used topically (Peterson et al., 2001; Tang et al., 1996).

The problems in vesicular system like chemical instability, drug leakage, aggregation and fusion during storage, solubility in physiological environment, lyzes of phospholipids, Purity of natural phospholipids, cost of production lack in dendritic system (Figure 7.2).

7.3 HOST GUEST INTERACTION

An area that has attracted great interest is the interaction between drugs and dendrimers. Several types of interactions have been explored, which can be broadly subdivided into the entrapment of drugs within the dendritic architecture (involving electrostatic, hydrophobic and hydrogen bond interactions) and the interaction between a drug and the surface of a dendrimer (electrostatic and covalent interactions).

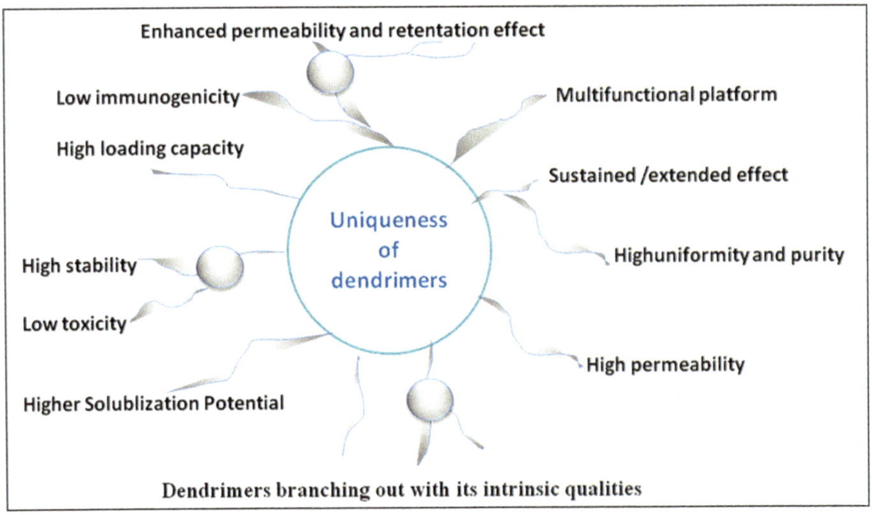

FIGURE 7.2 Dendrimers branching out with its intrinsic qualities.

Closer look at different motifs which are useful for complexation between dendrimers and various guest molecules, the so-called host–guest complexes. Present article explores the various complex interaction mechanisms between the dendrimer and a bioactive and their possible contribution in drug research (Choi et al., 2004; Peterson et al., 2001; Yellepeddi et al., 2008).

The mechanisms for the host-guest interaction can be broadly grouped into two main classes: (a) covalent binding, in which the guest molecule forms a chemically bonded conjugate (involving hydrophobic interactions, physical entrapment, hydrogen bonding, or electrostatic bonding either alone or in combination with these methods), and (b) noncovalent binding, in which the guest physically interacts with the dendritic architecture (Figure 7.3) (Davis et al., 2002; Kim et al., 2008).

Jansen et al. (1994) conceives the mechanism of the physical entrapment of a guest molecule within the internal architecture of the dendritic box, and this has been explored during their formulation process, while entrapping the guest molecules in these boxes. Newkome (1991) was the first to report this type of interaction in dendrimers, to solubilize hydrophobic solutes in aqueous media. Wallimann et al. (2000) reported Dendrimers specifically tailored to bind hydrophobic guests to the core have been created by the Diederich group under the name ‘dendrophanes. Newkome et al. (1996) reported hydrogen-bonding interactions between the dendritic host and the guest molecules, such as, glutarimide and barbituric acid.

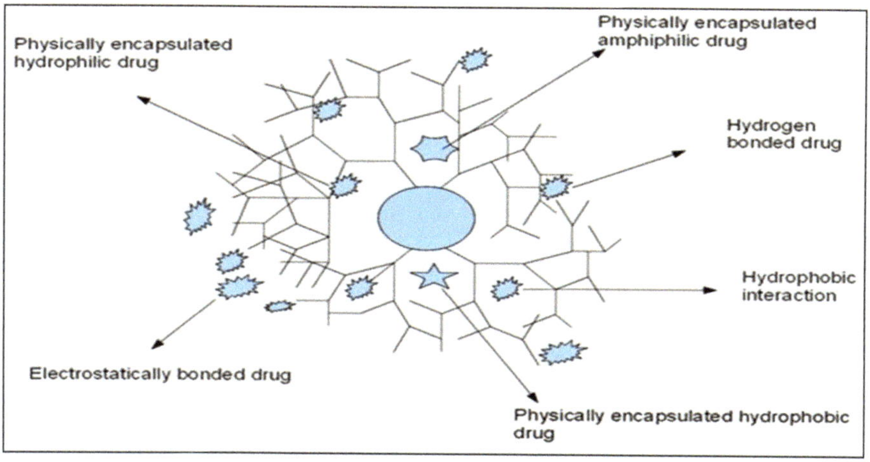

FIGURE 7.3 Dendrimers as drug carrier and mechanism of drug entrapment.

Milhem et al. (2000) showed that electrostatic interaction between hydrophobe and dendrimer was the major responsible mechanism for solubility enhancement.

Endoreceptors:

- Physical encapsulation (Cheng et al., 2007)
- Hydrophobic bondin (Gupta et al., 2006)
- Hydrogen bonding (Yang et al., 2004)

Exoreceptors:

- Electrostatic interaction (Zhang et al., 2010)

7.4 DEEP INSIGHT INTO HISTORY AND CURRENT PROGRESS

The first dendrimers were made by divergent synthesis approaches by Vögtle in 1978. In 1990 a convergent synthetic approach was introduced by Jean Fréchet (Figure 7.4) (Table 7.1) (Banerjee et al., 2004; Bosman et al., 1999; Caminade et al., 2010; Davis et al., 2003; Felder et al., 2005; Furer et al., 2004; Guoping et al., 2005; Hummelen et al., 1987; Karlos et al., 2008; Khopade et al., 2002; Kolev et al., 2005; Lajos et al., 2005; Patri et al., 2002; Park et al., 1999; Pesak et al., 1999; Popescu et al., 2006; Ronald et al., 2002; Singhai et al., 1997; Xiangyang et al., 2006; Zeng et al., 1997, 2001).

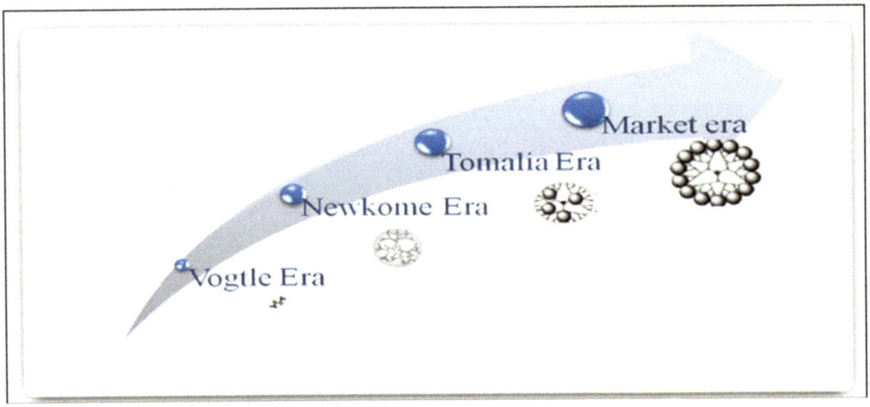

FIGURE 7.4 Dendrimers growth track.

TABLE 7.1 Timeline Milestones of Dendrimers

S. No.	Dendrimers	Year
1	Evidence for Branched-Chain Macromolecules	1941–1942
2	Stepwise Strategies for Synthesis of Marcocyclic Rings	1973
3	Cascade Synthesis	1978
4	Lysine-dendrimer Patent	1981
5	First Modular Dendrimer Synthesis, First Reference to Unimolecular Micelles	1985
6	First High-Generation Dendrimers-based on Linear Monomers	1985
7	First Convergent Synthesis	1990
8	First Theoretical Study on Dendrimers	1983
9	Phenylacetylene Dendrimers	1991
10	Improved Vogtle Procedure and Later Developed the Dendritic Box	1993
11	Silicone-based Dendrimers	1990
12	Phosphorous-based Dendrimers	1994
13	Self-Assembly of Dendrimers	1995
14	Dendronized-Polymers	1995
15	Metallodendrimers	1989
16	Chiral Dendrimers	1993
17	DNA-based Dendrimers	1993
18	Polyamidoaminedendrimers no viral vectors for gene transfer	1999
19	Starpharma is in phase II trials of dendrimers	1999
20	Treatment of various viral diseases	2001
21	US FDA for human trials of their dendrimer-based anti-HIV product	2003
22	Vaginal gel, SPL7013 human trials of their dendrimer-based anti-HIV product	2004
23	Pegylated Dendrimers	2010

Patents:

Tremendous rise in dendrimers-based patents attracts attentions of researchers to keep abreast in this important area. Credit for Very first patents goes to Donald Tomalia and Dow Chemical. Tomalia and Dow Chemical obtained most of the early dendrimers patents, but by the mid-1990's, many other companies had begun patenting dendrimers as well. For example, Xerox began developing a patent portfolio on dendrimers in ink/toner applications, Dow chemical and DSM also began pursuing dendrimers patents (Furer et al., 2004)

1. U.S. Patent No. 6,232,378 disclosed use of dendrimers as additives in commodity plastics.
2. DSM advocates, large-scale dendrimer production in their ASTRAMOL™ technology.
3. In biotechnology applications, dendrimers have become useful in many ways. For example, dendrimers have become transfection reagents, mediating transport across cell membranes.
4. University of Michigan disclosed a myriad of studies for antiviral application. In addition, U. S. Patent No. 6,190,650 disclosed ionic dendrimers useful for their antiviral properties (Table 7.2).

Classification of dendrimers: The classifications of dendrimers are given in Table 7.3.

7.5 SYNTHESIS OF DENDRIMERS

Dendrimer synthesis is a relatively new field of polymer chemistry defined by regular, highly branched monomers leading to a Monodisper, tree-like or generational structure. Synthesizing Monodisper polymers demands a high

TABLE 7.2 U.S. Patent in the Field of Dendrimers

U.S. Patent No.	Disclosures
U.S. Patent No. 6,232,378	Dendrimers as additives in commodity plastics.
U.S. Patent No. 6,084,030	Low cost plastics such as radiator Hoses, weather stripping, and gasket seals.
U.S. Patent No. 5,714,166	Transfection reagents, mediating transport across cell membranes.
U.S. Patent No. 6,190,650	Antiviral properties.
U.S. Patent No. 6,274,723	Dendritic nucleic acids which function as hybridization reagents for the detection of nucleic acid sequences.
U.S. Patent No. 4,507,466	Viscosity modifying agents.
U.S. Patent No. 5,714,166,	Antibody-labeled dendrimers can be localized in a mouse tumor.
U.S. Patent No. 6, 190,650.	Starpharma prepared an investigational New Drug application (IND) for phase I testing of dendrimers gels in humans.

TABLE 7.3 Classification of Dendrimers

Chemical classification	Physical classification	Miscellaneous
1. Poly amidoamine (PAMAM) Dendrimers	Simple Dendrimers	Dendrophanes
2. Polypropyleneimine (PPI) Dendrimers	Liquidcrystalline Dendrimers	Metallodendrimers
3. polyether(PE) Dendrimers	Chiral Dendrimers	Polyamino phosphine
4. L-lysine-based Dendrimers	Micellar Dendrimers	Dendritic box
5. Phenyl acetylene Dendrimers.	Hybrid Dendrimers	Carbohydrate vaccine Dendrimers

level of synthetic control which is achieved through stepwise reactions, building the dendrimer up one monomer layer, or "generation," at a time (Gardikis et al., 2006). Each dendrimer consists of a multifunctional core molecule with a dendritic wedge attached to each functional site (Figure 7.5).

7.6 POLYAMIDOAMINE (PAMAM) DENDRIMERS

Tomalia synthesized the first PAMAM dendrimers in 1985. Followed by Frechet's convergent approach in 1989. PAMAM dendrimers have

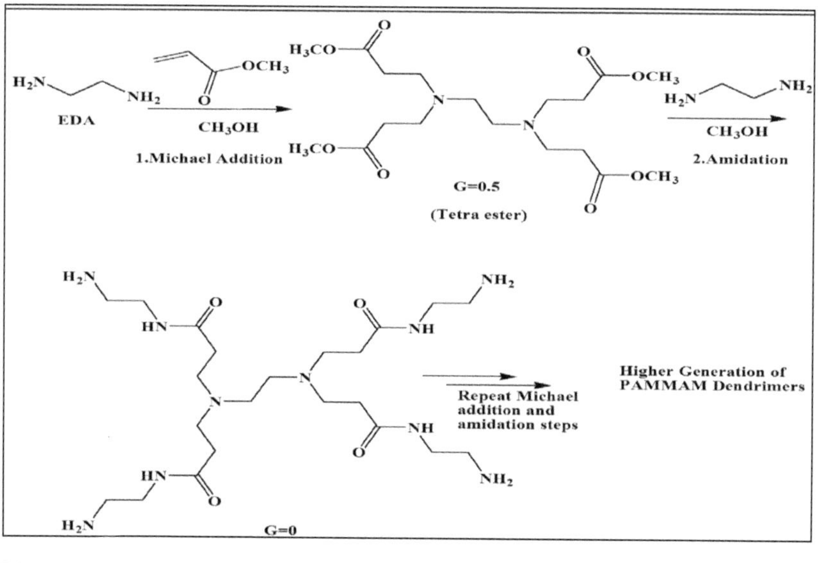

FIGURE 7.5 Schematic representation of PAMAM Dendrimers synthesis by divergent approach.

a unique tree like branching architecture that confers them a compact spherical shape in solution and a controlled incremental increase in size molecular weight and number of surface amine groups. PAMAM dendrimers have been shown to exhibits minimum cytotoxicity up to generation 5 (Figure 7.6).

Size and Shape:

PAMAM dendrimers ranges in diameter from about 2 nm (G-1) upto 13 nm (G-10). The conformation of dendrimers adjust to this steric crowding by developing a three dimensional structure for example G-1 or G-2 PAMAM dendrimers has open, flat structure, but by G-4 the 64 peripheral groups can only be accommodated if the 4.5 nm diameter dendrimer becomes spheroid. By G-8 the number of terminal groups has increased to 1024, but diameter has only got doubled, and therefore the periphery is very densely packed (Yoshimura et al., 2004).

7.7 CURRENT STATUS

Several products using dendrimers as platform have been developed and commercialized upon the approval of the FDA. Among them, VivaGel™ is a topical microbicide to prevent the transmission of HIV and other

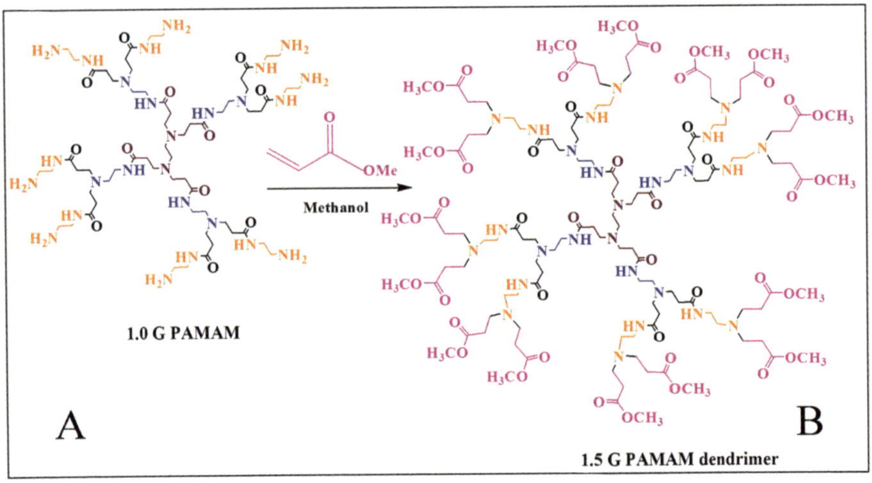

FIGURE 7.6 (a) 1.0 G PAMAM (Amine Terminated); (b) 1.5 G PAMAM (Carboxylic Acid Terminated).

sexually transmitted diseases. SuperFect® is used for gene transfection of a broad range of cell lines. Alert Ticket™ is an anthrax-detection agent. Stratus® CS is used for cardiac marker diagnostic (Table 7.4).

Currently Dendrimers are produced on larger scale and can be ordered through the Sigma-Aldrich catalog. Dow Chemical and dendritech maintain their dendrimer commercial position. Dendritic Nanotechnologies is providing on-going dendrimer leadership in all markets including exploiting traditional poly (amidoamine) PAMAM dendrimers. At DSM, Hybrane® hyper branched polyester amides are being developed, and are called Astramol. Dade Behring has patented and sells cardiac diagnostic technology-based on dendrimer linked monoclonal antibodies (Stratus® CS). Starpharma's dendrimer-doxorubicin construct (SPL8181) achieved the same inhibition of human breast-cancer tissue as doxorubicin alone. VivaGel vaginal microbicide is in Phase II clinical trials, and is the subject of a license agreement with Durex(R) condoms for use as a condom coating (Bielinska et al., 1996).

Priostar® Additives were specifically designed to improve adhesion, promote cross-linking, and enhance dispersion in formulations at low additive levels. DNT's Priostar® Additives is the latest innovation in polymeric dendrimer chemistry.

Other Dendrimer Products and Applications:

- Priostar99® dendrimers was First commercial product – EMD Chemicals launches Nanojuice™ DNA transfection kit.

TABLE 7.4 Dendrimers-Based Products

Dendrimers products	Application	Company
Stratus CS®	Cardiac Marker diagnostic	Dade Behring
SuperFect®	Gene Transfection technology	Qiagen
Starburst®	Dendrimers commercial	Sigma Aldrich
Priofect®	SiRNA & DNA transfection reagents	MERCK
VivaGel®	HIV and genital herpes Condom coating	Starpharmassl
SPL7013	Arthritis and cosmetic treatment	STARPHARMA
NanoJuice™	DNA transfection agent kit	EMD Chemicals
Targeted	MRI imaging Contrast agent	Baker
Drug Delivery	Cancer, Dermatological	Stiefel

- Priostar99® dendrimers are available as imaging agents to analyze structure of food.
- Agreement with Stiefel Laboratories, Inc., to apply SPL's 99 Dendrimer technology to delivery of drugs through the skin.
- Defense Department contract to DNT and Central Michigan University Research Corporation to develop water purification technology using Priostar.

Dendrimers commercialization can be loosely categorized into four groupings:

- Dow Chemical-related commercialization including Dow Corning;
- Tomalia-related commercialization including Dendritic Sciences, Inc. (with Central Michigan University) and Dendritic Nanotechnologies, Ltd. (with Starpharma);
- DSM commercialization (Buhleier et al., 1978);
- Others commercial efforts including but not limited to Perstorp, Xerox, Qiagen, Polyprobe, and Life Technologies.

7.8 APPLICATIONS

The potential for dendrimers in drug delivery certainly remains substantial (Figure 7.7). New biomaterials are being created, based on nanofibers, nanocrystalline metals, nanoporous silicon and new composites. Applications range from new implant materials to support matrices for growing new tissue, including nerve tissue. Here dendrimers are currently looking to be a minor player but their compatibility with other technologies, especially composites, does hold promise.

7.8.1 GENE THERAPY

Dendrimers can act as carriers, called vectors, in gene therapy. Vectors transfer genes through the cell membrane into the nucleus. Currently liposome and genetically engineered viruses have been mainly used for this. PAMAM dendrimers have also been tested as genetic material carriers. A transfection reagent called SuperFect™ consisting of activated dendrimers is commercially available (Yuan et al., 2008).

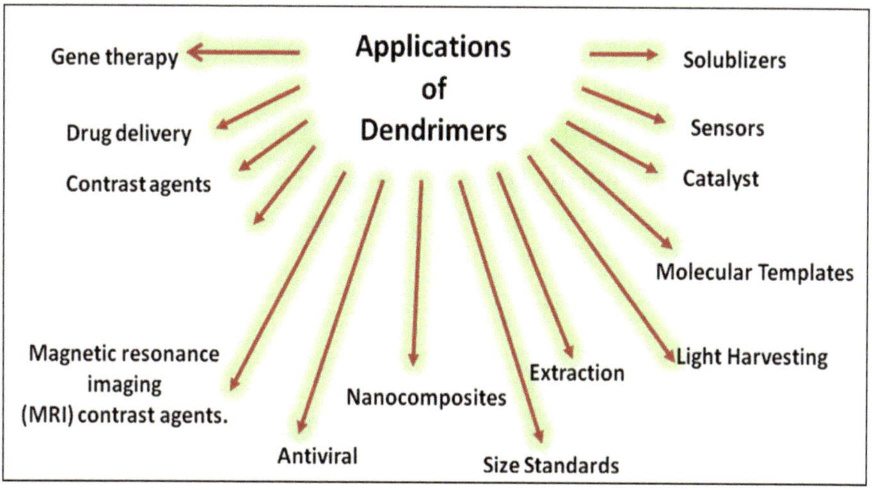

FIGURE 7.7 Application of dendrimers.

7.8.2 DRUG DELIVERY

Highly branched macromolecule complexed with gadolinium (III) ions is attracting a lot of attention. Gadomer-17 dendrimers is a promising candidate for magnetic resonance angiography and is in clinical trials. Gadomer-17 has a molecular weight of 17 kilodaltons. Surface Modified dendrimers can be used to carry drug molecules to a specific location and release them in a controlled way.

PAMAM dendrimers after acetylation can form dendrimer-5FU conjugates. Dendrimers can be used as coating agents to protect or deliver drugs to specific sites in the body or as time-release vehicles for biologically active agents (Pan et al., 2009).

7.8.3 CONTRAST AGENTS

Dendrimers have been tested in preclinical studies as contrast agents for magnetic resonance. Magnetic resonance imaging (MRI) is a diagnostic method producing anatomical images of organs and blood vessels.

Advantages of dendritic contrast agents versus Gd (III)/DTPA complexes:

1. Enhanced relaxivity.
2. Higher generation's dendrimers prevent diffusion of contrast agent into the interstitial space.
3. Introduction of target-specific sites is possible.

Gadolinium salt of diethylenetriaminepentaacetic acid (DTPA) is used clinically but it diffuses into the extravenous area due to its low molecular mass (Caroline et al., 1999).

7.8.4 SOLUBLIZERS

The therapeutic effectiveness of any drug is related with its good solubility in the body aqueous environment (Jansen et al., 1995).

7.8.5 CATALYST

Catalytic dendrimers were first synthesized by incorporating catalytic functionalities at either the dendrimer core or surface. Dendrimers are directly and covalently attached to solid supports via siloxybonds. The supported dendrimers were stable and easily separated from the reaction mixture. Sterically hindered dendrimer-metalloporphyrins was synthesized for use as shape-selective oxidation catalysts. These complexes have been examined as regioselective oxidation catalysts for both intra and intermolecular cases. Palladium complex was employed as catalyst in hydrogenation of organic compounds. The high activity of the complex was probably due to the formation of the coordinatively unsaturated palladium. Catalytic peptide dendrimers were first synthesized by Reymond and co-workers. They synthesized a series of peptide dendrimers having serine, aspartate and histidine residues at different positions of the dendrimer (Uchida et al., 1990).

7.8.6 LIGHT HARVESTING

Fréchet and co-workers at Berkeley are also investigating the use of dendrimers to harvest broadband light and convert the energy into monochromatic light with amplification, into electricity through charge separation, or into chemical energy. The systems consist of light-harvesting dendrons with numerous laser dye chromophores, such as coumarins, at the periphery. A single chromophore–an oligothiophene, for example lies at the core of the dendrimer. Dendrimer containing six ruthenium (II) polypyridine-type units undergoes as many as 26 reversible ligand-centered reduction processes. Since the same compound also exhibits six reversible metal-centered oxidation processes, it is capable of exchanging as many as 32 electrons altogether. This is the most extensive redox series reported so far (Launay et al., 1994).

7.8.7 EXTRACTION

Fluorinated dendrimers shows good solubility in supercritical CO_2 and can be used to extract strongly hydrophilic compounds from water into liquid CO_2. This may help develop technologies in which hazardous organic solvents are replaced by liquid CO_2 (Campagna et al., 1989).

7.8.8 ANTIVIRAL

Sialylated dendrimers have been shown to be potent inhibitors of the haemagglutination of human erythrocytes by influenza viruses. Sialodendrimers bind to haemagglutinin and thus prevent the attachment of the virus to cells (George et al., 1985).

7.9 CURRENT APPLICATIONS

7.9.1 IMAGING

PAMAM dendrimer conjugates with paramagnetic ions are used as magnetic resonance imaging (MRI) contrast agents (Hudson et al., 1993).

7.9.2 SENSORS

Due to their organized structure, ease of modification, and strong adsorption behavior to a variety of substrates, PAMAM dendrimers can be used to produce monolayers or stacked film layers, which can be used as sensors to detect hazardous chemical materials.

7.9.3 NANOCOMPOSITES

PAMAM dendrimers can form stable interior molecular nanocomposites with metal cations, zero-valent metals, other electrophilic ligands, and semiconductor particles. These materials are actively being investigated in electronics, optoelectronics and catalysis.

7.9.4 ORGANOSILANE COATINGS

PAMAM dendrimers are the basis of poly (amidoamine) organosilicon (PAMAMOS) coating technology. PAMAMOS coatings are tough, transparent, flexible coatings, which have many of the same attributes of PAMAM dendrimers in coating form (Buhleier et al., 1978).

7.9.5 SIZE STANDARDS/MOLECULAR TEMPLATES

Uniform molecular size of the various PAMAM dendrimer generations enable their use as calibration standards or precise scaffolds and templates to make organized thin films and stacked layers (Bielinska et al., 1996).

7.10 CONCLUSION

"Potted seeds have strengthened the roots and had started to give fruits." Even though dendritic polymers only have a short history of nearly three decades, but dendritic polymers have proved themselves to be promising material. Research findings would be helpful for the exploitation of dendrimers as future drug delivery carriers. Targeting moieties attached with surface groups will deliver drug at the site of intrust. Dendrimers-based drug delivery and targeting can offer impressive resolutions, when applied to medical challenges like cancer. Therefore, in the future, more attention should be paid to improving the synthesis of novel dendritic polymers and exploring the novel possible applications.

KEYWORDS

- **architecture**
- **dendrimer conjugates**
- **drug delivery**
- **drug targeting**
- **multifunctional platforms**
- **nanotechnology**
- **permeability**
- **polyamidoamine (PAMAM)**

REFERENCES

Bae, Y.; Nishiyama, N.; Fukushima, H.; Koyama, M.; Yasuhiro, K. Preparation and biological characterization of polymeric micelle drug carriers with intracellular pH-triggered drug release property: tumor permeability, controlled subcellular drug distribution, and enhanced in vivo antitumor efficacy. *Bioconjug. Chem.* **2005**, *16*, 122–130.

Banerjee, D.; Broeren, C.; Genderen, M.; Meijer, E. W.; Rinaldi, P. L. Multicomponent host-guest chemistry of carboxylic acid and phosphonic acid-based guests with dendritic hosts: An NMR study. *Macromol.* **2004**, *37*, 8313–8318.

Bielinska, A. U.; Kukowska-Latallo, J. F.; Johnson, J.; Tomalia, D. A.; Baker, J. R. Regulation of *in-vitro* gene expression using antisense oligonucleotides or antisense expression plasmids transfected using starburst PAMAM dendrimers. *Nucleic Acids Res.* **1996**, *24*, 2176–2182.

Bosman, A. W.; Janssen, H. M.; Meijer, E. W. About dendrimers: structure, physical properties, and applications. *Chem. Rev.* **1999**, *99*, 1665–1688.

Buhleier, E.; Winfried, W.; Fritz, V. Cascade and nonskid chain like syntheses of molecular cavity topologies. *Synthesis.* **1978**, 155–158.

Caminade, A.; Laurent, R.; Turrin, C.; Rebout, C.; Nicot, B.; Ouali, A.; Zablocka, M.; Majoral, J. Phosphorus dendrimers as viewed by 31P NMR spectroscopy; Synthesis and characterization. *Comp. Ren. Chim.* **2010**, *13*, 1006–1027.

Campagna, S.; Denti, G.; Sabatino, L.; Serroni, S.; Ciano, M.; Balzani, V. A new hetero-tetrametallic complex of ruthenium and osmium: Absorption spectrum, luminescence properties, and electrochemical behavior. *J. Chem. Soc. Chem. Commun.* **1989**, 1500–1501.

Caroline, D.; Mohamed, M. C.; Lipskier, J. F.; Caminade, Majoral, J. P. Characterization of dendrimers by X-ray photoelectron spectroscopy. *App. Spect.* **1999**, *53*, 1277–1281.

Chauhan, A. S.; Jain, N. K.; Diwan, P. V.; Khopade, A. J. Solubility enhancement of indomethacin with poly (amidoamine) dendrimers and targeting to inflammatory regions of arthritic rats. *J. Drug Target*. **2004**, *12*, 575–583.

Chauhan, A. S.; Sridevi, S.; Chalasani, K. B. Dendrimer-mediated transdermal delivery: enhanced bioavailability of indomethacin. *J. Cont. Rel.* **2003**, *90*, 335–343.

Cheng, Y.; Qu, H.; Ma, M.; Xu, Z.; Xu, P.; Fang, Y.; Xu, T. Polyamidoamine (PAMAM) dendrimers as biocompatible carriers of quinolone antimicrobials: An in vitro study. *Eur. J. Med. Che.* **2007**, *42*(7), 1032–1038.

Choi, J. S.; Nam, K.; Park, J.; Kim, J.; Lee, J. Enhanced transfection efficiency of PAMAM dendrimer by surface modification with l-arginine. *J. Cont. Rel.* **2004**, *99*, 445–456.

Davis, A. P. A. Thesis of synthetic modification of poly(amido)amine starburst dendrimers by acylation and a study of their NMR and vibrational spectra. *The Ohio State Univ.* **2002**, 1–147.

Davis, A. P.; Ma, G.; Allen, H. C. Surface vibrational sum frequency and raman studies of PAMAM G0, G1 and acylated PAMAM G0 dendrimers. *Anal. Chim. Acta.* **2003**, *496*, 117–131.

Devarakonda, B.; Hill, R. A.; DeVilliers, M. M. The effect of PAMAM dendrimer generation size and surface functional groups on the aqueous solubility of nifedipine. *Int. J. Pharm.* **2004**, *284*, 133–140.

Devarakonda, B.; Li, N.; Villiers, M. M. Effect of polyamidoamine (PAMAM) dendrimers on the in-vitro release of water-insolublenifedipine from aqueous gels. *A.A.P.S. Pharm. Sci. Tech.* **2005**, *6*(3), 504–511.

Duncan, R.; Dimitrijevic, S.; Evagorou, E. The role of polymer conjugates in the diagnosis and treatment of cancer. *S. T. P. Pharm. Sci.* **1996**, *6*, 237–263.

El-Sayed, M.; Ginski, M.; Rhodes, C.; Ghandehari, H. Transepithelial transport of poly (amidoamine) dendrimers across Caco2 cell monolayers. *J. Cont. Rel.* **2002**, *8*, 355–365.

El-Sayed, M.; Kiani, M. F.; Naimark, M. D.; Hikal, A. H.; Gandhehari, H. Extravasation of poly(amidoamine) (PAMAM) dendrimers across microvascular network endothelium. *Pharm. Res.* **2001**, *18*, 23–28.

Erk, N. Application of first derivative UV-spectrophotometry and ratio derivative spectrophotometry for the simultaneous determination of candesartan cilexetil and hydrochlorothiazide. *Phmz.* **2003**, *58*(11), 796–800.

Esfand, R.; Tomalia, D. Poly (amidoamine) (PAMAM) dendrimers: from biomimicry to drug delivery and biomedical applications. *Drug Discov. Today.* **2001**, *6*, 427–436.

Felder, T.; Schalley, C. A.; Fakhrnabavi, H.; Lukin, O. A combined ESI and MALDI-MS(/MS) study of peripherally persulfonylated dendrimers: False negative results by MALDI-MS and analysis of defects. *Chem. Eur. J.* **2005**, *11*, 1–13.

Furer, V. L.; Majoral, J. P.; Caminade, A. M.; Kovalenko, V. I. Elementoorganic dendrimer characterization by raman spectroscopy. *Poly.* **2004**, *45*, 5889–5895.

Furer, V. L.; Vandyukov, A. E.; Majoral, J. P.; Caminade, A. M.; Kovalenko, V. I. Fourier-transform infrared and Raman difference spectroscopy studies of the phosphorus-containing dendrimers. *Acta. Mol. Biomol. Spectrosc.* **2004**, *60*, 1649–1657.

Galia, E.; Nicolaides, E.; Horter, D.; Lobenberg, R.; Reppas, C.; Dressman, B. Evaluation of various dissolution media for predicting in vivo performance of class I and II drugs. *Pharm. Res.* **1998**, *15*, 698–705.

Gardikis, K.; Hatziantoniou, S.; Viras, K.; Wagner, M.; Demetzos, C. A DSC and raman spectroscopy study on the effect of PAMAM dendrimer on DPPC model lipid membranes. *Int. J. Pharm.* **2006**, *318*, 118–123.

George, R.; Yao, N. Z.; Gregory, R.; Baker.; Gupta V. K. Micelles. Part 1. Cascade molecules: a new approach to micelles. *A [27]-arborol, J. Org. Chem.* **1985**, *50*, 2003.

Greenwald, R.; Choe, Y.; McGuire, J.; Conover, C. Effective drug delivery by PEGylated drug conjugates. *Adv. Drug Deliv. Rev.* **2003**, *55*, 217–250.

Guoping, L.; Yunjun, L.; Huimin, T. PVP and G1.5 PAMAM dendrimer co-mediated synthesis of silver nanoparticles. *J. Solid State Chem.* **2005**, *178*, 1038–1043.

Gupta, U.; Agashe. B.; Asthana, H. A.; Jain, N. K. Dendrimers: Novel polymeric nanoarchitectures for solubility enhancement. *Biomacromol.* **2006**, *7*(3), 649–658.

Hudson, R. H. E.; Damha, M. J. Nucleic acid dendrimers: Novel biopolymer structures. *J. Am. Chem. Soc.* **1993**, *115*(6), 2119–2124.

Hummelen, J. C.; Dongen, J. L.; Meijer, E. W. Multivalency in the gas phase: The study of dendritic aggregates by mass spectrometry multivalency in the gas phase. *Chem. Eur. J.* **1987**, *3*, 1489– 1493.

Hummeln, J. C.; Van Dongen, J. L. J.; Meijer, E. W. Electrospray mass spectrometry of poly(proplyeneimine) dendrimers-The issue of dendritic purity or polydispersity. *Chem. Eur. J.* **1997**, *3*, 1489–1493.

Ihre, H. R.; De Jesus, O. L. P.; Szoka, F. C. J.; Frechet, J. M. J. Polyester dendritic systems for drug delivery applications: design, synthesis, and characterization. *Bioconjug. Chem.* **2002**, *13*, 443–452.

Jani, R.; Jani, K.; Setty, C.; Mallikarjuna.; Patel, D. Preparation and evaluation of solid dispersions of aceclofenac. *Int. J. Pharm. Sci. Drug Res.* **2009**, *1*(1), 32–35.

Jansen, J. F. G. A.; de Brabander-van den Berg, E. M. M.; Meijer, E. W. The dendritic box and bengal rose, Polym. *Mater. Sci. Eng.* **1995**, *73*, 123–124.

Jansen, J. F. G. A.; de Brabander-van den Berg, E. M. M.; Meijer, E. W. Encapsulation of guest molecules into a dendritic box. *Sci.* **1994**, *266*, 1226–1229.

Jevprasesphant, R.; Penny, R.; Jalal, D.; Attwood, N.; McKeown, A. The influence of surface modification on the cytotoxicity of PAMAM dendrimers. *Int. J. Pharm.* **2003**, *252*, 263–266.

Kallos, G.; Tomalia, D. A.; Hedstrand, D. M.; Lewis, S.; Zhou, J. Molecular weight determination of a polyamidoamine starburst polymer by electrospray ionization mass spectrometry. *Rapid Commun. Mass Spectrom.* **1991**, *5*, 383–386.

Karlos, X.; Simanek, E. Conformational analysis of triazine dendrimers: Using NMR spectroscopy to probe the choreography of a dendrimer's dance. *Macromol.* **2008**, *41*, 4108–4114.

Khopade, A. J.; Caruso, F.; Tripathi, P.; Nagaich, S.; Jain, N. K. Effect of dendrimer on entrapment and release of bioactive from liposomes. *Int. J. Pharm.* **2002**, *232*, 157–162.

Kim, Y.; Klutz, A. M.; Jacobson, K. A. Systematic investigation of polyamidoamine dendrimers surface-modified with poly(ethylene glycol) for drug delivery applications: Synthesis, characterization, and evaluation of cytotoxicity. *Bioconjug. Chem.* **2008**, *19*(8), 1660–1672.

Kojima, C.; Kono, K.; Maruyama, K.; Takagishi, T. Synthesis of polyamidoamine dendrimers having poly (ethylene glycol) grafts and their ability to encapsulate anticancer drugs. *Bioconjug. Chem.* **2000**, *11*, 910–917.

Kolev, T. M.; Velcheva, E. A.; Stamboliyska, B. A.; Spiteller, M. DFT and experimental studies of the structure and vibrational spectra of curcumin. *Int. J. Quant. Chem.* **2005**, *102*, 1069–1079.

Kolhe, P.; Khandare, J.; Pillai, O.; Kannan, S.; Lieh, M.; Kannan, R. M. Preparation, cellular transport, and activity of polyamidoamine-based dendritic nanodevices with a high drug payload. *Biomat.* **2006**, *27*, 660–669.

Konda, S.; Brechbiel, M.; Wiener, D.; Aref, M.; Wang. Targeting of folate-dendrimer Specific MRI contrast agents to the high affinity folate receptor expressed in ovarian tumor xenogafts, Magn. Reson. *Mater. Phys. Biol. Med.* **2001**, *12*, 104–113.

Kono, K.; Akiyama, H.; Takahashi, T.; Takahaschi, T.; Harada, A. Transfection activity of polyamidoamine dendrimers having hydrophobic amino acid residues in the periphery. *Bioconjug. Chem.* **2005**, *16*, 208–214.

Lajos, P. B.; Ganser, R. T.; Xiangyang, S. Characterization of dendrimer-gold nanocomposite materials. *Mater. Res. Soc. Symp. Proc.* **2005**, 847.

Launay, N.; Caminade, A. M.; Lahana, R.; Majoral, J. P. A general synthetic strategy for neutral phosphorus-containing dendrimers. *Angew. Chem. Int. Ed. Engl*, **1994**, *33*, 1589–1592.

Liu, M.; Kono, K.; Frechet, J. M. Water soluble dendritic unimolecular micelles: Their potential as drug delivery agents. *J. Control Release.* **2000**, *65*, 121–131.

Maiti, P. K.; Tahir, C.; Wang, G.; Goddard, W. A. I. Structure of PAMAM dendrimers: generations 1 through 11 macromolecules. **2004**, *37*, 6236–6254.

Malik, N.; Wiwattanapatapee, R.; Klopsch, R.; Lorenz, K.; Frey, H.; Weener, J. W.; Meijer, E. W.; Paulus, W.; Duncan, R. Dendrimers: relationship between structure and biocompatibility in vitro, and preliminary studies on the biodistribution of 125I labeled polyamidoamine dendrimers in vivo. *J. Controlled Release.* **2002**, *65*, 133–148.

Michaels, A. S.; Chandrasekaran, S. K.; Shaw, J. E. Drug permeation through human skin: theory and in vitro experimemtal measurement. *J. American Institute Chemical engineers. AIChE.* **1975**, *21*, 985–996.

Milhem, O. M.; Myles, C.; McKeown, N. B.; Attwood, D.; Emanuele, A. D. Polyamidoamine Starburst dendrimers as solubility enhancers. *Int. J. Pharm.* **2000**, *197*, 239–241.

Namazi, H.; Adeli, M. Dendrimers of citric acid and poly (ethylene glycol) as the new drug-delivery agents. *Biomat.* **2005**, *26*, 1175– 1183.

Nekkanti, V.; Karatgi, P.; Prabhu, R.; Pillai, R. Solid self-micro emulsifying formulation for candesartan cilexetil. *A.A.P.S. Pharm. Sci. Tech.* **2009**, *122*, 49–47.

Nekkanti, V.; Pillai, R.; Venkateshwarlu, V.; Harisudhan, T. Development and characterization of solid oral dosage form incorporating candesartan nanoparticles. *Pharm. Dev. Tech.* **2009**, *14*(3), 290–298.

Newkombe, G. R.; Moorefield, C. N.; Voegtle, F. Dendrimers and dendrons: Concepts, syntheses, applications; Wiley-VCH. *Weinheim.* **2001**.

Newkome, G. R.; Moorefield, C. N.; Baker, G. R.; Saunders, M. J.; Grossman, S. H. Unimolecular micelles. *Angew Chem. Int. Ed. Engler.* **1991**, *30*, 1178–1180.

Newkome, G. R.; Woosley, B. D.; He, E.; Morefield, C. N.; Guther, G.; Baker, R. Supromolecular chemistry of flexible, dendritic-based structure employing molecular recognition. *Chemical Comm.* **1996**, 2737–2738.

Pan, B.; Cui, D.; Xu, P.; Ozkan, C.; Feng, G.; Ozkan, M.; Huang, T.; Chu, B.; Li, Q.; He, R.; Hu, G. Synthesis and characterization of polyamidoamine dendrimer-coated multiwalled carbon nanotubes and their application in gene delivery systems. *Nanotech.* **2009**, *20*, 125101.

Park, K. M.; Lee, M. K.; Hwang, K. J.; Kim, C. K. Phospholipid-based microemulsions of flurbiprofen by the spontaneous emulsification process. *Int. J. Pharm.* **1999**, *183*, 145–154.

Patri, A. K.; Kukowska-Latallo, J. F.; Baker, J. Targeted drug delivery with dendrimers: Comparison of the releaser kinetics of covalently conjugated drug and noncovalent drug inclusion complex. *Adv. Drug Deliv. Rev.* **2005**, *57*, 2203–2214.

Patri, A. K.; Majoros, J. J.; Baker, J. R. J. Dendritic polymer macromolecular carriers for drug delivery. *Curr. Opin. Chem. Biol.* **2002**, *6*, 466–471.

Pesak, D. J.; Moore, J. S. Columnar liquid crystals from shape persistent dendritic molecules. *Angew. Chem. Int. Ed. Engl.* **1999**, *36*, 1636–1639.

Peterson, J.; Ebber, A.; Allikmaa, V.; Lopp, M. Synthesis and CZE analysis of PAMAM dendrimers with an ethylenediamine core. *Proc. Estonian Acad. Sci. Chem.* **2001**, *50*(3), 156–166.

Popescu, M. C.; Filip, D.; Vasile, C.; Cruz, C.; Rueff, J. M.; Marcos, M.; Serrano, J. L.; Singurel, G. Characterization by Fourier transform Infrared Spectroscopy (FT-IR) and 2D IR correlation spectroscopy of PAMAM dendrimer. *J. Phys. Chem.* **2006**, *110*, 14198–14211.

Purohit, G.; Sakthivel, T.; Florence, A. T. Interaction of cationic partial dendrimers with charged and neutral liposomes. *Int. J. Pharm.* **2001**, *214*, 71–76.

Ronald, C. H.; Bauer, B. J.; Paul, S. A.; Franziska, G.; Eric, A. Templating of inorganic nanoparticles by PAMAM/PEG dendrimer–star polymers, Polymer. **2002**, *43*, 5473–5481.

Santo, M.; Fox, M. A. Hydrogen bonding interactions between starburst dendrimers and several molecules of biological interest. *J. Physical Org. Chemistry*. **1999**, *12*, 293.

Sathesh, P. R.; Subrahmarnyam, C. V. S.; Thimmasetty, J.; Manavalan, R.; Valliappan, K. Solubilization of rofecoxib through Cosolvency. *J.pharm.sci.* **2008**, *7*(2), 119–126.

Singhai, A. K.; Jain, S.; Jain, N. K. Evaluation of an aqueous injection of ketoprofen. *Pharmazie*. **1997**, *52*, 149–151.

Smith, P. B.; Martin, S. J.; Hall, M. J.; Tomalia, D. A. A characterization of the structure and synthetic reactions of polyamidoamine "Starburst" polymers. In applied polymer analysis and characterization, (Mitchell, J., Jr., ed.). Hanser, München/New York, **1987**, 357–385.

Stevemans, S.; Van Hest, J. C. M.; Jansen, J. F. G. A.; VanBoxte, D. A. F. J.; De Breabander van den Berg, E. M. M.; Meijer, E. W. J.; *Am. Chem. Soc.* **1996**, *118*, 73–98.

Stiriba, S. E.; Frey, H.; Haag, R. Dendritic polymers in biomedical applications: From potential to clinical use in diagnostics and therapy. *Angew. Chem. Int. Ed.* **2002**, *41*, 1329–1334.

Tang, M. X.; Redemann, C. T.; Szoka, F. C. In-vitro gene delivery by degraded polyamidoamine dendrimers. *Bioconjug. Chem.* **1996**, *7*, 703–714.

Tomalia, D. A.; Baker, H.; Dewald, J. R.; Hall, M.; Kallos, G.; Martin, S.; Roeck, J.; Ryder, J. A new class of polymers: Starburst-dendritic macromolecules. *J. Polym.* **1985**, *17*, 117–132.

Tomalia, D. A.; Baker, H.; Dewald, J.; Hall, M.; Kallos, G.; Martin, S.; Roeck, J.; Ryder, J.; Smith, P. Dendritic molecules: synthesis of starbrust dendrimers. *Macromol.* **1986**, *19*, 2466–2468.

Tomalia, D. A.; Naylor, A. M.; Goddard, W. A. Starburst dendrimers: Molecular-level control of size, shape, surface chemistry, topology, and flexibility from atoms to macroscopic matter. *Chem, Int. Edn.* **1990**, *29*, 138–175.

Twyman, L. J.; Beezer, A. E.; Esfand, R.; Hardy, M. J.; Mitchell, J. C. The synthesis of water-soluble dendrimers, and their application as possible drug delivery systems. *Tetrahedron Letters.* **1999**, *40*, 1743–1746.

Uchida, H.; Kabe, Y.; Yoshino, K.; Kawamata, A.; Tsumuraya, T.; Masamune, S. "General strategy for the systematic synthesis of oligosiloxanes. silicone dendrimers," *J. Am. Chem. Soc.* **1990**, *112*(19), 7077–7079.

Xiangyang, S.; Ganser, R. T.; Kai, S.; Lajos, P. B.; Baker, J. R. Characterization of crystalline dendrimer-stabilized gold nanoparticles. *J. Nanotech.* **2006**, *17*, 1072.

Yang, H.; Morris, J. J.; Lopina, S. T. Polyethylene glycol–polyamidoamine dendritic micelle as solubility enhancer and the effect of the length of polyethylene glycol arms on the solubility of pyrene in water. *J. Colloid and Interface Sci.* **2004**, *273*(1), 148–154.

Yang, J.; Morris, S.; Lopina, T. Polyethylene glycolpolyamidoamine dendritic micelle as solubility enhancer and the effect of the length of polyethylene glycol arms on the solubility of pyrene in water. *J. Colloid Interface Sci.* **2004**, *273*, 148–154.

Yellepeddi, V. K.; Pisal, D. S.; Kumar, A.; Kaushik, R. S.; Michael, B.; Hildreth; Guan, X.; Palakurthi, S. Permeability of surface modified polyamidoamine (PAMAM) dendrimers across $CaCo_2$ cell monolayers. *Int. J. Pharm.* **2008**, *28*, *350*(1–2), 113–121.

Yiyun, C.; Tongwen, X. Dendrimers as potential drug carriers part I solubilization of non-steroidal antiinflammatory drugs in the presence of polyamidoamine dendrimers. *Eur. J. Med. Chem.* **2005**, *40*, 1188–1192.

Yoshimura, T.; Abe, S.; Esumi, K. Characterization of quaternizedpoly(amidoamine) dendrimers of generation 1 with multiple octyl chains. Colloids and Surfaces A: Physicochemical and Engineering Aspects. **2004**, *251*, 141–144.

Yuan, W.; Jinying, Y.; Mi, Z.; Caiyuan, P. Synthesis, characterization, and fluorescence of pyrene-containing eight-arm star-shaped dendrimer-like copolymer with pentaerythritol core. *J. Poly. Sci.* **2008**, *46*, 2788–2798.

Zeng, F.; Zimmerman, S. C. Dendrimers in supramolecular chemistry: from molecular recognition to self-assembly. *Chem. Rev.* **1997**, *97*, 1681–1712.

Zhang, C.; Tomalia, D. A.; Frechet, J. M. J.; Tomalia, D. A. Dendrimers and other Dendritic Polymers. *John Wiley & Sons.* **2001**, 239–253.

Zhuo, R. X.; Du, B.; Lu, Z. R. In vitro release of 5-fluorouracil with cyclic core dendritic polymer. *J. Cont Rel.* **1999**, *57*, 244–257.

CHAPTER 8

NANOFIBERS: PRODUCTION TECHNIQUES AND APPLICATIONS

HEMANT K. S. YADAV,* NOUR A. H. ALHALABI, and GHUFRAN A. R. ALSALLOUM

Department of Pharmaceutics, RAK College of Pharmaceutical Sciences, RAK Medical and Health Sciences University, UAE,
**E-mail: haisunny2@yahoo.co.in*

CONTENTS

ABSTRACT

The utilization of nano-sized structures has become a field of great interest in the recent years. Nanofibers are ultra-fine filaments that have a diameter in nano scale. They can be classified mainly into core-shell, hollow and porous nanofibers. Plenty of advantages associated with nanofibers are introduced. A variety of these properties including mechanical, electrical, thermal, optical and magnetic are discussed. Diverse techniques are used for production of nanofibers, such as electrospinning, self-assembly, phase separation, template synthesis. Different alignments can be obtained by minor changes in electrospinning set-up. Characterization and evaluation of nanofibers can be done via a number of methods which involve X-ray diffraction, light spectroscopy, TEM, SEM, AFM, DSC as well as tests for thickness and tensile strength. Utilization of nanofibers as a drug delivery system is a promising field of applications. Numerous approaches demonstrate drug loading to reduce burst release of other conventional dosage forms and provide sustained release over prolonged period of time. They are successfully employed in wound dressing, cancer therapy, gene and growth factor delivery. They are also involved in tissue engineering. Many studies elucidate bone, cartilage, tendons, dental, cardiovascular, neural and skin tissue engineering. This chapter mainly focuses on methods of preparation and applications that can come in handy for the reader.

8.1 INTRODUCTION

The development of particles at the nano scale has become a field of great interest in the recent years. Research about nanotechnology has been going on since the second half of the 1980 s. As a result of ambiguity of the term "nanotechnology" and the uncertainty of the time span of the early stages of nanotechnology development, it is actually rather difficult to describe the history of nanotechnology. The possibility of using atoms as building particles to create nanosized products was first introduced in the well-known lecture of Mr. R Feynman, the Professor of Californian institute of technology, "there is a lot of space down there" delivered in 1959 at the session of the American Physical Society. However, the introduction of the term "nanotechnology" into the world of scientific research was at the hands of N. Taniguchi at the International Conference on industrial production in Tokyo in 1974, Taniguchi discussed the super-thin processing of materials with nanometer accuracy and the creation of nano-sized mechanisms. Generally, bringing nanotechnology into practice and producing nanosized objects as well as using related nano-level processes, happened spontaneously and without deep understanding of the actual properties of these nanosized materials (Tolochko, 2009).

In the international system of units, the "nano" is exactly 1 billionth of a meter. To visualize how small a nano is, a sheet of paper is 100,000 nanometers thick, while a strand of human DNA is 2.5 nanometers in diameter. It is important to understand that nanosized materials are found in abundance in nature. For example, hemoglobin, the oxygen-carrying protein in red blood cells, is 5.5 nanometers in diameter. Understanding of the various types and dimensions of nanoscale materials is essential for working at the nano scale (Nikalje, 2015).

The various shapes and dimensions of nanomaterials give rise to highly versatile nanoproducts. Generally, these products include particles, tubes, wires, films, flakes, fibers, or shells.

As the name indicates, nanofibers can be considered as strings, filaments or threads that have a diameter in the nanometer range. Many manufacturing methods have been developed for the production process of nanofibers. However, in most cases, nanofibers are manufactured through a method called electrospinning, in which continuous polymer fibers are produced through the action of an external electric field applied on a polymeric solution, or melt using various materials and polymers such as proteins, lipids and carbohydrates. Nanofibers can be classified according to a variety of parameters, for example, they can be classified according to their structure into four types: core-shell, bi- component, hollow and porous nanofibers (Khajavi and Abbasipour, 2012). The small particle size, large surface area and small paore size are the main properties that contribute to nanofibers being widely used in several different applications (Blanco-Padilla et al., 2014).

Health and medicine, electronics, transportation, energy, environment and space exploration are fields that find potential application of nanotechnology (Nikalje, 2015). The application of nanotechnology in medicine can be termed as nanomedicine. The aim of nanomedicine is the comprehensive monitoring, control, construction, repair, defense and improvement of the human biological systems, working from the molecular level using nanostructures, to ultimately achieve medical benefits. These nanostructures are used in medical diagnostics such as imaging, implants and sensors, biomarkers and nano-biopsy. Another application is nano-pharmaceutical, where nanocarriers are used to aim therapies directly and selectively at diseased tissues or cells, thereby preventing toxicity and unwanted side effects. This is greatly useful in therapy of cancer and inflammation. Regenerative medicine is one of the most promising applications of nanotechnology. It is defined as the process of creating living, functional tissues, to repair or replace tissue or organ function lost due to age, disease, damage, or congenital defects. This field holds the promise of regenerating damaged tissues and organs in the body by stimulating previously irreparable organs to heal by themselves. By stimulating irreparable organs to heal, damaged tissue can be regenerated or even new tissue can be grown in a laboratory and safely implanted when the body fails to heal by itself (Boisseau and Loubaton, 2011). When it comes to nanofibers, applications in health and medicine include being used in tissue engineering, tissue repair, drug delivery and a huge variety of biomedical applications (Ramakrishna et al., 2006).

8.2 DEFINITION

Nanostructures are present in the form of particulate materials, layered materials and fibrous materials. Nanofibers have two dimensions in the nanoscale whereas nanoparticles have all three dimensions in the nano scale and nanoplates have only one dimension in the nano scale (Figure 8.1).

To construct an appropriate definition of nanofibers, we need to understand the terms that came before, like micron. A microsized structure is 1 millionth of a meter that is 10^{-6} of a meter. However, a nano-sized structure is 1 billionth of a meter. The transitioning from micro scale to nano scale will result in dramatic changes in properties and characteristics of the fibers.

Nanofibers can be defined as ultra-fine threads or filaments that have a diameter in nano scale, with a length that is not necessarily as small. In other words, two dimensions are less than 100 nanometers or smaller than one micron, while the length is determined according to the manufacturer's purpose, which results in the ultra-thin structure of nanofibers (Alubaidy et al., 2013).

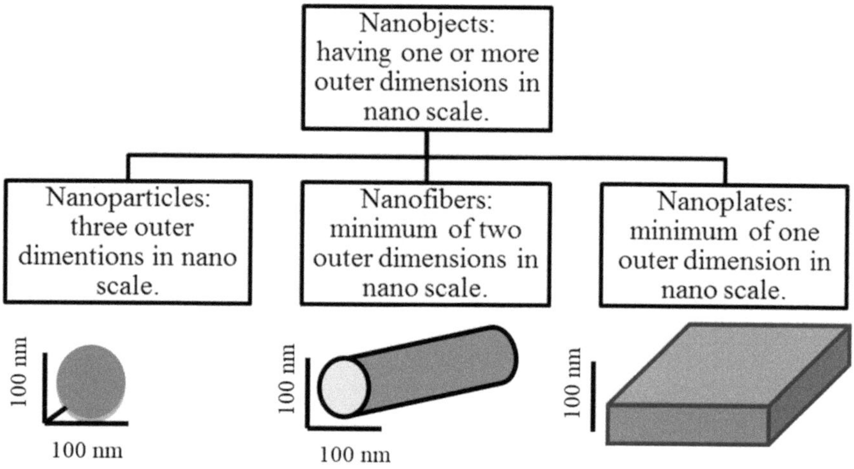

FIGURE 8.1 Different nano objects and their dimensions.

8.3 CLASSIFICATION OF NANOFIBERS

Nanoparticles are simply classified according to their diameter since it's the only general parameter to consider upon classification. However, when compared to classifying nanofibers, it is relatively considered more complex as there are plenty of parameters to consider (Unrau et al., 2007). Figure 8.2 demonstrates the order of classification of nanofibers on different basis.

8.3.1 ON BASIS OF SIZE

Nanofibers' size is determined by miscellaneous instruments and devices to measure diameter, length, or both. Those, which assort nanofibers-based on diameter, take advantage of their aerodynamic properties (Tolochko, 2009). A good example can be given on using inertial impaction and centrifugation. In inertial impactors, the fibers with large diameter will have relatively larger size (larger inertia), hence, get precipitated. Where the ones with smaller diameter (lesser inertia) will be swept away with the flow of fluids introduced to the impactor. By this way they are segregated.

Regarding the length classification, it is ranked through screen penetration, gravitational settling and majorly electrostatic classification.

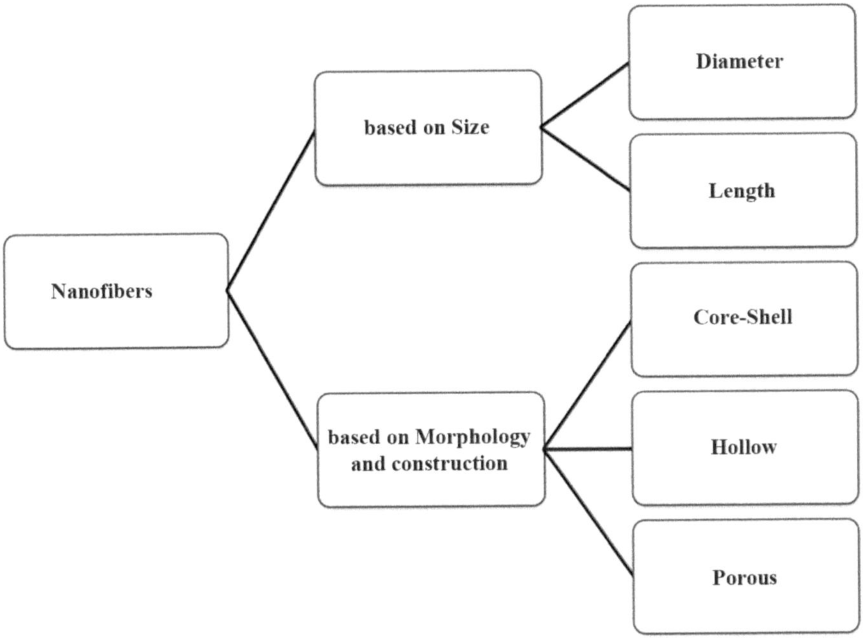

FIGURE 8.2 Classification of nanofibers on different basis.

It's executed in by means of electrophoresis and dielectrophoresis. In electrophoresis, the fibers are charged by bi- or uni-polar ions. Then, an electrical field is applied and the fiber's motility is observed. Afterwards, electrical motility is correlated with charge distribution that can correspond to a certain length. On contrary with dielectrophoresis, the uncharged neutral fibers are placed under a nonunifrom electrical filed in which polarization is induced and will be responsible for the motility within the electric field. In this case, the degree of motion is corresponding to the fiber length (Unrau et al., 2007).

8.3.2 ON BASIS OF MORPHOLOGY AND CONSTRUCTION

An enormous sum of studies was done in this field, but it came down to three types of structures that are found: core-shell, hollow and porous nanofibers (Khajavi and Abbasipour, 2012) (Figure 8.3).

FIGURE 8.3 Core-shell nanofibers.

8.3.2.1 Core-Shell Nanofibers

The name itself reveals the structure comprises of two parts; core and shell. Plenty of methods where exploited to fabricate such fibers, e.g., template synthesis, surface initiated atom transfer radical polymerization. However, it's coaxial electrospinning whom displayed great efficiency and versatility for this purpose. The set-up of the coaxial spinneret is accustomed specially for accommodation of two solutions or fluids in internal and external capillaries held together and connected with high power source. Then, the polymer solutions are released out simultaneously and as the solvent evaporates they solidify and possess a nano-scale fibrous texture. A formulator must be very careful dealing with this type, because many factors influence their uniformity and deformation. The core may readily deform if the electric field is weak or stretching occurs quickly.

A slightly different conformation of this type of nanofibers can be obtained known as side-by-side nanofibers. Those employ an exclusive type of coaxial spinneret in which the capillaries are adjacent to each other rather than being one surrounding another. This conformation offers two sided fibers; each side will have its own characteristics (Khajavi and Abbasipour, 2012).

8.3.2.2 Hollow Nanofibers

Those are known for being empty at core just like a tube which is why there referred to as nanotubes. This void can be filled with several materials and agents which make it favorable in many fields. They can be prepared

FIGURE 8.4 Hollow nanofiber.

using two techniques either chemical vapor deposition (CVD) or coaxial electrospinning. By CVD, a template must be ready-made from the precursor polymer. A metal or a polymeric coating is administered. A typical electrospinning is then followed and the outcome of fibers is dissolved to eliminate the template and dried (Figure 8.4).

Coaxial electrospinning for hollow nanofibers is the same as pursued for core-shell. Only one further step is added to dissolve the core and give rise to hollow void. Depending on which, appropriate selection of solvent and heating rate become crucial for successful synthesis of nanofibers (Khajavi and Abbasipour, 2012).

8.3.2.3 Porous Nanofibers

Presence of pores throughout the fibers provides relatively the largest surface area and capacity to intake various materials. This qualifies such fibers to a large domain of applications. So, it's not a surprise to find porous nanofibers employed in filtration, membrane and tissue engineering and most importantly in drug delivery and release. To gain this morphology, one should be keen regarding choice of suitable solvents/solvent mixtures or polymer mixtures. Phase separation is the main technique used. Multiple subdivisions can be incorporated like vapor-induced, nonsolvent-induced, thermally induced and rapid phase separation (Figure 8.5).

Controlling the porosity, pore size, depth, shape and distribution are dependent on the ratio of polymers used and the technique of solvent removal (vacuum or temperature) (Khajavi and Abbasipour, 2012).

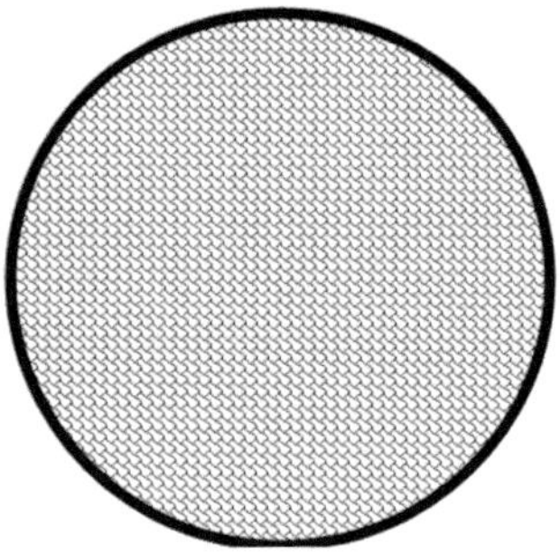

FIGURE 8.5 Porous nanofibers.

8.4 ADVANTAGES AND LIMITATIONS

In the modern world, nanofibers have become a great excitement due to their special properties that have guaranteed their wide range of applications from medical, industrial to information technology.

8.4.1 ADVANTAGES OF NANOFIBERS

- High surface area to volume ratio and porosity.
  This is particularly useful in applications where large surface area is required like sensors. The high porosity of nanofibers will result in high liquid or gas permeability, which gives rise to industrial applications like filters (Blanco Padilla et al., 2014).
- Capable of incorporating a variety of polymers and materials with ease.
  Even though polymeric nanofibers are the most used, ceramic and metal nanofibers are also quite useful.
- Constructed of a variety of structures.
  Depending on the setup of the electrospinning process, various nanofibrous structures are produced, such as tubes, coatings and 3D structures.
- Easily deposited onto other surfaces, including metal, glass and water.
- Wide range of applications.
- Large scale production is possible.
- Capable of carrying heat sensitive materials.

8.4.2 LIMITATIONS OF NANOFIBERS

- Recycling and disposal.
  Nanofibers are relatively new; therefore, clear policies about their recycling and disposal are yet to be developed. In addition, results of exposure to nanofibers are not available; hence, their toxicity is still under question.
- Expensive.
  When compared to conventional fibers, the manufacture of nanofibers is quite costly, due to high cost of the technology and low rate of production.
- Environmental and Health hazards.
  Electrospinning solutions emit vapors during manufacture, which need to be disposed of in an environmental friendly way, which involves the addition of cost and equipment. Moreover, the inhalation of emitted vapor and the very fine fibers, may lead to possible health hazard.
- Handling of nanofibers.
  The ultra-fine structure, as well as other properties of nanofibers, has caused some difficulty in packaging, handling and shipping.
  Nanofibers have been proven to display exceptional qualities and fascinating applications, therefore, manufacturers are trying to economically deal with their limitations (Blanco-Padilla et al., 2014).

8.5 PROPERTIES OF NANOFIBERS

From a manufacturer's point of view, different scales are considered when producing nanofibers.

- The macroscopic scale: which has a length of 10^{-2}
- The mesoscopic scale: which has a length of 10^{-4}
- The microscopic scale: which has a length of 10^{-6}
- The nanoscopic scale: which has a length of 10^{-9}.

While designing a nanofiber structure, it is important to consider properties on the macroscopic scale, including properties like stiffness and tensile strength. On the mesoscopic scale, fiber architecture and large defects are described. When looking on the microscopic scale, single fiber characteristics are determined such as fiber diameter and length. However, it is the smallest scale that determines the behavior of materials. The nanoscale describes how atoms and molecules are arranged and how they interact among each other. Also it is on the nanoscale where polymer interaction with nanoparticles is described (Wierach, 2012).

When a specific material is reduced to the size of a micron, its properties will remain without any significant changes. Whereas, the manufacturing of nano-sized particles of the same material will produce a drastic difference from the bulk. This difference is mainly due to the exceptionally large surface area of the nano-sized parts. Nanostructures will also exhibit spatial confinement effect and reduced imperfection of the material due to the very small particle size. When the size of a material is reduced to the nano scale, the energy band structure as well as the charge carrier density will be changed, which in turn will modify the electronic and optical properties of the material. In addition, while reducing the size of a material, impurities and intrinsic defects will rise to the surface of the particles, thereby increasing the perfection and purity of this material. Enhancing the perfection of materials will lead to enhanced chemical stability and mechanical properties. Generally, nano-sized materials are known to show novel properties, which in turn have led to the emergence of new applications (Alagarasi, 2009).

Main properties of nanofibers include: mechanical properties, optical properties, electrical properties, magnetic properties, thermal properties, and surface properties.

8.5.1 MECHANICAL PROPERTIES

Nanofibers have found many areas of application, including tissue engineering, drug delivery and composite reinforcement, due to their special properties including small fiber diameter (20 to 100 nm), high specific surface area (tens to hundreds m^2/g), high porosity and small pore size. However, many applications require high mechanical properties as well. Even though, nanofibers have enhanced orientation of polymer backbones along the longitudinal axis, the increase in tensile strength is not significant.

When comparing nanofibers of a specific polymer with corresponding textile fibers made of the same polymer, nanofibers are found to have poor mechanical properties. Tensile strength and Young's moduli are used as measures of mechanical properties. Nanofibers exhibit tensile strengths less than 300 MPa and Young's moduli below 3 GPa.

This is mainly due to low degree of orientation and extension of the polymer chains along the fiber axis. Nanofibers ultra-thin structure will display insufficient overlap between chains, and poor stress transfer leading to poor tensile strength (Yao et al., 2014).

In addition, hardness and elastic modulus can be determined as measures of mechanical properties. Hardness is defined as the resistance of a material to local surface deformation. The elastic modulus is a measure of the overall stiffness of polymer network. The thread-like structure of nanofibers and their large surface area have a great influence on their mechanical properties, for example, the interaction of polymer chains with nanofibers is enhanced due to the high area available for interaction. Also, there is an increase in hardness due to nucleation of crystalline phases, because of the nanofiber structure that provides an abundance of sites for nucleation. The flow strength of a material is directly related to its hardness (Alubaidy et al., 2013).

It is possible to manufacture high performance nanofibers that display good mechanical properties. However, most of modern research is going on mechanical properties of nanofiber mats or bundles rather than single fibers. Therefore, future research should focus more on single fiber properties and characterization in order to develop a clear concept of improving the mechanical properties of nanofibers. The degree of dispersion, processing history and the type of polymer incorporated into the nanofiber has shown a great influence on various mechanical properties. For example, carbon nanofibers display improved tensile strength, compression strength, Young's modulus, shear strength and fracture toughness of the base polymer (Nagasaka et al., 2014).

8.5.2 OPTICAL PROPERTIES

While reducing the size of material from macro into nano scale, random surface structures can be developed to control the optical properties of this material. With increase in the capability of structuring materials at a very small scale, more precise control of light-material interaction can be achieved. This has caused the appearance of useful applications such as optical detectors, sensors, imaging and solar cells. Generally, the refractive index of nanofibers is found to be smaller than that of the macrosized material (Flory et al., 2011).

8.5.3 ELECTRICAL PROPERTIES

When nanofibers are dispersed in a polymer, contact conditions between neighboring nanofibers will result in the formation of a conduction network resulting in good electrical conductivity of these fibers.

The electrical conductivity of various polymers, which are incorporated into nanofibers, is the basis of many applications, for example, pressure sensors, actuators and electromagnetic interference shielding. Nanofibers are commonly used as conductive composition materials (Alagarasi, 2009). These materials exhibit an ability of charge storage, electric current passage and redox activity. Nanofibers are known to be used in insulating polymers to improve their electrical properties. In addition, by the incorporation of conducting fillers, such as metal powders, almost any common polymer can be modified to conduct an electric current. Interesting electrical properties of nanofibers, like electrical conductivity, electrochromism and electroluminescence, have brought about several practical uses, including diodes and light emitters, memory storage and rechargeable batteries. When it comes to electrical conducting, nanofibers have an advantage over other nanostructures, due to their properties like anisotropy and high surface area. When nanofibers have electroactive polymers incorporated into them, new and unrelated properties will arise, which may be useful in electronic devices, optics and biomedical materials (Picciani et al., 2011).

8.5.4 MAGNETIC PROPERTIES

The properties of materials change as they transition from their normal bulk size into the nano-sized structures. For example, the color of gold when reduced to the size of a nano becomes red, or even blue, depending on the size and the distance between its particles. Moreover, bulk gold is not magnetic, but at the nano size, it is. Magnetism occurs at the nano range due to different reasons. The change in size can give rise to structural changes of materials causing them to display magnetic properties; examples of these materials are Pt and Pd. However, gold nanoparticles exhibit magnetism when capped with a specific material due to the charge localization at the surface (Alagarasi, 2009). Clinical applications of magnetic nanostructures include targeted drug delivery, magnetic resonance imaging and magnetic fluid hyperthermia, which are a novel biomedical application for the treatment of cancer (Tolochko, 2009).

8.5.5 THERMAL PROPERTIES

When a material conducts heat, it is considered thermally conductive. Heat is conducted throughout a solid via electrons, which is the case in metals, phonons, that is the thermal vibrations of atoms and is the case in nonmetal materials, and photons. Since atoms of a solid are physically bound, heat will travel from one end to the other due to the vibrations of the atoms. However, when it comes to nanofibers, not only the shape and volume of the incorporated polymer are taken into consideration, but also interface resistance and phonon scattering become increasingly significant. If there is a mismatch in thermal expansion between the polymer and the fiber surface, transport of heat energy will not be efficient. Moreover, poor chemical adhesion of the polymer to the fiber surface will result in discontinuity of the temperature at the particle-polymer interface; this scenario is called interfacial thermal resistance. These cases do not seem to occur when the particle size is more than 100 nm. Therefore, it is understood that microfibers have a much better thermal conductivity than nanofibers (Kochetov, 2012).

8.5.6 SURFACE PROPERTIES

Porosity of a nanofiber has a direct effect on it wettability and ability of drug delivery. It can be measured for polymer nanofibers assembly or for a single nanofiber. Porosity is the unfilled three-dimensional voids present between nanofibers in a nonwoven assembly. The size and shape of the pores greatly influences the action of nanofibers. Even if a single nanofiber was observed to be smooth, a significant internal porosity can be present. Whereas, surface porosity will be apparent on the surface of the nanofiber, making it appear rough. High surface to weight ratio, high pore volume and tight pore size give nanofibers an advantage over other nanostructures. Surface features can vary in size from one nanometer to hundreds of nanometers, depending on their nature and the fabrication method. Wettability of nanofibers is a very important property as it has given rise to nanofibers being used as coating to control the degree of water adhesion and spreading (Pisignano, 2013).

8.6 PRODUCTION OF NANOFIBERS

Over years there were various methods which are used to synthesize fibers. However, limited techniques are suitable for producing fibers on nano-scale. Commonly, production takes place by electrospinning, self-assembly, phase separation and template method. Electrospinning seems to capture more attention as it displays high efficiency and versatility in comparison to the rest techniques which can be used on specific or narrow range of studies. Each technique is described in the following subsections.

8.6.1 ELECTROSPINNING

Electrospinning is the most versatile and used technique on large-scale production, depending on electrical field as the driving force stimulating the fabrication of nanofibers of defined dimensions (Goonoo et al., 2014).

The simplest regular set-up requires four major components: a spinneret with a pointed tip, a syringe pump which is the reservoir of polymeric solution ejecting it at a particular rate, direct current representing the high-voltage source or electromotive force applied and grounded collector (usually an aluminum foil) which collects the freshly synthesized nanofibers (Figure 8.6 displays the conventional electrospinning set-up) (Liu et al., 2011).

The chief principle of this process is-based on the electrostatic repulsion of the polymer solution to be fabricated. The procedure pursued is as follows: First of all, the droplet of the polymeric solution is allowed to emerge from the spinneret tip. Due to surface tension, it maintains a

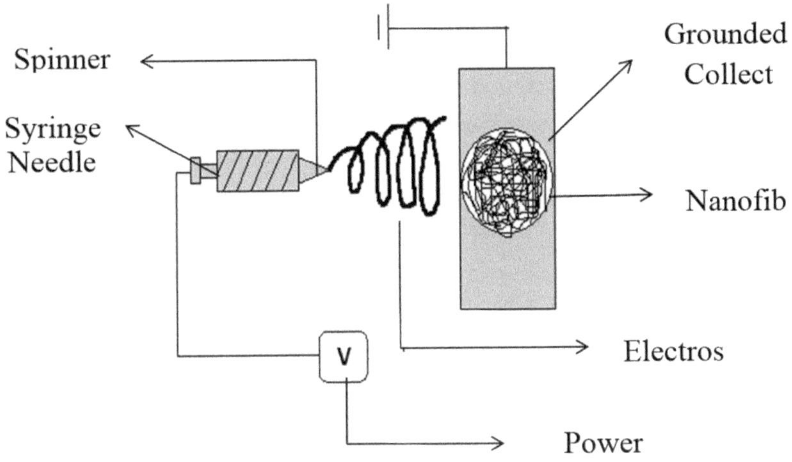

FIGURE 8.6 Equipment requirement for electrospun nanofibers.

droplet shape until an electrical voltage, usually ranging between 1–30 kV, is applied (Hu et al., 2014). This prompts the droplet to electrify and induce charges thus give rise to repulsion in the polymer. The established repulsion exerts an opposite action to surface tension which results in change of conformation into bent hemispherical-surfaced jet. As time flows, elongation continues and a conical shape called Taylor cone is created. When the length is considered sufficient, potential is increased in a way it overcomes the attraction forces of surface tension permitting the jet to erupt from Taylor cone's tip. Owing to elongation and solvent evaporation along with air and moisture the fibers are subjected to, nanofibers undergo a kind of instability where fibrous texture perishes. These instabilities associated with electrospinning fall under three categories: Axisymmetric instability, Non-axisymmetric instabilities and Rayleigh instability. The latter is manifested by jet breakage into droplets. This occurs due to the forces applied by means of viscosity of fluids involved or surface tension which are opposing the ones applied by the electric field/ voltage (Goonoo et al., 2014).

Therefore, crosslinking the polymer used with another one is advisable to overcome these instabilities and improve mechanical strength (Liu et al., 2011). Eventually, fibers are directed towards grounded collector constructing a fibrous polymer mesh (Figure 8.6).

8.6.1.1 Merits and Demerits of Electrospinning

As mentioned earlier, electrospinning is the favorable technique among all by virtue of several reasons:

- Being a practical and economical approach;
- The setup that isn't time consuming nor expensive;
- The ability to control many variables as in diameter, length, orientation and composition to give the desired properties that correspond to the intended use and route of administration.

Limitations and drawbacks:

- Use of organic solvents;
- Broad spectrum of thickness;
- Casual and irregular orientation;
- Poor mechanical properties;
- Limited control of pore structure (Dahlin et al., 2011).

8.6.1.2 Influential Factors

There are a variety of parameters that can be manipulated to obtain a final product possessing certain properties. They can be categorized into two groups: Systemic (solution) parameters and processing parameters as shown in Figure 8.7.

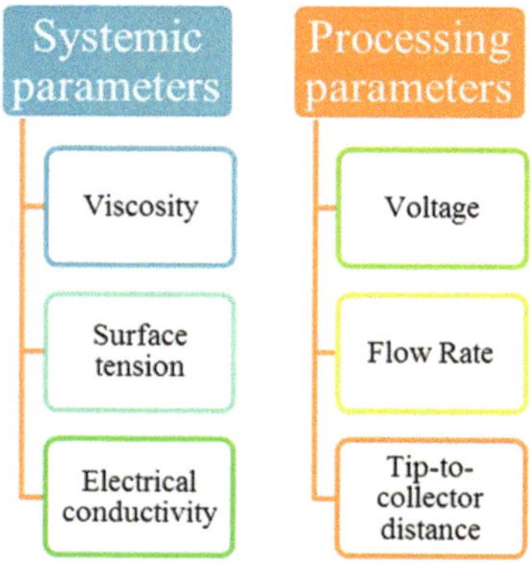

FIGURE 8.7 Parameters influencing electrospun nanofibers.

8.6.1.2.1 Systemic Parameters

They are precisely determined by the polymer's molecular weight and concentration, solvents and additives incorporated. Each parameter has its own impact (Sharma et al., 2015).

- Viscosity:
  When the system has low viscosity, chances of electrospraying outweigh electrospinning. On the other hand, if it has exceedingly high viscosity, jet ejection becomes difficult.
- Surface tension:
  If the surface tension is high, bending instability is reinforced and round/beaded fibers are obtained.
- Conductivity:
  This parameter can be adjusted by addition of ionic salts. High conductivity systems produce thinner fibers (smaller diameters). However, very high conductivity becomes mischievous as it results in round/beaded fibers along with irregular diameter distribution.

8.6.1.2.2 Processing Parameters

- Voltage:
  When low voltage is applied, surface tension forces prevail and Taylor's cone isn't launched, but extremely high voltages cause round/beaded fibers.
- Flow rate:
  It manipulates the fibers diameter in direct proportion. Any increase in flow rate is subsequently followed by an increase in diameter. However, it should be carefully observed as excessively high flow rate causes accumulation of beaded structures on the fiber. This occurs in correspondence to short evaporating time.
- Tip-to-collector distance:
  The distance should be optimal. If collector was set at short distance, there won't be enough space to permit elongation or enough time to permit solvent evaporation. At the same time, very long distance makes the surface tension forces dominate over electric field forces (Sharma et al., 2015).

Voltage and flow rate can alter the degradation rate by changing polymer molecular weight and distribution, or decide nanofibers diameter. It can determine probability of bead formation as well by modifying one or all of the following: viscosity, temperature, surface tension or conductivity. The latter parameter, presented as the distance between the tip and grounded collector, specifies the extent of solvent evaporation, while the collector's movement in special set-ups lays out the sedimentation pattern (Vasita and Katti, 2006).

Based on these factors many set-ups can be made to deliver an exclusive type of nanofibers. Both coaxial electrospinning and emulsion electrospinning are employed to draw out core-shell nanofibers with high loading capacity for hydrophilic drugs. In case of emulsion set-up, the hydrophilic molecule is enclosed inside an organic phase then electrospinning takes place (Heunis and Dicks, 2010).

8.6.1.3 Alignment of Nanofibers

Different alignment textures of nanofibers highly influence mechanical properties and biological significant roles related to cell adhesion, proliferation and migration. Controlling what seems to be, at first glance, a minor detail contributes to significant effects like decreasing the chances of tissue scarring or even stimulating faster wound healing (Liu et al., 2011). Since this is the case, knowing how these alignments are made is extremely important for drug delivery and design. Those alignments are produced because of one of three forces: Mechanical, electrostatic or magnetic. Table 8.1 points out the main differences between alignments obtained by each of these forces.

8.6.1.3.1 Mechanical Forces-Induced Alignment

The rotation velocity given by the mandrel must be optimal to yield uniform uniaxial nanofibers. That is to say, it should be high enough to prevent random spontaneous alignment and low enough to avoid fiber's necking. Some studies have proven the fact that mandrel width assists the generation of electrostatic forces from both ends. This steers the nanofibers to

TABLE 8.1 Comparison Between Alignments Introduced by Three Different Forces

	Mechanical forces	Electrostatic forces	Magnetic forces
Equipment used	Metallic Rotating mandrel	• Metallic staple • Metallic ring with central metallic pin • Array of metallic bead	Two permanent magnets
Morphology of fibers	Uniaxial	• Uniaxial • Radial • Microwells array	Ranges from uniaxial to wavy (depending on flow rate of electrospun jet)
Parameter determining alignment	Rotation velocity	External Electric field	External Magnetic field
Special requirements	Using an alternating current rather than conventional direct current	Electrostatically charged jet	Magnetic nanoparticles (or less the magnetic field is useless)

twist in a condensed fashion on the edges. This sense of direction supports uniform alignment. However, it hinders formation of large scaffolds due to small collecting area (Liu et al., 2011).

8.6.1.3.2 Electrostatic Forces-Induced Alignment

The presence of electrostatically charged jet in an electrical field pushes the jet to align in certain orientation. There are three equipments, which rely on this principle:

1. Metallic staple for uniaxial alignment: just like a staple it contains a gap separating between two collecting pieces (which act like electrodes). Once the jet starts descending, it will be subjected to two types of forces. One of which pulls the jet towards the electrodes or the collecting pieces, the other will stretch or expand it straight across the gap.
2. Metallic ring with a central pin to achieve radial alignment: the needle is positioned on the upper surface of the ring. It, then, ejects the electrostatically charged jet and targets the pin and overall

radially aligned scaffolds are established. This morphology is very useful for wound healing by drawing a particular path for cells to migrate and seal the wound.

3. Array of metallic bead to formulate microwells array: the metallic beads in an array are organized beside each other separated by gaps which are intended to accommodate uniaxial fibers. Therefore, they can be tailored to fit the application they are intended for. The procedure proceeds until a concave nonwoven mat is produced (Liu et al., 2011).

8.6.1.3.3 Magnetic Forces-Induced Alignment

Those alignments are prepared on basis of application of magnetic field on a magnetized polymer solution to lead them to extraordinary oriented fibers. Hereby, two paralleled magnets are placed generating a magnetic field that stretches the fibers to take a uniaxial morphology. It was noticed, by some researches, that the flow rate of the jet influences the conformation of these fibers. If the flow rate increases, wavy fibers are synthesized. While other researches showed that magnetization of the polymer solution and solvents had nothing to do with the final product architecture. So, this area of magnetic field-assisted electrospinning is still under research and study to investigate this technique (Liu et al., 2011).

8.6.2 SELF-ASSEMBLY

Self-assembly is a naturally occurring process in which various building blocks present in our body, like nucleic acids and proteins, are arranged into a definite format. This gave inspiration to many approaches to take advantage of this fundamental in synthesizing nanofibers. This technique is-based on unprompted organization of uncovalently bonded molecules through weak intermolecular forces.

Hartgerink et al. (2001) and others have studied peptide-amphiphiles (PAs) which was found to be of quiet the resemblance to human extracellular matrix (Vasita, 2006; Hartgerink, 2001).

In essence, PAs are composed of amide linked hydrophobic tail and hydrophilic peptide sequence of numerous amino acids (Sharma et al., 2015).

Once the ion content of PAs is settled, the assembly begins to give an initial gel-like structure and develops electrostatic repulsion. Upon electrical neutrality, peptides agglomerate spontaneously and turns into cylindrical, micelle-like framework in which hydrophobic tails bundle towards the core and hydrophilic component projects out.

Prominent features of self-assembly:

- Smaller diameter nanofibers in comparison to electrospun nanofibers (5–8 nm).
- Short nanofibers can be produced by this technique (1 to several μm).
- Suitable for injection purposes in tissue repair due to similarity with in vivo peptides. However, this similarity gives rise to some sort of competition with physiological amphiphiles which complicates application.

Drawbacks and limitations:

- Only few limited polymers can be involved.
- Poor mechanical strength.
- Undefined pore structure.
- Liable to enzymatic degradation which may result in unpredictable rate of scaffold degradation.

It's noteworthy to mention that further studies made in this area showed three types of self-assembly methods: pH-controlled self-assembly, drying on surface-induced self-assembly and divalent ion-induced self-assembly in which diverse motives can initiate self-assembly (Vasita and Katti, 2006).

8.6.3 PHASE SEPARATION

This method relies on the fact that once a polymer uniform phase is irrigated with a solvent, it urges a state of thermodynamic instability provoking phase separation into polymer-rich and solvent-rich phases (Gupte and Ma, 2011).

Phase separation encompasses of 5 steps as follows (Figure 8.8) (Vasita and Katti, 2006):

1. Polymer dissolution into single uniform phase.
2. Liquid-Liquid phase separation (quenching polymer solution).

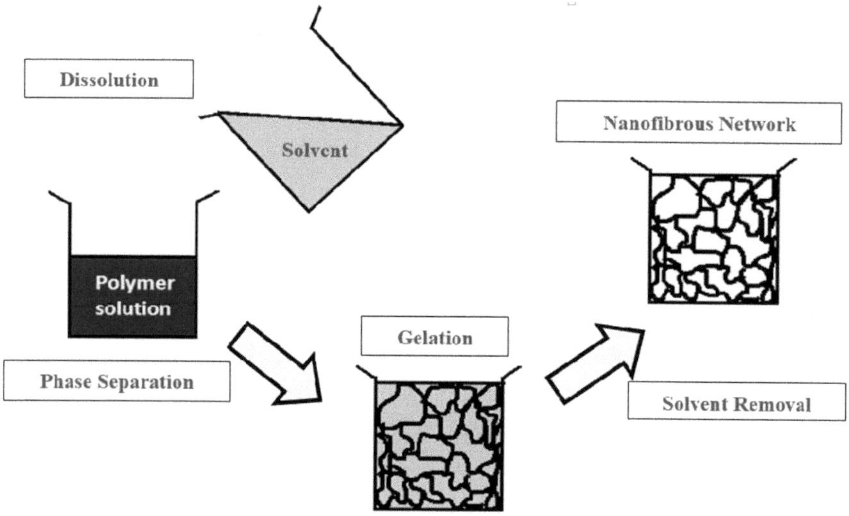

FIGURE 8.8 Steps of phase separation.

3. Gelation which is a very crucial step to determine the porosity of scaffolds depending on choice of gelation temperature. Low temperatures favor the formation of nanofibrous network, whereas high temperatures lead to crystal growth which establishes a platelet-like morphology.
4. Extraction and removal of solvent.
5. Freezing and Freeze-Drying under vacuum conditions.

This technique features many outstanding characteristics that distinguish it from previously discussed techniques. For instance:

- It's capable of synthesizing nanofibers with a diameter ranging between 50–500 nm with 98% to 98.5% porosity which cannot be obtained by electrospinning or self-assembly (Sharma et al., 2015; Vasita and Katti, 2006).
- It specifically offers internal macropores and sophisticated scaffold geometries.
- The increase of polymer concentration doesn't quiet affect the nanofibers diameter; instead, it increases the tensile strength and enhances the mechanical properties to fit the desired application.
- It's a simple procedure that doesn't demand much equipment.

- It rarely introduces batch variations.
- On biological aspect, it promotes mass transport, cell distribution and tissue organization.(Vasita and Katti, 2006)

Limitations and drawbacks:

- Narrow range of polymers can be used
- Difficulty to scale it up for commercial purposes.

Phase separation is usually referred to when synthesizing porous nanofibers. However, when volatile solvents are involved it becomes a bit problematic as it results in filling of pore with solvent-rich phase reducing the tremendously large surface area. So, to overcome this, electrospinning is chosen in which a blend of two polymers of different solubility is spun then dissolved in a solvent in which the intended polymer is insoluble and the auxiliary polymer is easily removed (Ramakrishna et al., 2006) (Figure 8.8).

8.6.4 TEMPLATE SYNTHESIS

As the name indicates, nanofibers are fabricated through a template-based synthesis in which a template or a mold is prepared from ceramic or a particular polymeric membrane. Then, it is used to guide the extrusion of nanofibers. This method is occasionally used for small scale production to fabricate nanofibers depending on some perforations of defined size present in the template that grant polymer passage in form of fibers once water pressure is applied. A good example of this method can be illustrated by Wang and co-workers' attempt to synthesize carbon nanofibers containing linear mesocage arrays. In his experiment, he synthesized anodized aluminum oxide (AAO) template to synthesize carbon tube (hollow nanofibers) (Wang et al., 2010). AAO template is one of the commonly used ceramic templates in production of diverse types of nanofibers such as; conducting polymeric, carbonic and metallic nanofibers. Silica templates are another example which is mainly involved in polymer nanofibers synthesis (Sharma et al., 2015). Many polymers have been studied to figure out if they are candidates for template design to be used in fabricating special nanofibers (Figure 8.9).

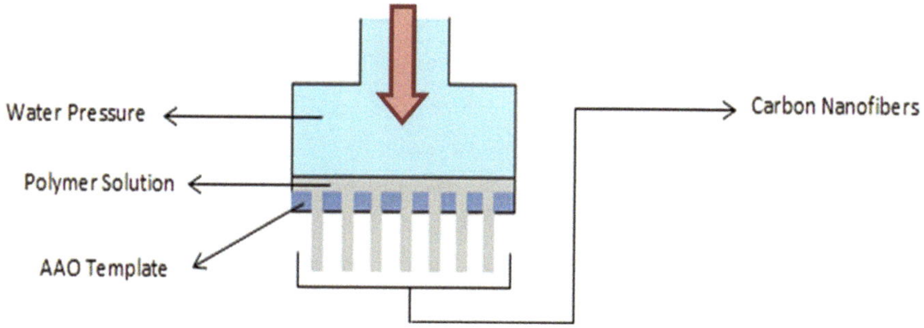

FIGURE 8.9 Template Synthesis of carbon nanofibers by AAO template.

8.7 CHARACTERIZATION AND EVALUATION OF NANOFIBERS

A variety of tests are conducted on nanofibers in order to characterize and evaluate their many properties. Some techniques that can be used to characterize most of nanomaterials are electron microscopy and scanning probe microscopies, which includes atomic force microscopy neutron diffraction, x-ray diffraction, x-ray scattering, x-ray fluorescence spectrometry, and various spectroscopies. What is really challenging about nanofibers is to achieve perfect control over their properties at the nano-scale, therefore, it has become a world-wide interest to synthesize and characterize new nanofibrous materials with well-controlled shape, diameter, porosity and structure (Ikhmayies, 2013).

Differential scanning colorimeter is used to characterize the mechanical properties, such as polymer crystallinity of nanofibers. Whereas, scanning electron microscopy is performed to obtain information about fiber diameter, diameter distribution and fiber alignment. In addition, thickness test and tensile strength test are performed to evaluate other properties of nanofibers (Zhang, 2009).

The following Table 8.2 lists a number of characterization techniques.

8.7.1 X-RAY DIFFRACTION (XRD)

Generally, the most extensively used method for the characterization of materials is X-ray diffraction, which was first introduced by the German physicist, Von Laue in 1921. XRD data can provide a lot of information

TABLE 8.2 Methods of Characterization

Types of characterization techniques	Fiber parameter measured
X-ray diffraction	Size
	Crystallinity
	Structure
	Strain
	Alignment or orientation
Scanning electron microscopy	Size
	Diameter
	Shape
	Composition
	Microstructure
	Porosity
	Pore size and distribution
	Topography
	Fiber Orientation
Transmission electron microscopy	Size
	Composition
	Crystal structure (inferred)
	Alignment or orientation
Optical Spectroscopy	Chemical composition
	Chemical bonding
Atomic force microscopy	Topography surface profile
	Surface mapping
	Mechanical properties
	Fiber geometry
	Diameter
	Alignment
Differential scanning colorimeter	Crystallinity
	Mechanical properties
Near field and confocal light microscopy	Size
	Shape
	Topography
	Morphology
	3D image reconstruction (confocal) of microarchitecture

TABLE 8.2 (Continued)

Types of characterization techniques	Fiber parameter measured
Differential scanning colorimeter	Crystallinity
	Mechanical properties
Near field and confocal light microscopy	Size
	Shape
	Topography
	Morphology
	3D image reconstruction (confocal) of microarchitecture

for all forms of samples, powder, bulk or thin fibers. The data, extracted from XRD, describes the nature of crystals, lattice parameters, and grain size. Unit cell parameters and microstructural parameters, such as grain size and microstrain, can also be obtained from the orientation and shape of the fibers. The nature of strain in a nanofibrous system can be inferred from the lattice parameters (Dahlin et al., 2011).

8.7.2 SCANNING ELECTRON MICROSCOPY (SEM)

This method involves the use of energetic electrons beam to examine the properties at a very fine scale. SEM can provide images that are more magnified than those obtained from light microscopy. It describes surface compositions and provides information about regions that are near the surface. The principle of this method is to scan the specimen surface with an electron beam. The electrons will generate a number of signals after striking the surface of the specimen. The signals are later on collected by detectors to form images. This technique is used to observe the nanofiber diameter, diameter distribution and fiber consistency. Moreover, the void area of nanofibrous scaffolds can be measured by modifying the images produced from SEM, the color of fibers and background is changed to black and white. Then, by using a specific software, the void area and single fiber surface area is calculated and later on used to analyze mechanical properties data (Zhang, 2009). SEM is easy to use, which is why it is commonly used in determining the morphology of fibers. However, the electron beam can cause damage to fibers with diameters less than 200 nanometers. In addition, it is not directly applicable for nonconductive samples, hence, the sample is coated with a conductive metal. This gives rise to questionable accuracy for the very thin nanofibers. This results in Transmission Electron Microscopy and Atomic Force Microscopy being better suited for characterizing the morphology of exceptionally fine nanofibers (Dahlin et al., 2011).

8.7.3 TRANSMISSION ELECTRON MICROSCOPY (TEM)

In this technique, the beam of electrons is transmitted through a very thin specimen, with whom it reacts, to form an image that is magnified and focused on a layer of photographic film or detected by a sensor. Basically, the TEM follows the same principle of operation as the light microscope. However, it uses electrons instead of light. Since electrons have much smaller wavelength than light, the resolution of the produced images is thousand times better than those produced with a light microscope. TEMs allow the user to examine extremely fine details like a single column of atoms. The images produced in this technique allow the observation of crystal orientation, chemical identity, and electron structure of a nanofiber. Other than structural and chemical characterization of nanofibers, TEM is applicable in other tests, for instance, melting point determination, in which the sample is put under direct contact with the electron beam, which heats up the nanofibers and the melting point is indicated by the disappearance of electron diffraction. Another application is to measure mechanical and electrical properties of a single nanofiber.

8.7.4 OPTICAL SPECTROSCOPY

Optical spectroscopy is categorized into two groups: absorption and emission spectroscopy, which determines electronic structures of different entities, like atoms, ions, molecules or crystals, by exciting electrons (absorption spectroscopy) or relaxing them (emission spectroscopy).

The second group is vibrational spectroscopy, which involves the interaction of photons with the sample resulting in energy transfer via vibrational excitation or de-excitation. The frequency of vibration will provide information about chemical bonds in the sample. Different types of spectroscopy are: Infrared spectroscopy, photoluminescence spectroscopy, and ultraviolet- visible spectroscopy (Cao, 2004).

8.7.5 ATOMIC FORCE MICROSCOPY (AFM)

AFM is gaining popularity because of its ability to make images of biomaterials without damaging the sample surface. However, its main advantage is that it is capable of imaging materials at the nano scale. The basic set up of an AFM consists of a micromachined cantilever, made of silicon and silicon nitride, with a needle at one end to detect any deflection of the cantilever tip. This deflection can be due to electrostatic, Vann der Waals repulsion, and attraction between the atoms at the tip and the sample. AFM is used for investigations about size, shape, structure, dispersion and aggregation of nanomaterials as well as nanomaterials' dynamic interactions with biological molecules in real time (Lin et al., 2014). In this technique, visualization of three dimensions is possible. However, it is limited on the vertical axis by the vibrational environment of the set up, and on the horizontal axis by the diameter of the tip used for scanning. The vertical resolution is typically less than 0.1 nanometer and horizontally around 1 nanometer. Size information and other physical properties including morphology and surface texture can be measured via AFM. Image analysis and data processing programs can be used to provide statistics of particle count, particle size distribution, surface area distribution, volume distribution and, if density of the material is known, mass distribution. The fact that AFM can be performed in solid or liquid medium, offers a great advantage when it comes to nanofibers characterization. AFM is considered a cost effective technique for nano scale imaging. When compared to SEM and TEM, AFM has a number of advantages over them, for example, AFM produces 3D data whereas the others produce 2D data. It is much simpler, requires less laboratory space, less processing time and less money than SEM and TEM (Scalf and West, 2006).

8.7.6 DIFFERENTIAL SCANNING COLORIMETER (DSC)

This technique is mainly used to describe material transitions like melting and crystallization (Lin et al., 2014). In a DSC, samples are sealed in aluminum pans with a sealed empty pan as reference and placed under nitrogen atmosphere. The sample is heated and the heat energy, required to melt the sample, is measured. The variations in the heat energy can indicate the variations of crystallinity of the measured samples. DSC also describes the melting behavior of a variety of polymers. The mechanical properties of single fibers can be assessed by microstructure properties like the degree of crystallinity in a polymer (Zhang, 2009).

8.7.7 TENSILE TEST

Mechanical testing of individual fibers is rarely performed, due to the difficulty in handling them and their extremely small size. However, tests for tensile strength, bending and indentation of nanofibers have been developed. Tensile tests have many set ups and systems. One method of tensile testing is the use AFM cantilever to hold one end of a nanofiber while the other end is positioned on a movable optical microscope stage. The microscope and camera are used to observe the fiber as the stage applies tensile force to the fiber. Another technique is a three-point bending test which measures the tensile modulus and fracture strength of single fibers. The elastic modulus of nanofibers can be determined by nanoindentation tests, in which an AFM probe is used to indent the nanofiber, the force applied is then measured and used in various calculations (Dahlin et al., 2011). Fiber diameter has an effect on the mechanical properties of nanofibers. For nonwoven mats, the breakage is due to weak cohesion among fibers, therefore, their mechanical properties are not affected by fiber diameter and crystallinity (fiber properties). However, breakage in aligned fibrous mats occurs due to failure of the fibers themselves (Zhang, 2009).

8.7.8 THICKNESS TEST

When measuring mechanical properties of nanofibers, a better representation of their characteristics is by measuring macroscopic fiber sheets.

The average mechanical properties are better assessed by stretching a nanofibrous mat instead of measuring and individual ultra-thin nanofiber (Chen Zhang et al., 2011). This test is used to study the thickness of nanofibrous mats. Sample mats are placed between an anvil and a pressure foot, and accurate measurements taken via a sensor. The average value of this test is used in tensile testing calculations (Zhang, 2009).

8.8 BIOMEDICAL APPLICATIONS

Nanomedicine is the branch of nanotechnology that is concerned with the biomedical applications of nanostructures.

Nanofibers are ideal for this application because they operate at the same size scale that biological molecules and structures function at. Several other properties of nanofibers have contributed to them being used in medical technologies; the surface/volume ratio is very large due to their small size, resulting in high surface area available for interaction with biomolecules. In actuality, as the particle size decreases, the biochemical reaction time falls sharply, it is due to this property that nanotechnology was used in biochemical analytical devices, making them faster and more sensitive. Moreover, the extremely small size of nanofibers has allowed the miniaturization of these devices. As the devices got smaller in size, fast, safe and noninvasive techniques were developed. Another advantage of miniaturizing the devices lies in making the sensing part smaller, which allows the measurement of really small, or rare biological samples, like some biopsies (Boisseau and Loubaton, 2011). Major applications of nanofibers in various areas of medical field are mentioned in the following subsections.

8.8.1 DRUG DELIVERY

Nanofibers can be introduced as drug delivery carriers in various routes of administration, usually, used to accomplish localized treatment or targeted, controlled release drugs. In order to present a successful drug delivery, it has to cover two major areas: loading drug cargo and providing controlled release over a prolonged duration (Sharma et al., 2015).

8.8.1.1 Loading Techniques

In any drug-loaded fiber, drug particles can be found in one of four modes. Either the drug is blended with polymer becoming an integral part of a single uniform fiber, held at the surface of the fiber by physical electrostatic forces or chemical covalent bonds, fabricated separately and braided with the polymer, or electrospun into hollow or porous form with capacity to encapsulate the drug within (Goonoo et al., 2014). This gives rise to various techniques that can be used to incorporate therapeutic agents within nanofibers. Figure 8.10 shows the different techniques performed for this purpose.

8.8.1.1.1 Blending

The predominant, simplest and most common technique in which the drug is dissolved or dispersed in a suitable polymer solution and mixed thoroughly to form a homogeneous blend which is later electrospun into nanofibers (Goonoo et al., 2014; Sharma et al., 2015). To ensure successful drug loading, major requirements must be fulfilled, which are even drug distribution in the fiber (Dahlin et al., 2011), suitable physicochemical

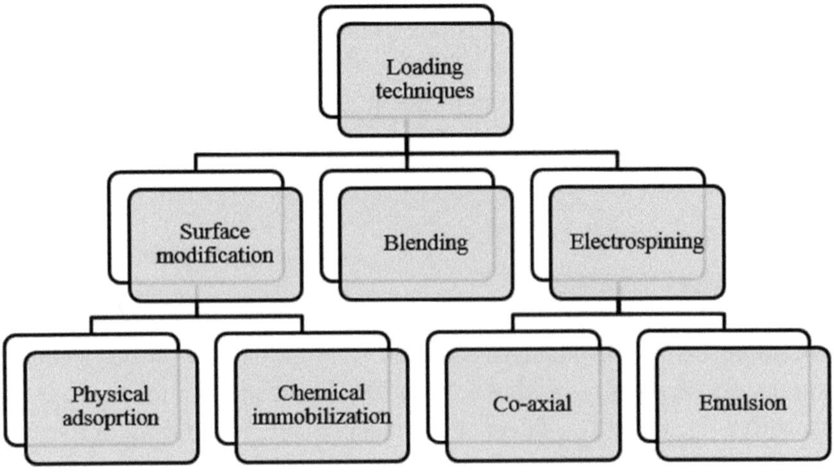

FIGURE 8.10 Classification of techniques involved in drug loading.

characteristics of both the polymer and drug as well as their interactions (Goonoo et al., 2014). Equal distribution can be visually confirmed if a cross-section or a surface-section was taken from the fiber. This feature can be most noticed when the drug has low solubility in the polymer solution. In this case they may be dispersed homogeneously and randomly across the fiber or even concentrated on the surface since they are prone to migration.(Dahlin et al., 2011) According to which, it will cause burst release, in other words, high initial release rate that cannot be extended for a prolonged period of time. This may seem beneficial when rapid delivery is desired. However, when sustained release delivery systems are prepared, this approach doesn't seem to be the best loading method to provide appropriate drug release rates and duration.

8.8.1.1.2 Surface Modification

8.8.1.1.2.1 Physical Adsorption

It is another simple method of drug loading, sometimes referred to as sorption method (Sharma et al., 2015). Here, the previously fabricated or electrospun nanofibers are drenched into the drug solution, or the drug solution is added drop wise until the maximum adsorption is reached. The drug will be attached to the surface by means of electrostatic forces, hydrogen bonding, hydrophobic and Vann der Waals interactions (Goonoo et al., 2014). Being adsorbed at the surface via those weak interactions also leads to burst release. However, coating the fiber with a polymer may delay the release and decrease the burst release (Dahlin et al., 2011).

Yoo et al. (2009) have demonstrated three modes through which the drug molecules are superficially adsorbed:

Simple physical adsorption

It refers to the process in which the driving force for adsorption is electrostatic weak interactions. What distinguishes this technique is inhibiting early degradation which helps in retaining biological activity intact (Dahlin et al., 2011). Therapeutic efficacy that is offered by the localized drug delivery provided accompanied with the nano-structures allowed these surface-functionalized nanofibers implementations in various health-related areas. It was found that this type of loaded nanofibers is successfully employed as postsurgical antiadhesion barriers. Usually after surgical incisions, the body tends to promote adhesion as a part of wound healing, which will cause inflammations unless suitable antibiotics are used. These can be best administered through nanofibers with antibiotics loaded by this technique (Yoo et al., 2009).

Nanoparticle assembly on the surface

Due to their low size, they acquire large surface area. Their large surface area can be further increased by assembly of nanoparticles on the surface of the fibrous system. The nanoparticles are assembled on the nanofibers in a distinctive hierarchical fashion through attractive forces between opposite charges gained through electrospinning. The pharmaceutical industry took a huge advantage of this design in formulating sustained release drug delivery system. The drug release obtained cannot be found in either nanoparticles or nanofibers alone (Sharma et al., 2015). Hereby, the drug is encapsulated within nanoparticles and side-by-side electrospun with the nanofibers to give a final product of stable, hierarchically structured, sustained release drug delivery system (Yoo et al., 2009).

Layer-by-Layer (LbL) multilayer assembly

This innovation is based on coating the nanofibers with LbL polyelectrolyte multilayers increasing the thickness from nanometers to few micrometers. The assembly is driven by major electrostatic forces of charged polyanions and polycations deposited on the layers (Yoo et al., 2009). Therefore, incorporation of charged therapeutic agents like DNA and heparin is easy and feasible. On the other hand, insoluble and electrically neutral therapeutic agents faced difficulty in encapsulation until recent studies acknowledged many forces like hydrogen bonding, covalent bonding, acid-base pairing, metal-ligand complexation and hydrophobic interactions to be an eligible driving force for such assembly (Zamani et al., 2013). It has been devised to provide a sustained release delivery via control of many parameters such as pH, temperature, polyelectrolyte multilayer coating and hydrophobic coating (which acts as a diffusion barrier) (Figure 8.11).

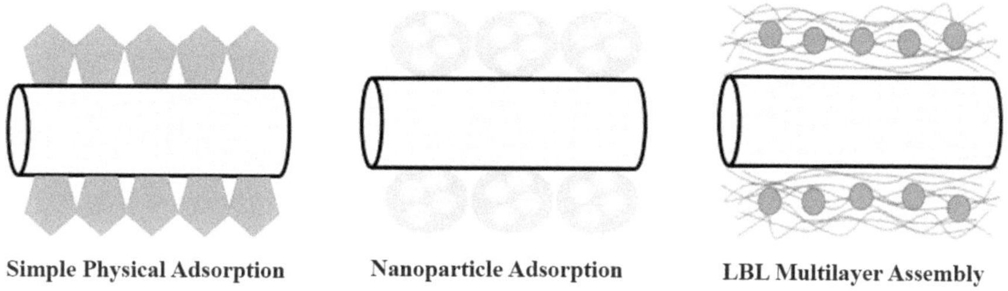

FIGURE 8.11 Physical adsorption methods.

8.8.1.1.2.2 Chemical Immobilization

It's sometimes called surface- covalent modification, as it introduces covalently bonded or conjugated functional groups on the surface of the fibers. Generally, this technique enhances the surface properties of the nanofibers. A model was presented to illustrate how do functionalities enhances solubility, targeting, and release, known as Ringsdorf's model which is elucidated in Figure 8.12. It is chosen for the sake of resembling the structure and biochemistry of the biological tissue, especially when synthetic polymers are involved (Zamani et al., 2013). Various functionalities can be immobilized such as; amine, carboxyl and hydroxyl groups which reduce the drug release (Sharma et al., 2015) and accurately control it by incorporating responsive materials to local external cues (Goonoo et al., 2014).

Surface modification can be attained through one of three options; plasma treatment, wet chemical method and surface graft polymerization.

Plasma treatment

Typically, the surface is treated with oxygen, ammonia or air to create and attach carboxyl or amine groups that permit immobilization of extracellular matrix protein-based components (like gelatin, collagen and laminin) (Yoo et al., 2009). Those alter cell adhesion and proliferation without compromising bulk properties. Another approach employs air or argon treatment to increase hydrophilicity of the surface as well as to remove contaminants.

Wet chemical method

In this method, the polymer's surface is partially hydrolyzed in acidic or basic medium to improve wettability or develop surface functional groups. Accomplishing such function is-based on cleavage of ester linkages present

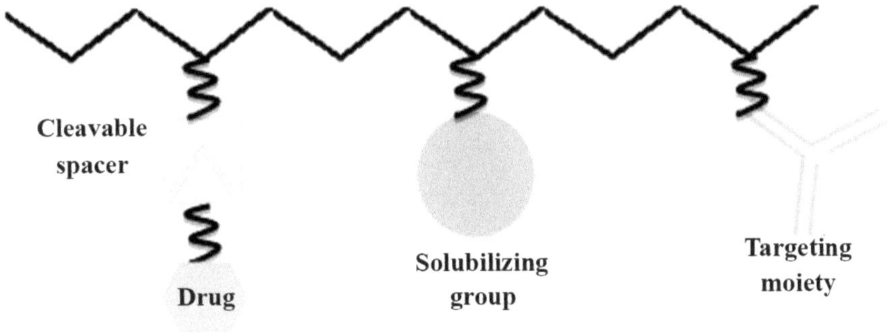

FIGURE 8.12 Ringsdrof's model.

on the surface of polymers skeleton structures (Goonoo et al., 2014). The hydrolysis, here, is considered kind of critical and should be closely monitored, especially regarding the duration and the concentration of hydrolyzing agent as well. If not properly optimized, functional groups are created with concomitant change in bulk properties. In comparison to the aforementioned method, when handling mesh with deeply situated nanofibers, it was found that wet chemical method is preferred for two reasons: (1) providing flexibility of surface modification across thick meshes. (2) The restricted depth penetrability of plasma through nanopores of the mesh (Yoo et al., 2009).

Surface graft polymerization

This method is more applicable for synthetic polymers as they possess a surface of hydrophobic nature. So, in order to seize a favorable response, hydrophilic modification becomes crucial (Goonoo et al., 2014). The principle of this process is dependent on provoking the formation of free radicals from treatment with plasma or UV radiations to polymerize them on the surface. Thus, better cell adhesion, proliferation and differentiation are achieved without any structural damage (Yoo et al., 2009).

8.8.1.1.3 Electrospinning

8.8.1.1.3.1 Co-Axial Electrospinning

This technique is very feasible when the drug requires protection from exposure to the organic solvent where the polymer is dissolved in.

This is because the nanofibers are fabricated from separate solutions which minimize the contact. Consequentially, it preserves the biological activity which in other techniques was compromised by that contact. Moreover; it provides a sustained release of the drugs or encapsulated molecules. All these factors qualify coaxial electrospinning for gene- and growth factor- delivery as well as antibiotics and antioxidant drugs. To ensure efficiency of drug incorporation some parameters must be accurately modulated, for example: concentration of core and shell polymer, their relative flow rate, drug molecular weight and concentration and sometimes the morphology of the fiber when low-molecular-weight drugs are involved (Zamani et al., 2013).

8.8.1.1.3.2 Emulsion Electrospinning

The drug or protein to be incorporated is emulsified with the polymer solution then electrospun to give one of two products: (1) if the drug possess low molecular weight it will form a uniformly distributed drug-polymer type of fibers, or (2) if a drug is macromolecular and is likely to be found in the aqueous phase, it will form core-shell fibers (Goonoo et al., 2014; Zamani et al., 2013). What differentiates this type of electrospinning from any other is that there is no need to find a common solvent for the drug and polymer. Each of which, can be dissolved in an appropriate solvent then emulsified together which allows the synthesis of various combinations of hydrophilic/ hydrophobic polymers. As long as their distribution is uniform, the release is sustained, the stability is maintained and bioactivity is preserved (Zamani et al., 2013).

8.8.1.2 Routes and Delivery Systems

8.8.1.2.1 Wound Dressing and Transdermal Drug Delivery

Whenever any part of the skin is wounded, natural body mechanisms tends to regenerate dermal and epidermal tissues from underlying source cells through orchestrated cascade to mend the damage. Inert wound healing process undergoes four stages: inflammatory, proliferative, remodeling and epithelialization. When the wound is deep, as in full thickness burns and deep ulcers, the healing process becomes a bit more problematic as there are no source cells other than the ones on the edges to regenerate the tissue (Fang et al., 2011). Thus, demanding longer period of time for healing. So, they need a boost to accelerate wound closure, which is nothing but appropriate wound dressing.

This pharmaceutical approach converts the passive, naturally occurring process into active, drug induced healing via involvement of several antimicrobials, growth factors or stimulants within nanofibers. In order to perform this role, ideal nanofibers for wound dressing must reflect the following features (Goonoo et al., 2014; Sharma et al., 2015; Zamani et al., 2013):

- Hemostasis (where blood coagulates and stops flowing in the injured area).
- Capable of absorbing exudate effusing from the area.
- Sufficient permeability to facilitate gas exchange.
- Establishing high sterile conditions.
- Maintaining acceptable moisture content.
- Biocompatible with tissues and nontoxic.
- Hindering full adhesion to the wound.
- Providing scar-free regeneration and improve cosmetic appearance.
- Accelerating wound-healing process.

There have been many forms of dressings administered for this purpose like films, hydrogels, sponges and foams. However, compared to nanofibers, the aforementioned don't seem to be as efficient. What distinguish nanofibers over others are their remarkable characteristics. For instance, their high surface area contributes to satisfactory absorption of exudates and adjustment of moisture content. Moreover, high porosity and small pores lead to smooth oxygen entry to the cell without the risk of microorganism evasion and subsequent infection (Fang et al., 2011). In addition, they allow incorporation of variety of therapeutic agents as well as different types of polymers. Natural biodegradable and synthetic non-biodegradable polymers may be used in formulation of nanofibers-based dressing to regulate the release of the drug through precise modulation of biodegradability and hydrophilicity of the fibrous system (Hu et al., 2014).

Generally, wound dressing can be categorized into three types: passive, interactive and bioactive. The passive dressing is basically acting as a cover to the wound. It doesn't interact or stimulate the healing process. It just prevents potential microbial infection from the surrounding environment. Traditional materials like guaze usually fall under this category (Fang et al., 2011). Interactive dressing is permeable and grants passage of oxygen and water vapor. Yet, it blocks the entry of microbes. All polymeric dressings are included in this category. Bioactive dressings are the ones exerting a biological activity and might be loaded with antibiotics, anti-inflammatory agents, growth promoting factors, vitamins, minerals, and inorganic ions like silver and iodine (Sharma et al., 2015). Release of dressing can be controlled to suit the application. So, in case of open wounds like surgical incisions, quick release is preferred and obtained. At the same time, if prolonged release is desired, it's also achievable.

Transdermal delivery system delivers drugs and bioactive factors through the skin. It provides an alternative route of administration for sensitive drugs; especially the ones subjected to large presystemic effect through oral route (Zamani et al., 2013). Not to mention how it comparatively improves release efficiency since it decreases the fluctuations in drug plasma levels that are usually observed in oral administrations. This route has also proven many advantages regarding patient compliance, gastric irritations and controlled therapeutic response (Goonoo et al., 2014). In spite of all these facts, the low permeability of the skin as well as restricted number of polymers that can be used for this purpose makes full dependence on this route kind of unfeasible. Only drugs with low molecular weight and proper hydrophobicity and concentration can be administered transdermally (Zamani et al., 2013). Good examples of these drugs are vitamins, antioxidant and anti-inflammatory agents. In addition, herbal compounds were also encapsulated as extracts as well as crude plants for topical, transdermal use and even wound dressing.

If transdermal delivery was provided through nanofibers, instead of conventional films, the release pattern is much slower and prolonged avoiding burst release given by transdermal films. Due to which, an optimal therapeutic effect can be readily obtained by accurate regulation of release amount and rate. In addition to that, few modifications on pH, temperature and electrical field can be applied if they reasonably influence the release (Goonoo et al., 2014).

8.8.1.2.2 Cancer Therapy and Implanted Drug Delivery

Cancer is one of the most demanding challenges in the medical field. Many approaches have been investigated to come up with a convenient treatment strategy. Currently, the procedures followed include: surgery, radiotherapy, chemotherapy, hyperthermia, immunotherapy, hormone therapy, stem cell therapy or a combination of them according to the severity and individual case (Yu et al., 2015). In addition, many anticancer agents were designed to exert the therapeutic effect via numerous mechanisms such as; metabolism revision, retarding mitosis, decreasing cell motility and hinder intracellular signal relay (Zamani et al., 2013). Despite all these attempts there were always some limitations which stood an obstacle in the way of treatment. Mostly, these drawbacks are related to poor solubility and instability of therapeutic agents within the body when administered in conventional dosage forms, especially if administered through oral and intravenous routes. Moreover, the therapeutic agent leakage from the tumor site leads to low efficacy and severe toxicity to other healthy tissues (Goonoo et al., 2014).

Although many formulations were successfully tailored to overcome these drawbacks offering efficient, safe, sustained and sometimes targeted drug delivery, the biphasic release (significant initial burst release with subsequent trivial release along incubation time) remained an issue compromising the efficacy of cancer therapy. So, in order to obtain a prolonged release with maximal profits and minimal risks, nanofibers were introduced as a sustained implanted postsurgical delivery (Goonoo et al., 2014; Zamani et al., 2013). This remarkable attention given to nanofibers is based on its characteristics of surface area, porosity, morphology and mechanical strength that can be easily manipulated to attain the desired release pattern (Yu et al., 2015). Amna et al. (2013) has demonstrated through her study on Camptothecin nanofibers that an initial burst release is in the matter of fact beneficial to attain an initial dosage, but it's supposed to provide consecutive sustained release for cancer cells which survived the initial dose. Anticancer loaded nanofibers can get implanted at the tumor site providing full coverage. This results in high local dosage with low toxicity that extends the duration of the effect reducing the frequency of administration and eventually contributes to better patient compliance (Zamani et al., 2013).

Many studies have investigated anticancer agents loaded in nanofibrous systems as mentioned below:

- Multilayer coated paclitaxel-loaded CS nanofibers with hyaluronic acid for treatment of pancreatic cancer.
- Green tea polyphenol (GTP) – loaded nanofibers to inhibit proliferation of hepatocytes in hepatocellular carcinoma.
- Titanocene dichloride in PLLA nanofibers to inhibit mitosis in lung tumor cells.
- Addition of solubilizer (HPCD) to enhance solubility and stability of (HCPT) nanofibers for breast cancer (Zamani et al., 2013).
- Paclitaxel core-shell fibers to produce antiproliferative effect which illustrated 3 phases of release. First phase (between the 1st and 10th week): release is merely occurring because of diffusion through shell. Second Phase (between 10th and 20th week): release has witnesses quiet a surge due to polymer degradation and reduced drug attachment. Third phase (between 21st and 38th week): release has increased moderately due to destruction of shell polymer (Goonoo et al., 2014).

8.8.1.2.3 Gene and Growth Factor Delivery

Many biomacromolecules have proved its significance and important role in a variety of applications including cancer therapy, tissue engineering and regenerative medicine. These molecules can interrupt, interfere or regulate cellular process by promoting or inhibiting particular intracellular signals and cues. They tend to modify genetic information or regulate exogenous stimulation of target cells depending on the type of the biomacromolecule. These types are categorized as:

8.8.1.2.3.1 Gene/DNA/siRNA Delivery

Before the discovery of nanofibers, the pursued method for gene delivery was direct administration of liquid formulation orally or intravenously. This classical approach has elucidated how it diffuses across the body without getting localized on the site of action resulting in some side effects related to toxicity or immune responses (Lee et al., 2014). On the other hand, the recent approach aims for the accomplishment of various biological functions like distinguishing and classifying target cells into specialized classes of cells, triggering apoptosis signals to initiate cell self-destruction in cancer cells, secreting bioactive factors that impact the tissue or even production of cellular therapeutics (Lee et al., 2014). The most critical step accompanying gene delivery is bypassing immune system effects, potential toxicity and reaching the nucleus for transfection in the safest journey possible.

Accordingly, gene delivery can be further classified into viral and non-viral delivery systems (Zamani et al., 2013). Viral delivery employs viral vectors which are nothing but viruses stripped off their DNA (or genomic sequences) and the remaining capsid acts as a carrier. The gene intended for administration is loaded within the viral capsid then electrospun to become an integral part of the fibrous network. This technique offers advanced delivery of genes and prolonged period of gene expression (Lee et al., 2014). Successful delivery is determined by the type of cells and viruses, structure of gene to be incorporated and the delivery technique chosen.

Non-viral delivery indicates incorporation of the gene biomacromolecules such as naked plasmid DNA (pDNA) or DNA/polyplexes directly to the electrospun nanofibers in presence of a suitable polymer. Compared to viral delivery, it's distinguished by the ease of production and capability to preserve intact properties of the carrier. In addition, it provides less toxicity levels and allows administration of DNA in various types and sizes. Small interfering RNA (siRNA) is also a biological macromolecule which inhibits the expression of some particular proteins. This ability was well-used in cases of prohibited tissue repair, inhibitory factors secretion and in cancer therapy as well as for the genes that aid tumor growth (Zamani et al., 2013). Some studies and investigations made in this field showed that co-encapsulation siRNA with a transfecting reagent or cell-permeating peptides, which facilitates its penetration to the nucleus, can yield a sustained release over a period of 28 days (Lee et al., 2014). Sustained release of gene delivery is tunable and can range from days to months by proper choice of polymer template and technique of drug loading. The main challenge in gene delivery was retaining the biological activity of the genes after electrospinning. Blend, emulsion, coaxial electrospinning were used. The former didn't give much satisfactory results regarding bioactivity due to the contact with organic solvents. The latter two methods incorporated the genes into core-shell fashioned fibers which maintain bioactivity and adequate release profiles (Lee et al., 2014; Vasita and Katti, 2006).

8.8.1.2.3.2 Growth Factor (GF) Delivery

These biomacromolecules regulate some natural biological mechanisms like proliferation, migration, and differentiation of cells required for tissue regeneration. This is established through signal transfer between cells and their Extracellular Matrix (ECM). GFs are essentially needed when wounded or damaged tissues lack the proteins needed for successful regeneration (Zamani et al., 2013). For example, there is a growth factor known as fibroblast growth factor-2 (FGF2) which is found in very low levels in case of chronic wounds. Therefore, it can be administered as a temporary, nanofibrous, dermal matrix to assist wound healing and closure (Jeon et al., 2010). Proteins, peptides and growth factors possess very short half-lives in vivo (Hu and Ma, 2011). They vanish from the body before reaching the effective concentration (threshold). For this purpose, many techniques as blending, surface modification, coaxial and emulsion electrospinning were employed in attempt to overcome this instability (Zamani et al., 2013). Other aspects must be carefully considered like potential denaturation due to exposure to organic vehicles or heat treatment, or possibility of partial or complete elution by vehicles. They influence the morphology, mechanical strength and degradation properties of scaffolds. Therefore, for optimal delivery many approaches suggested conjugation with heparin which simulates the body ECM and interacts with GF by the sulfate group on its backbone. The heparinized GF serves a protective property against proteolytic and chemical inactivation as well as providing sustained release. Recently, some studies proposed the combination between hydrogels and nanofibers into sandwich architecture which improves GF delivery and retain its bioactivity intact (Goonoo et al., 2014; Zamani et al., 2013).

8.8.2 TISSUE ENGINEERING

One of the most innovative applications of nanofibers is tissue engineering, which brings together a variety of sciences including biology, chemistry, medicine and engineering. The overall aim of this multidisciplinary field is to improve the quality of life for thousands of people by restoring, maintaining and enhancing the functions of organs and tissues in the body. The concept of tissue engineering is-based on applying the foundations of medicine, biology and engineering to develop and manipulate three dimensional physiological substitutes, with the purpose of recovering or sustaining a variety of bodily functions (Kanani and Bahrami, 2010). Tissue engineering can be broadly divided into three major approaches: (1) transplantation of isolated cells to an injured tissue; (2) delivery of tissue-inducing biomolecules to a targeted tissue; (3) growth of specific cells onto three-dimensional scaffolds (Chen et al., 2013). Scaffold-based tissue engineering represents a new and improved replacement of other tissue repair techniques. Tissue engineering is-based on three components: (1) stem cells; (2) signaling molecules; and (3) scaffold or extracellular matrix. Tissue repair techniques involved autografts and allografts, both of which have displayed a number of limitations. Autografts are associated with disadvantages such as limited availability and donor site morbidity. Although allografts are not limited in supply, they have shown potential of disease transfer and generation of immune response. Therefore, tissue engineering has emerged as an alternative approach for regeneration and repairing damaged tissue (Vasita and Katti, 2006).

To explain why nanofibers are ideal for tissue engineering, the anatomical aspect of the human body should be explored. The human body contains four major tissues: epithelial tissue, covering the body surfaces, connective tissue, supporting and binding other tissues of the body, nervous tissue, specializing in conducting impulses, and muscle tissue, providing movement. Every type of tissue contains fibers, present in the form of bundle structures, in both micro and nano scale. They serve the purpose of providing strength enforcement and elasticity, conducting nervous impulses, and moving body parts. Nanofibers exhibit structural characteristics similar to those of the natural tissue fibers, thereby making them highly acceptable for tissue engineering purposes (Sharma et al., 2015).

Moreover, nanofibers are used as biomimetic scaffolds due to their small diameter which matches the size scale of ECM fiber. The ECM is composed of interwoven protein fibers, like fibrillary collagen and elastin, and adhesion proteins that serve as specific binding sites for cell adhesion.

The alignment of the nanofibers in a scaffold greatly influences the mechanical properties of the scaffolds. This influence is represented as an increase in scaffold anisotropy and stiffness, which is often in the direction of the alignment. This has a major impact on engineering anisotropic, load bearing tissues, like tendons, myocardium and annulus fibrosis. In tendons, the collagen fibrils are aligned parallel to each other, therefore an aligned nanofibers scaffold should be used. Whereas, the myocardium in the cardiac tissue is composed of perpendicularly interwoven collagen stripes. Another influence of nanofibers alignment in a scaffold is on the migration and extension of cells. This has a direct impact on the process of wound closure. Since the cells will migrate according to the alignment of the nanofibers, a shorter time will be needed for the cells to migrate on aligned nanofibers scaffolds, compared to random nanofibers scaffolds (Liu et al., 2011). The high surface area to volume ratio of nanofibers is ideal for cell attachment. Nanofibers have also shown high rates of protein adsorption, which is a key factor in tissue engineering. In addition, the unique mechanical properties of nanofibers, specifically tensile strength, tensile modulus, and shear modulus, are useful for modulating cell behavior and providing adequate strength and tension to the tissue (Dahlin et al., 2011). The porosity of the scaffold greatly influences its degradation rate. When comparing a scaffold made of electrospun fibers with a thin film cast from the same polymer, the scaffold displays higher porosity which, in turn, leads to higher degradation rate. The degradation rates of the scaffold should be matching with the rate of tissue regeneration. The degradation of individual nanofiber occurs once implanted in the body and differs from one scaffold to another according to the polymer used, since the prevailing mechanism of degradation is-based on the hydrolysis of the polymer backbone. This mechanism involves a number of steps: firstly, water penetrates the surface of a nanofiber and attacks the amorphous regions of a polymer, converting its chains into shorter, more water-soluble ones, and eventually, the whole nanofiber disintegrates and disappears. If enzymes are present, rapid loss of the nanofiber mass might occur due to enzymatic digestion (Liu et al., 2011).

The basic approach of using nanofibers in tissue engineering involves three key elements: cells, scaffolds, and biochemical and/or mechanical stimuli (Dahlin et al., 2011). Generally, the tissue engineering strategy involves isolation of healthy cells, expanding them in vitro, and then seeding them on a scaffold that is implanted and, later on, degraded and replaced by newly grown tissue from the seeded cells (Vasita and Katti, 2006). The core technologies applied in tissue engineering are: cell technology, scaffold construct technology, and in vivo integration technology. The integration of nanofibers into the extracellular matrix of the body will affect signaling pathways that alter cellular responses including adhesion, proliferation, differentiation, and tissue neogenesis. Nanofibers scaffolds promote adhesion of cells to produce morphology similar to that in-vivo. Cells cannot survive without sufficient adhesion to surroundings. Some studies suggest that initial adhesion of cells to the nanofibers scaffold is due to increased adsorption of ECM components like fibronectin, vitronectin, and laminin. This, in turn, leads to increased expression of integrins, thereby mediating cell-scaffold integration. Later on, proliferation of cells should occur after adhesion for a successful tissue engineering process. Differentiation of human stem cells has been proved to happen more efficiently on nanofibers scaffolds when compared to microfibers or other materials (Gupte and Ma, 2011).

Nanofibers scaffolds, especially those produced through electrospinning, are structures composed of continuous nanofibers. The unique properties of nanofibers scaffolds have guaranteed their wide utilization in the field of tissue engineering. A scaffold should meet specific requirement in order to be fully functional, including:

- High degree of porosity.
- Appropriate pore size distribution.
- Large surface area.
- Biodegradability, with an appropriate degradation rate that matches the neo-tissue formation rate.
- Appropriate mechanical properties, dimensional stability and structural integrity in order to prevent the scaffold from collapsing onto itself.
- Nontoxicity and biocompatibility.
- Easy processability, malleability and sterilizability.

Nanofibers scaffold have been proven to meet all the previously mentioned requirements, resulting in positive interaction between them and the cells and promoting cell adhesion, proliferation, and penetration (Fang et al., 2008). The function of scaffolds is very critical as they are the managers and directors of the growth of cells either seeded within the porous structure of the scaffold or migrating from the tissue surrounding it. Basically, they serve a function of supporting and delivering cells, inducing, differentiating and channeling tissue growth and stimulating cellular response (Chen et al., 2013). To perform a successful tissue engineering process, close simulation of the natural extracellular matrix should be achieved. Therefore, the scaffold architecture and morphology should be similar to the components of the native extracellular matrix. The electrospun nanofibers for tissue engineering use a number of biopolymers, which include synthetic polymers, such as poly(α-hyroxyl acid) and poly(hydroxyalkanoate) like poly(hydroxybutyrate)(PHB), and natural polymers, such as gelatin, chitosan, silk and collagen. Blends of synthetic and natural polymers have been successfully used in engineering a number of tissues in the human body (Gupta et al., 2014). Natural polymers show greater biocompatibility in comparison to synthetic polymers. This is due to the fact that natural polymers are identical to macromolecules present in the ECM. However, synthetic polymers have an advantage of being easy to prepare, having good mechanical strength, and giving reproducible results that are not observed in the case of natural polymers (Sharma et al., 2015).

Blood vessels, bones, cartilages, tendons and ligaments, muscles, skins, and neural tissues are examples of body tissues than can be engineered using nanofibers.

8.8.2.1 Cardiovascular Tissue Engineering

The function of blood vessels in the body is to carry and transport blood from and to the heart. Depending on their location and function, blood vessels have various sizes, mechanical and biochemical properties, cellular content, and structural organization. Tissue of blood vessels and the heart are delicate and complex in structure, and any damage to them can lead to serious health problems. The vessel wall consists of three layers, tunica intima, tunica media, and tunica adventitia. The innermost layer is the tunica intima, with non-thrombogenic monolayer endothelial cells. Separated from the tunica intima, by an internal elastic lamina, is the tunica media, which is composed of concentrically organized smooth muscle cells. The tunica adventitia is the outermost layer and is composed of collagenous extracellular matrix and fibroblast cells. Generally, the ECM surrounding the vascular cells contains collagen (type 1 and type 3) elastin, proteoglycans and glycoproteins (Kanani and Bahrami, 2010). The ECM of cardiac tissue has cells in fiber-like bundles, which allows mechanical coupling of adjacent fibrils.

When engineering vascular tissues, the major mechanical properties to take into consideration are tensile stiffness, for the resistance against rupture, elasticity and compressibility. Collagen provides the tensile stiffness, elastin gives the elastic properties, proteoglycans contribute to compressibility, and a combination of collagen and elastin prevents the deformation of vessels against pulsatile blood flow. Electrospun nanofibers made with collagen are proved to promote cell growth and penetration of cells into the engineered matrix. When aligned nanofibers are used, structural integrity, vasoactivity, and mechanical strength are greatly improved in comparison with nonaligned nanofibers. To engineer a successful vascular graft, the fabricated nanofibers scaffolds should resemble the dimensions of natural ECM, possess mechanical properties comparable to those of blood vessels, and support the adhesion and proliferation of smooth muscle cell. It is interesting to note that endothelial cells display better growth and enhanced cell penetration with the increase of the diameter of the nanofibers used. To fabricate a nanofibers scaffold that mimics the morphological and mechanical properties of a native blood vessel scaffold, a bilayered scaffold, consisting of a stiff and oriented poly(lactic acid) (PLA) outer nanofiber layer, and a randomly oriented poly(ε-caprolactone) (PCL) inner nanofiber layer, was electrospun. It was reported to be successful in supporting the attachment, spread and growth of human myofibroblasts. Another scaffold was prepared, with a highly porous poly(ester-urethane) urea (PEUU) inner layer and an external layer of electrospun nanofibers, for small diameter vascular grafts. The bilayered scaffold has displayed mechanical properties that are comparable with native vessels, and the combination of the two layers ensured better cell integration and growth. Modifying the surface of electrospun nanofibers with natural proteins, like collagen and gelatin, is an effective way to enhance the endothelial cells spreading and proliferation. Another way to enhance not only cell growth but also cell differentiation, migration and survival, is by binding fibroblast growth factor (FGF-2) to the electrospun nanofibers matrix. The latest approach to match the mechanical properties of vascular ECM and to mimic the ratio of collagen and elastin in blood vessels is by electrospinning nanofibers with blends of collagen, elastin and synthetic polymer such as poly(lactic-co-glycolic acid) (PLGA) and poly L-lactic acid (PLLA) in a ratio of 45:15:40 w/w/w. Scaffolds made with these blends have shown dimensional stability, mechanical properties similar to the native blood vessels, and no cytotoxicity (Fang et al., 2008).

Vascular graft failure occurs majorly due to partial growth of endothelial cells on the surface of the graft so that it remains uncovered. To overcome this issue, gelatin-modified polyethylene terephthalate (PET) nanofibers as well as collagen-grafted PCL nanofibers were fabricated, resulting in maintenance of cell phenotype and enhancement of endothelial cells spreading and proliferation (Chen et al., 2013).

When it comes to heart tissue engineering, biodegradable, nonwoven poly(lactide) and poly(glycolide) based scaffolds were fabricated. Cardiomyocytes (CMs) were cultured on these scaffolds, and the result was that, mature contractile machinery (sarcomeres) was developed, morphological and electrical communication was established and synchronized excitability (beating) was observed (Kanani and Bahrami, 2010).

8.8.2.2 Bone Tissue Engineering

Basically, bone tissue has unique anisotropic structures and fibrous architectures (Ma et al., 2013). Bone is a bio-composite consisting of inorganic components, mainly hydroxyapatite (HA) crystals contributing with 65–70% of bone weight, and organic components, including glycoproteins, proteoglycans, sialoproteins and bone 'gla' proteins (Ma et al., 2013). The organic components of bone matrix are 90% collagen, out of which 95% is type 1. The fibers in bone matrix are approximately 50 nanometers in diameter and are not organized in a specific pattern, some fibers are aligned while some are irregular, thereby making it exceptionally hard to engineer scaffolds that fit to be used in bone tissue engineering

(Dahlin et al., 2011). The fabrication of nanofibers scaffolds for bone tissue engineering is-based on the physical properties of bones like mechanical strength, hardness, porosity, pore size, and overall three dimensional structure. To fabricate a successful scaffold for this purpose, the scaffold should have pore size in the range of 100–350 μm, porosity greater than 90%, and with fibers ranging in diameter from 20 nm to 5 μm (Vasita and Katti, 2006). Nanofibers scaffolds fabricated for bone tissue engineering should meet three requirements: (1) easy for vascularization by making them highly porous; (2) mimicking ECM components and mechanical properties by incorporating hydroxyapatite and collagen; (3) incorporating signaling molecules (growth factors) (Ma et al., 2013). Naturally, the collagen (Col) present in the bone matrix provides the resilient nature, and the inorganic minerals, including hydroxyapatite (HA), are the reason behind bone hardness. Hydroxyapatite is present in bones in the form of calcium complexes. Therefore, when fabricating scaffolds for bone tissue engineering, nHA/Col biocomposite, blended with PCL, nanofibers are being used, in which PCL provides mechanical stability, collagen supports cell proliferation and nHA aids in the mineralization of oseteoblast cells. The addition of 50% gelatin to the PCL enhanced cell attachment, growth and migration on the scaffold. In addition, blending bioactive glass (BG) with PCL composite resulted in improved differentiation of osteoblast cells in comparison to pure PCL. Another approach to stimulate growth, proliferation and differentiation of cells is by applying mechanical stresses, induced by embedded magnetic nanoparticles, onto the PCL scaffold. Mechanical stimulation has been proven to enhance fracture healing due to anabolic osteogenic effects (Fang et al., 2008).

8.8.2.3 Cartilage Tissue Engineering

Anatomically, the human cartilage, with all three types of it, has four different zones: superficial, transitional (middle), radial (deep) and calcified zones, each of which has special morphology, matrix composition, cellular, mechanical and metabolic properties. Collagen fibrils in the cartilage are arranged in specific organizations; in deep and middle zones, fibrils display the surface periodicity property of collagen and are oriented, in large bundles approximately 55 μm thick, towards the articular surface. Fibrils in superficial zone are oriented in a manner that is parallel to the surface. Surface fibrils are randomly oriented. Therefore, multilayerd nanofibers should be fabricated to fit with the varying compositions and orientations of fibrils in the native cartilage matrix (Ma et al., 2013).

Articular cartilage is a type of connective tissue consisting of collagen and chondrocytes, or yellow elastic fibers. The cartilage matrix is firm, gel-like, rich in mucopolysaccharides and exhibiting elasticity and flexibility. To examine the potential of nanofibers in cartilage tissue engineering, two types of nanofibers were electrospun from PLGA, the first one was a solid cylindrical type and the second was a cannulated tubular type. The tubular scaffolds displayed better performance in cartilage tissue regeneration and had higher histology scores than solid ones. In combination with stem cells and chondrocytes, electrospun poly(vinyl alcohol) (PVA)/ poly(ε-caprolactone) (PCL) nanofibers scaffolds were applied to regenerate cartilage. The observed result was that the cartilage defects were repaired, after 12 weeks, with chondrocyte-like cells exhibiting a rounded morphology with lacunae. Moreover, the expression of type 2 collagen was increased in the neocartilage. Healing of defects was improved due to seeding of stem cells in the scaffold. This conclusion was supported by another study, in which mesenchymal stem cells (MSCs), derived from human bone marrow, were seeded in electrospun PCL electrospun nanofibers scaffolds. The results stated that these scaffolds, in presence of transforming growth factor-β1 (TGF-β1), displayed good cell differentiation, no inductive properties and mechanical properties suitable for cartilage tissue engineering. Self-assembling peptide hydrogel scaffold were investigated to estimate their potential for cartilage generation. Bovine chondrocytes were combined with the peptide KDK-12, which has a sequence of (AcN-KLDLKLDLKLDL-CNH2) (where K is lysine, D is aspartic acid, and L is leucine), and allowed to self-assemble into a hydrogel. The chondrocytes proliferated well and maintained a chondrocytic phenotype. In addition, cartilage-like ECM, rich in type 2 collagen and proteoglycans, was produced by the cells (Chen et al., 2013).

8.8.2.4 Tendons and Ligaments Tissue Engineering

The structure of a tendons and ligaments consists of densely packed collagen fiber bundles oriented parallel to the longitudinal axis. While tendons connect muscle to bone, ligaments join one bone to another. Tendons and ligaments exhibit a high degree of ECM fiber alignment because mechanical loading is restricted to one direction, leading to high degree of anisotropy. As a result, MSCs and fibroblasts are seeded on aligned nanofibers in order to engineer highly anisotropic and tensile scaffolds. Braided nanofibers scaffolds can be used; however, their mass transfer, cell seeding, cell infiltration and mechanical strength properties are poor. As an alternative, knitted microfibers were used in combination with nanofibers to provide good mechanical strength and high surface area for tendon and ligament tissue engineering. Human ligament fibroblasts were seeded on aligned polyurethane electrospun nanofibers scaffolds, resulting in increased collagen synthesis and spindle morphology in fibroblasts; however, the alignment showed no effect on cell proliferation. Another approach is culturing MSCs into electrospun PLGA nanofibers on top of PLGA microfibers, leading to enhanced cell seeding and proliferation and higher levels of tendon/ligament-specific gene expression (Dahlin et al., 2011).

8.8.2.5 Tendon-To-Bone Insertions Tissue Engineering

There are four different zones in a direct tendon-to-bone insertion: tendon, nonmineralized fibrocartilage, mineralized fibrocartilage and bone. The mismatch between tendon and bone at the insertion site is one of the biggest mechanical mismatches in nature; while tendons are soft tissues with a modulus of 200 MPa, bones have a modulus of 20 GPa.

The gradual transition from tendon to bone has two features: (1) gradual organization in the orientation of collagen fibers; and (2) linear gradient in mineral content. Considerable efforts have been made to fabricate a nanofibers scaffold that mimics these features; however, this was achieved only recently. An electrospun PCL nanofibers scaffold was fabricated with gradations in fiber organization and controlled biomineralization, which was achieved by using polydopamine as a mediator. The study demonstrated a nanofiber design with dual gradation in both fiber orientation and mineral content as well as a performance of implanting such a scaffold to the rotator cuff injury site. The nanofibers scaffold was sutured in a way that makes it connect between the bone and the tendon, provide mechanical stability, regulate cellular activity and improve the healing process.

A modern approach to healing tendon-to-bone insertion injury is to combine stem cells with the nanofibers scaffold, as stem cells can differentiate into tendon fibroblasts at tendon site, osteoblasts at bone site and fibrochondrocytes in between (Ma et al., 2013).

8.8.2.6 Meniscus Tissue Engineering

Menisci are fibrocartilaginous, load-bearing tissues, present in the knees. They have a critical role in the functioning of knees, and any damage to them results in high contact stresses and may lead to the development of osteoarthritis. The collagen fibers in meniscus tissue are circumferentially aligned, making the meniscus tissue exceptionally complex and hard to engineer or regenerate. Initial studies in the field of menisci tissue engineering have confirmed that aligned PCL nanofibers scaffold with MSCs seeded into them have a great potential for this application. The same study was successful in fabricating circumferentially aligned nanofibers, which can mimic the collagen fibers in menisci, by developing a novel electrospinning method that uses a rotated plate as a collector. However, no in-vivo studies have been conducted so far. Thus, future studies should be devoted to designing nanofibers scaffolds capable of mimicking the structure and the function of meniscus tissue (Ma et al., 2013).

8.8.2.7 Intervertebral Discs Tissue Engineering

The intervertebral discs (IVDs) hold the vertebrae of the spine together, absorb shock and allow the spine to rotate, bend and twist. IVDs are composed of three tissue components: the annulus fibrosus (AF) surrounding the nucleus pulposus (NP), both of which are sandwiched between cartilage end plates and vertebral bodies. The AF is a fibrous structure, consisting of approximately 15 to 25 concentric collagen type 1 and 2 sheets called the lamellae that contain the pressurized NP. The lamellae provide structural support for proteoglycan synthesis and maintain the tensile strength of IVDs. It is composed of a number of layers, each of which displays different collagen fibrils orientation. The fibrils are oriented concentrically in the outer AF but as it moves towards the NP, the angle of orientation changes from 62° to 45°. Inability to mimic this organization is a major obstacle in intervertebral discs tissue engineering. In order to overcome this obstacle, nanofibers scaffolds were fabricated, stacked using a layer-by-layer strategy and seeded with MSCs. Firstly, aligned nanofibers scaffolds, seeded with MSCs were used to form lamellar tissues then shape them into bilayers. Bilayers were then oriented into parallel or opposing alignments. Electrospun nanofibers mats, 250 μm thick, were used to match the natural lamellar thickness of the AF. The scaffold was able to successfully replicate the structure of AF after 10 weeks of in-vitro culture. Other studies on this application were able to engineer IVDs that demonstrated a resemblance of gross, histological, biosynthetic, architectural and biochemical properties of the native IVD, by incorporating a novel biomaterial amalgam into nanofibers scaffold with seeded MSCs and a center made of hyaluronic acid hydrogel (Ma et al., 2013).

8.8.2.8 Dental Tissue Engineering

Although, currently, limited studies, about the application of nanofibers scaffolds in dental tissue engineering, exist, nanofibers hold a great potential in supporting the formation of dental composite tissues enamel, dentin, and periodontium, as well as the mandible. One recent study suggests that the utilization of nanofibers self-assembling peptide amphiphile could support enamel formation and initial tooth development. Dentin, the mineralized layer below enamel, is formed from dental pulp stem cells (DPSCs) because they can differentiate into odontoblasts. Both phase-separated nanofibers PLLA scaffolds and electrospun nanofibers polymer scaffolds have been proved to promote the attachment and proliferation of DPSCs; however, the differentiation of DPSCs to odontoblasts was enhanced on the phase-separated scaffold when compared to the solid-walled scaffold. Odontogenic differentiation, similar to the osteogenic differentiation, is enhanced when nano-hydroxyapatite is incorporated into electrospun nanofibers. For defects in periodontal ligament, cementum, and alveolar bone, electrospun nanofibers scaffolds were fabricated as a potential repair method. Periodontal ligament cells have displayed good attachment, proliferation and differentiation on electrospun gelatin and electrospun PLGA scaffolds. In another study, collagen type 1 and type 3, which are main periodontal ligament ECM components, were deposited when human periodontal ligament cells were cultured on self-assembled peptide nanofibers scaffolds. Due to the unique features, complex shapes and special geometry of craniofacial bones of each individual, it is necessary to design patient-specific solutions. Fabricating a scaffold that is unique to each individual is the basic principle in regenerating the mandible shape. After creating a wax mold of the patient's mandible, it is, then, used to form phase-separated nanofibers scaffolds with a macroporous structure and patient specific geometry (Gupte and Ma, 2011).

8.8.2.9 Neural Tissue Engineering

The nervous system, composed of central nervous system and peripheral nervous system, is responsible for transmitting signals between different parts of the body and coordinating motor and sensory functions. Any damage done to nervous tissue is irreversible and cannot be repaired, hence, regeneration of this tissue is a huge challenge. Many therapeutic approaches have been attempted in order to repair damages nerves, however, the most promising approach is the adaption of neural tissue engineering strategy, which employs biological tools like normal or genetically engineered cells and ECM-like biomaterials for scaffold design. Nanofibers are ideal for this application because they exhibit a number of properties that greatly enhance cell proliferation and differentiation, deliver neurotrophic factors to the site of injury and direct the growth of the neural stem cells (Dahlin et al., 2011).

Aligned and random PLLA electrospun nanofibers scaffolds were tested for the purpose of neural tissue engineering. The results indicated that, in the aligned nanofibers scaffold, the neural stem cells (NSCs) displayed high rates of adhesion and differentiation, and their neurites outgrew in a direction parallel to that of the fiber alignment (Fang et al., 2008). Furthermore, increased rates of proliferation and differentiation were observed with the decrease in nanofiber diameter. Conductivity of the nanofibers scaffold is a critical property when it comes to neural issue engineering. Conductive electrospun polyaniline/PCL/gelatin nanofibers scaffolds with an average diameter of 112 to 189 nm were fabricated and studied. The results demonstrated that cell proliferation and neurite outgrowth was enhanced upon electrical stimulation. The incorporation of growth factor into the nanofibers scaffold also increased proliferation and differentiation of the NSCs into neurons. Furthermore, immobilizing the epidermal growth factor (EGF) and basic fibroblast growth factor (bFGF) was proved to greatly promote the axon growth (Dahlin et al., 2011).

8.8.2.10 Skin Tissue Engineering

The skin functions as a barrier between the body and the external surroundings. Besides the subcutaneous tissue, the skin consists of two layers: (1) Epidermis, which is divided into five layers starting with Stratum corneum, the outermost layer, made of dead keratinocytes and ending with Stratum basale, the innermost layer, made of proliferating keratinocytes, melanocytes, and Merkel cells. (2) Dermis, which is next to the subcutaneous tissue, is composed of collagen, elastin, glycosaminoglycans, and fibroblasts, as well as sweat glands, leucocytes, adipocytes and mast cells (Beck et al., 2011). Although skin has the ability to regenerate, low generation rates occur in cases of severe skin damage as in burns, infections and inflammation. Normally, skin wounds do not require full skin regeneration to heal but only require the formation of epithelialized scar tissue. While the dermis has a great capacity to regenerate, the epidermis has a low capacity to heal, therefore, when large areas of the epidermis are damaged, the tissues need to be replaced (Vasita and Katti, 2006). Therefore, epidermal nanofibers scaffolds have to enhance the growth of keratinocytes, while dermal nanofibers scaffolds should promote the attachment, diffusion, and growth of fibroblasts. Collagen/PCL electrospun nanofibers were fabricated for the purpose of regenerating skin issues. The collagen had an effect of promoting cell proliferation and enhancing cell migration inside the scaffold (Fang et al., 2008). In another study, electrospun nanofibers were made of poly(vinyl acetate) (PVA) and poly(3-hydroxy butyrate) (PHB) for the same purpose and gave positive results. When tested with human keratinocyte cell line (HaCaT) and fibroblasts, the PHB promoted cell adhesion and proliferation of both, while PVA promoted only HaCaT growth and inhibited fibroblasts. This bio-selectivity can be altered by changing the ratio of PVA and PHB in the polymer blends.

Cell attachment tests were conducted on bovine serum albumin (BSA) coated electrospun nanofibers made of chitin, which is similar to glycosaminoglycan (GAG), and poly(glycolic acid) (PGA) blends, using normal human epidermal cells. Among a variety of blends made, the best cell adhesion was observed in the blend with 25% PGA with BSA coating (Sharma et al., 2015).

Increase in the wettability and hydrophilicity of a nanofiber, causes improvement of cell attachment and proliferation. To fabricate a highly hydrophilic nanofibers scaffold, a small fraction of low molecular weight poly(ethylene glycol) (PEG), blended with poly(L-lactic acid) (PLLA), was incorporated into the nanofibers. The presence of hydrophilic nanofibers scaffolds, such as chitosan/poly(vinyl alcohol) (PVA) scaffolds, promoted the absorption of nutrients during cell culture, thereby enhancing cell attachment, infiltration, proliferation, and migration in the nanofibers scaffold matrix. In addition, the alignment of nanofibers has an impact on cell adhesion and proliferation, for example, aligned collagen nanofibers scaffolds demonstrated lower cell adhesion but higher cell proliferation in comparison to the nonaligned nanofibers scaffolds of the same polymer, however, better cell infiltration was observed in aligned PLLA nanofibers scaffolds (Fang et al., 2008).

Due to their low toxicity, a number of natural polymers, other than collagen, were explored. Silk nanofibers scaffolds have shown a potential to be used in skin tissue engineering. The nonwoven silk electrospun nanofibers, coated with type 1 collagen, have demonstrated great keratinocyte/fibroblast adhesion and spreading, due to their high porosity and high surface area to volume ratio (Vasita and Katti, 2006). The large pores, present in the 3D structure of silk nanofibers, provide good excess for cells to infiltrate into the nanofibers scaffold. Chitosan (CECS), another natural polymer known for its antibacterial properties, was used in combination with PVA to fabricate electrospun nanofibers scaffolds, which demonstrated uniform tissue formation and no cytotoxicity (Sharma et al., 2015).

Nanofibers have proven to be a promising platform upon which biomedical applications, like tissue engineering, occur; however, more studies are required in order to, actually, perform an ideal tissue engineering process.

8.8.3 NANOCOSMETICS

Nanofibers are promising tools for, not only delivery of drugs, therapeutics, and molecular medicines, but also for the delivery of body-care supplements. Generally, nanofibers, in cosmetic application, are used for applying skin wellness agents via a technique similar to the drug delivery. For example, electrospun nanofibers were fabricated in combination with a hydrophilic polymer, a skin wellness agent and vitamins like vitamin A, C and E. Some polymers that are used for this application are polyether, poly(alkylene oxide) and poly(vinyl alcohol). Skin wellness agents are substances that serve a purpose of making the skin healthier, and they include glycerin, hyaluronic acid, collagen, ammo acid derivatives, lactic acid, ceramide and lipid mix, dimethicone, polyphenols, natural bisabolol-herbalia green tea, bentonite clay, peptides, hydrolyzed wheat gluten, and ceratonia siliqua gum mix.

After combining vitamins, skin wellness agents, with a hydrophilic polymer, the system was electrospun into nonwoven nanofibers scaffolds for delivery of beneficial components into the skin. Nanofibers scaffolds display a number of advantages, such as prolonged release of ingredients into the skin, increased contact time and increased surface area, which in turn increased their potential in cosmetics applications. The most cosmetic application of nanofibers is for skin such as healing, cleansing, and care mask. Nanofibers can be incorporated into masks, which then can be impregnated with skin-revitalizing factors for skin health and renewal. Cellulose acetate electrospun nanofibers containing retinoic acid and vitamin E were fabricated for the purpose of delivering vitamins into the skin. When compared with conventional delivery methods of vitamins, using nanofibers resulted in the release of vitamin E over 24 h and retinoic acid over 6 h, while conventional film formulations exerted burst release of these ingredients. The matrix system of such nanofibers formulations is suitable for cell proliferation and growth because of its hyperbranched polymeric structure. This application does not only cover vitamins delivery but also includes colorants, sunscreens, emollients, exfoliants, humectants, nutrients and antioxidants. Furthermore, active ingredients, derived from compounds like aloe Vera, zinc oxide, titanium oxide, alpha, beta or poly hydroxy acid, vitamins, retinol, retinal, retinoic acid and tocopherol, can be used in the nanofibers matrix for treatment of skin conditions such as acne, eczema, contact dermatitis, rosacea and psoriasis.

A promising approach of nanocosmetics is the formulation of antiwrinkle nanofibers masks. The fabrication of nanofibers for the production of these masks is-based on combining ascorbic acid, retinoic acid, and collagen with gold nanoparticles. Using nanofibers increases the contact area between the skin and the mask, due to the high surface area to volume ratio of nanofibers, thereby enhancing the effect of the mask. This novel product is manufactured to overcome the problems of conventional preparations that contain ascorbic acid. These problems occur because the preparation is premoisturized, and taking in consideration the instability of ascorbic acid, this could lead to problems for the long shelf life of the formulation. 10% of polyvinyl alcohol and 20% of randomly methylated β-cyclodextrin were used to fabricate nanofibers containing collagen, ascorbic acid, retinoic acid and gold nanoparticles. Loading these nanofibers with vitamins was proved to enhance skin permeation of ascorbic acid and retinoic acid. Moreover, incorporating β-cyclodextrin into the formulation slowed down the degradation rate of the formulation.

Generally, many cosmetic aimed ingredients can be incorporated into a nanofibers matrix. In a variety of novel formulations, not only cosmetic ingredients, but also antibacterial agents, can be incorporated into the nanofibers matrix in order to enhance their clinical outcome (Ulubayram et al., 2015).

8.9 CONCLUSION

Nanofibers have been undergoing research all over the globe for various potential applications in field of medicine and health care, they have wide variety of applications which make them highly suitable for efficient drug delivery. Currently they are being investigated for various uses like, in field of biomedical applications, cancer, periodontal regeneration, delivery of poorly water soluble drugs, site specific and controlled release of drugs. In future they will be highly potential material for improving public health word wide.

KEYWORDS

- **classification**
- **drug delivery**
- **electrospinning**
- **nanofibers**
- **properties**
- **tissue engineering**

REFERENCES

Alagarasi, A.; Introduction to Nanomaterials. In: Viswanathan, B. (ed.), *Nanomaterials*. Narosa Publishing House, **2009**.

Alubaidy, A.; Venkatakrishnan, K.; Tan, B. Nanofibers Reinforced Polymer Composite Microstructures. Advances in Nanofibers. Maguire, R. (Ed). (pp. 165–184). DOI: 10.5772/57101.2013.

Amna, T.; Barakat, N.; Hassan, M.; Khil, M.; Kim, H. Camptothecin loaded poly(ε-caprolactone) nanofibers via one-step electrospinning and their cytotoxicity impact. *Colloids Surf. A Physicochem. Eng. Asp.* **2013**, *431*, 1–8.

Beck, R.; Guterres, S.; Pohlmann, A. *Nanocosmetics and Nanomedicines: New Approaches for Skin Care*. Verlag Berlin Heidelberg, Germany: Springer, **2011**.

Blanco-Padilla, A.; Soto, K.; Hernández Iturriaga, M.; Mendoza, S. Food Antimicrobials Nanocarriers. *Scientific World J.* **2014**, 1–11.

Boisseau, P.; Loubaton, B. Nanomedicine, Nanotechnology in Medicine. *C.R. Acad. Sci.* **2011**, 12, 620–636.

Cao, G. *Nanostructures & Nanomaterials*. London, UK: Imperial College Press, **2004**.

Chen Zhang; Sizhu Wu; Xuejia Ding. The microstructure characterization and the mechanical properties of electrospun polyacrylonitrile-based nanofibers. Lin, T. (ed.) *Nanotechnology and Nanomaterials*. InTech. **2011**.

Chen, H.; Truckenmüller, R.; Van Blitterswijk, C.; Moroni, L. Fabrication of nanofibrous scaffolds for tissue engineering applications. *Nanomaterials in Tissue Engineering*. **2013**, *6*, 158–183

Dahlin, R.; Kasper, F.; Mikos, A. Polymeric Nanofibers In Tissue Engineering. *Tissue Eng. Part B: Reviews* **2011**, *17*, 349–364.

Fang, J.; Niu, H.; Lin, T.; Wang, X. Applications of electrospun nanofibers. *Chin. Sci. Bull.* **2008**, *53*, 2265–2286.

Fang, J.; Wang, X.; and Lin, T. Functional applications of electrospun nanofibers. Lin, T. (ed.). *Nanotechnology and Nanomaterials*. InTech. **2011**, 287–326.

Flory, F.; Escoubas, L.; Berginc, G. Optical Properties of Nanostructured Materials: A Review. *J. Nanophoton.* **2011**, *5*, 052502.

Goonoo, N.; Bhaw-Luximon, A.; Jhurry, D. Drug Loading and Release From Electrospun Biodegradable Nanofibers. *J. Biomed. Nanotechnol.* **2014**, *10*, 2173–2199.

Gupta, K.; Haider, A.; Choi, Y.; Kang, I. Nanofibrous scaffolds in biomedical applications. *Biomater. Res.* **2014**, *18*, 5.

Gupte, M.; Ma, P. Nanofibrous scaffolds for dental and craniofacial applications. *J. Dent. Res.* **2011**, *91*, 227–234.

Hartgerink, J. D. Self-assembly and mineralization of peptide-amphiphile nanofibers. *Science.* **2001**, *294*, 1684–1688.

Heunis, T. D. J.; Dicks, L. M. T. Nanofibers Offer Alternative Ways to the Treatment of Skin Infections. *BioMed Res. Int.* **2010**, 1–10.

Hu, J.; Ma, P. Nano-fibrous tissue engineering scaffolds capable of growth factor delivery. *Pharm. Res.* **2011**, *28*, 1273–1281.

Hu, X.; Liu, S.; Zhou, G.; Huang, Y.; Xie, Z.; Jing, X. Electrospinning of Polymeric Nanofibers For Drug Delivery Applications. *J. Control. Release.* **2014**, *185*, 12–21.

Ikhmayies, S. Characterization of Nanomaterials. *JOM.* **2013**, *66*, 28–29.

Jeon, O.; Powell, C.; Ahmed, S.; Alsberg, E. Biodegradable, photocross-linked alginate hydrogels with independently tailorable physical properties and cell adhesivity. *Tissue Eng. Part A.* **2010**, *16*, 2915–2925.

Kanani, G.; Bahrami H. Review on electrospun nanofibers scaffold and biomedical applications. *Trends. Biomater. Artif. Organs.* **2010**, *24*, 93–115.

Khajavi, R.; Abbasipour, M. Electrospinning as a versatile method for fabricating coreshell, hollow and porous nanofibers. *Scientia Iranica*, **2012**, *19*, 2029–2034.

Khan, K., Rehman, S.; Rahman, H. U.; Khan, Q. Synthesis and application of magnetic nanoparticles.

Kochetov, R. *Thermal and Electrical Properties of Nanocomposites, Including Material Processing.* (Thesis) Lappeenranta University of Technology, Finland. Wöhrmann Print Service, Zutphen, the Netherlands, **2012**.

Lee, S.; Jin, G.; Jang, J. Electrospun nanofibers as versatile interfaces for efficient gene delivery. *J. Biol. Eng.* **2014**, *8*, 30.

Lin, P.; Lin, S.; Wang, P.; Sridhar, R. Techniques for physicochemical characterization of nanomaterials. *Biotechnol. Adv.* **2014**, *32*, 711–726.

Liu, W.; Thomopoulos, S.; Xia, Y. Electrospun Nanofibers For Regenerative Medicine. *Adv. Healthcare Mater.* **2011**, *1*, 10–25.

Ma, B.; Xie, J.; Jiang, J.; Shuler, F.; Bartlett, D. Rational design of nanofiber scaffolds for orthopedic tissue repair and regeneration. *Nanomedicine.* **2013**, *8*, 1459–1481.

Nagasaka, A.; Miyawaki, T.; Machida, T.; Oshida, K.; Kawamura, W.; Yanagisawa, K.; Momose, N. The Effect Of Vapor-Grown Carbon Fiber (VGCF) Content On The Mechanical Properties And Abrasion Resistance Of VGCF/PTFE Composites. *Carbon* **2014**, *66*, 742.

Nikalje, A. P.; Nanotechnology and its Applications in Medicine. *J. Med. Chem.* **2015**, *5*, 81–89.

Picciani, P.; Medeiros, E.; Mattoso, L.; Orts, W. Advances in Electroactive Electrospun Nanofibers. In: Lin, T. (Ed.). *Nanofibers – Production, Properties and Functional Applications.* INTECH Open Access Publisher. DOI: 10.5772/23229. **2011**.

Pisignano, D. *Polymer Nanofibers: Building Blocks for Nanotechnology;* Royal society of Chemistry Publications. Cambridge, **2013**.

Ramakrishna, S.; Fujihara, K.; Teo, W.; Yong, T.; Ma, Z.; Ramaseshan, R. Electrospun nanofibers: solving global issues. *Mater. Today*, **2006**, *9*, 40–50.

Scalf, J.; West, P. Introduction to nanoparticle characterization with AFM. *Pacific Nanotechnology*, **2006**, *16*, 1–8.

Sharma, J.; Lizu, M.; Stewart, M.; Zygula, K.; Lu, Y.; Chauhan, R.; Yan, X.; Guo, Z.; Wujcik, E.; Wei, S. Multifunctional Nanofibers Towards Active Biomedical Therapeutics. *Polym. J.* **2015**, *7*, 186–219

Tolochko, N.; History of nanotechnology. In: Valeri, N.; Chunli, B.; Sae-Chul, K. (eds) Nanoscience and nanotechnologies. *Encyclopaedia of Life Support Systems (EOLSS)*. Developed under the auspices of the UNISCO, Eolss Publishers, Oxford, **2009**.

Ulubayram, K.; Calamak, S.; Shahbazi, R.; Eroglu, I. Nanofibers-based antibacterial drug design, delivery and applications. *Curr. Pharm. Des.* **2015**, *21*, 1930–1943.

Unrau, C. J.; Axelbaum, R. L.; Biswas, P.; Fraundorf, P. Online size characterization of nanofibers and nanotubes. In: Mansoori, G. A.; George, T. F.; Assoufid, L.; Zhang, G. (Eds.). *Molecular Building Blocks for Nanotechnology: From Diamondoids to Nanoscale Materials and Applications*. New York: Springer; **2007** (pp. 212–245).

Vasita, R.; Katti, S. Nanofibers and their applications in tissue engineering. *Int. J. Nanomedicine*, **2006**, *1*, 15–30.

Wang, Y.; Zheng, M.; Lu, H.; Feng, S.; Ji, G.; Cao, J. Template Synthesis Of Carbon Nanofibers Containing Linear Mesocage Arrays. *Nanoscale Res Lett.* **2010**, *5*, 913–916

Wierach, P. Nano-Micro-Macro. In: Wiedemann, M.; Sinapius, M. (eds.). *Adaptive, Tolerant and Efficient Composite Structures* (pp. 17–27). **2012**. Germany, Berlin: Springer-Verlag.

Yao, J.; Bastiaansen, C.; Peijs, T. High Strength and High Modulus Electrospun Nanofibers. *Fibers* **2014**, *2*, 158–186.

Yoo, H.; Kim, T.; Park, T. Surface-functionalized electrospun nanofibers for tissue engineering and drug delivery. *Adv. Drug Deliv. Rev.* **2009**, *61*, 1033–1042.

Zamani, M.; Prabhakaran, M. P.; Ramakrishna, S. Advances in drug delivery via electrospun and electrosprayed nanomaterials. *Int. J. Nanomedicine.* **2013**, *8*, 2997–3017.

Zhang, S. Mechanical and physical properties of electrospun nanofibers. A thesis submitted to the Graduate Faculty of North Carolina State University in partial fulfillment of the requirements for the degree of Master of Science, Textile Chemistry, **2009**.

CHAPTER 9

DRUG AND FOOD APPLICATIONS OF LIPOSOMES AND NANOPARTICLES: FROM BENCHMARK TO BEDSIDE?

MARCUS VINÍCIUS DIAS-SOUZA[1,3,*] and
RENAN MARTINS DOS SANTOS[2,3]

[1]*Biological Sciences Institute, Federal University of Minas Gerais, Belo Horizonte, Brazil, *E-mail: souzamv@ufmg.br*

[2]*Faculty of Pharmacy, Federal University of Minas Gerais, Belo Horizonte, Brazil*

[3]*Integrated Pharmacology and Drug Interactions Research Group (GPqFAR), Brazil*

CONTENTS

ABSTRACT

Liposomes and nanoparticles are currently among the most investigated technologies for drug delivery, diagnostic and treatment of diseases, and for food preservation, processing and optimization of nutritional properties. These nanodevices provide controlled release, improvement of kinetic and diffusion properties, and also vectorized delivery of loaded compounds. In general, in order to guarantee the safety for human or veterinarian use, liposomes and nanoparticles must not induce immunogenicity or toxicity. The aim of this chapter is to review the recent advances in the fields of liposomes and nanoparticles regarding their use in improving drugs and food features. Moreover, the directions that these nanotechnologies are taking in pharmaceutical and food sciences and markets are discussed.

9.1 INTRODUCTION

Controlled and vectorized release of different compounds is among the main features that stimulates research on the use of liposomes and nanoparticles in health sciences. Beyond the sustainable delivery of substances, liposome entrapment and nanoparticle loading of different compounds present several advantages when compared to the same nonentrapped substances, which include: protection against reactive species, resistance to extreme pH and temperature, protection against high ion concentrations, improvement of stability, pharmacokinetics properties and of bioavailability (Gregoriadis, 2007; Koçer, 2010; Holzschuh et al., 2015).

Liposomes can be defined as thermodynamically stable lipid-based vesicles formed by self-assembly of phospholipids in aqueous systems, resulting in entrapment of a determined compound. These systems were designed for studying the behavior of biological membranes. Their biotechnological advantages due to controlled release and entrapment protection were firstly explored only in the 1970s by Gregoriadis. Since then, they have been widely studied in different applications, being pharmaceutical, cosmetics and food sciences the most advanced fields regarding their exploration. Hydrophilic and hydrophobic molecules can be localized in the aqueous core and in the membranes, respectively. Ideal liposomes are biocompatible with biological membranes and should lack immunogenicity and toxicity (Gregoriadis, 2007; Holzschuh et al., 2015; Koçer, 2010; Wang et al., 2015).

Nanoparticles, on the other hand, can be made of varied inorganic and/or organic materials, such as polyesters, metals and proteins. Nanoparticles present enhanced characteristics like optical, magnetic and electric behaviors (in specific conditions) when contrasted to the same materials in greater dimensions. These modified behaviors can be relevant for optimizing several drug and food features, beyond controlled release of substances to their targets. Nanoparticles should also lack immunogenicity, toxicity and present adequate biocompatibility as well as liposomes (Markoutsa et al., 2011; Liu et al., 2014).

The encapsulation of natural or synthetic compounds is of great potential to the treatment of several diseases, and to improve food preservation and nutritional properties. These effects encourage research on the development of new liposome and nanoparticle-based products, although the basic compounds for their preparation are still very expensive. Nevertheless, it is expected that the growing number of manufacturers may provide less expensive high quality raw material for liposomes and nanoparticles research in a near future (Holzschuh et al., 2015; Markoutsa et al., 2011).

Synthetic drugs have been successfully encapsulated by several research groups, and a strong improvement in drug effects is generally experimentally observed, due to the increased surface contact amidst the drug and its target. A number of nutrients and phytomolecules can be associated with both drugs and other food components when freely available on the intestinal tract, causing unwanted effects like aggregation or precipitation (what can trigger or even abrogate immune responses), and quantity and functional loss of these molecules (Holzschuh et al., 2015). Different nutrients have been investigated aiming their use as nutraceuticals or as dietary supplements, but many of them present poor water solubility, low oral bioavailability, minimum residence time in gastrointestinal tract and poor physicochemical stability (Dos Santos et al., 2015). Taken together, these characteristics can impair the use of these nutrients in food technology. Entrapment in liposomes or nanoparticles loading can provide solutions to overcome these issues (Gregoriadis, 2007).

By reviewing the recent advances in this field, this chapter aims to present to the audience relevant data on how liposomes and nanoparticles have been explored regarding applications for improving drugs and food features of interest, such as administration/intake, kinetics parameters, general preservation, bioavailability, and interactions with cells or specific physical or chemical interactions of interest, for the development of products that might reach the market in a near future.

9.2 LIPOSOME AND NANOPARTICLES IN DRUG DELIVERY

9.2.1 OVERVIEW

The engineering of drug delivery systems aiming to provide controlled release of therapeutic molecules in a given space-time frame is among the most striking challenges of nanotechnology in health sciences. Controlled release systems can be defined as devices that are able to reach therapeutically effective concentrations of drugs or any other molecule of interest in a specific target over a determined period of time. Some examples include therapeutic contact lenses, drug-covered orthopedic implants and transdermic patches. And why are these devices better than the traditional dosage forms? Different problems can be identified here. Because of drugs exposure to biological events that may lead to their inactivation, therapeutic regimens are often complex, require multiple administrations and long time duration. This condition generally decreases patients' compliance to therapy. Systemic accumulation, a common event drug use, may lead to toxicity. *In situ* release of therapeutic molecules by these devices can increase the efficiency of drugs and decrease the risks of possible systemic side effects (Hatakeyama et al., 2007).

We direct now our attention to nanodevices for drug delivery. Liposomes and nanoparticles are the most explored technologies in this context. Liposomes are the most investigated and commercialized drug delivery nanosystem nowadays. Examples of liposome-entrapped drugs currently available on the market include Ambisome® (amphotericin B), Doxil® (doxorubicin hydrochloride) and Epaxal® (inactivated hepatitis A virus vaccine). Drug entrapment in liposomes can improve pharmacokinetics features; and, as stated, the localized delivery of therapeutic molecules can preclude toxic effects (Wang et al., 2015). Until the time of preparation of this chapter, no formulation of nanoparticles-loaded drugs is available at the pharmaceutical market, although several researches, as the ones described in the next paragraphs, have been published.

Nanotechnologies in drug delivery are not free of limitations. The main problems in this sense include:

- Risk of idiosyncratic behavior in body fluids flow;
- Sensitivity to pH, electrolytes and adsorption by plasma proteins;
- Structural instabilities such as erosions;
- Problems on drug release (diffusion) speed, such as unwanted burst release;
- Uptake by nontargeted cells, decreasing the concentration of the active molecule and increasing the risk of toxic effects;
- Complex steps to scale up the manufacturing process.

To overcome part of these problems, some strategies have been explored to improve the use of liposomes and nanoparticles for drug delivery. To increase a desired bioaccumulation on the targeted tissue, antibodies, generally of the IgG class, can be attached by covalent binding to polymeric molecules (nanoparticles) or to phospholipids (liposomes). This provides vectorization of drugs and does not interfere in stability or physicochemical properties of liposomes and nanoparticles (Markoutsa et al., 2011). In addition, to provide long circulation and increased bioavailability, polymers like polyethylene glycol can be attached to the phospholipids, increasing their flexibility. PEG forms a protective layer over the liposomes and delays (or also avoid) they recognition by opsonins, and the subsequent clearance (Hatakeyama et al., 2007; Wang et al., 2015).

Also in this context, Osawa and Erickson (2013) developed a system in which Z rings were incorporated inside large unilamellar liposomes, and structures such as narrow gaps between invaginating edges of the vesicles were formed. The authors demonstrated that the incorporated Z rings could divide the liposomes by forming a clear septum between the daughter vesicles. This technology can be useful as a strategy to maintain stable kinetics parameters such as distribution and bioavailability. Therefore, it is expected that in a near future, the number of liposome-entrapped drugs available at the pharmaceutical market will increase, allowing the planning of new and less complicated therapeutic regimens, which may improve patients' compliance to drug therapy.

In the following sections, we show in brief some of the most striking advancements recently published in the field of pharmaceutical nanotechnologies for drug delivery in the treatment or diagnostic of different diseases. By discussing these advancements, we wish not only to provide insights regarding nanotechnologies trends for drug delivery, but to arouse questions concerning points that these technologies might be lacking to reach the pharmaceutical market in spite of everything that has been investigated so far.

9.2.2 LIPOSOMES AND NANOPARTICLES IN CANCER TREATMENT

Among the main benefits of exploring liposomes and nanoparticles in cancer treatments is their ability to provide vectorized delivery of anticancer drugs, which are often highly citotoxic and may trigger several adverse effects. In addition to this, the controlled release of drugs can contribute to overcome an eventual resistance developed by the tumoral cells to the drugs, as in the work of Gharib et al. (2014). The authors developed artemisinin and transferrin-loaded magnetic nanoliposomes for treating breast cancer. The thermosensitive formulations (drug release at 42°C) presented high antiproliferative activity against cell lines MCF-7 and MDA-MB-231, when exposed to magnetic fields.

Also aiming breast cancer treatment, Kong et al. (2015) investigated the pharmacological potential of an advanced platform composed of porous silicon nanoparticles and giant liposomes encapsulating doxorubicin. This system was assembled on a microfluidic chip against human doxorubicin-resistant MCF-7 cancer cells, and also M28 cells. The system was also loaded with DNA nanostructures, gold nanorods, and magnetic nanoparticles, and presented photothermal and magnetic responsiveness. It was able to release all its entrapped components, being effective against breast cancer cell lines.

Boks et al. (2015) proposed the use of antigen-carrying liposomes as a noninvasive skin vaccine. The authors explored a human skin model to assess the efficiency of intradermal delivery of liposomes against melanoma. Liposomes were prepared with gp100280–288 peptide, a melanoma-associated antigen, and monophosphoryl lipid A (MPLA), a Toll-like receptor 4 ligand that was used as an adjuvant. The uptake of liposomes by dendritic cells was superior to the uptake of conventional liposomes, and antigen presentation to $CD8^+$ T cells was significantly improved with MPLA-liposomes. These modified liposomes induced upregulation of CD83, CD86, TNF-α and IL-6 by dendritic cells, and effective maturation and cytokine production by these cells was observed after the treatment with the modified liposomes.

Xiao et al. (2011) formulated silicon nanoparticles via electrochemical etching from a single porous silicon crystal against HeLa and NIH-3T3 cells. The nanoparticles were toxic for upto 45% of cells. Also exploring synthetic compounds, Floyd et al. (2015) developed aerosolized nanoparticles of rhodamine and poly(lactic-co-glycolic acid) (PLGA) spherical shape polymeric particles for glioma therapy. The authors demonstrated in an *ex vivo* model that rhodamine-PLGA spheres presented a high retention rate in brain tissues, suggesting its potential for glioma therapy.

Cui et al. (2015) developed RNA nanoparticles for gastric cancer therapy using the thermostable three-way junction of bacteriophage phi29 motor pRNA. The nanoparticles consisted in three fragments, which were functionalized with folate as a targeting ligand. The authors observed that the nanoparticles could bind specifically to tumoral gastric cells and induce apoptosis. Also, the breast cancer-associated antigen 1 (BRCAA1) gene was inactivated.

9.2.3 LIPOSOMES AND NANOPARTICLES IN GASTROHEPATIC AND CARDIOMETABOLIC DISEASES

The fast uptake of superparamagnetic iron oxide nanoparticles by Kupffer cells has encouraged their exploration for liver imaging for diagnosis and treatment of carcinoma. Although the biodistribution of these nanoparticles to vascular lesions in the liver cab be impaired because of cirrhosis, their use still provide advantages when compared to conventional methods, like the increased contrast between tumoral cells and healthy tissues (Barthia et al., 2003; Tanimoto et al., 2006).

Nanodevices have also been developed against liver fibrosis. IFN-α and IFN-1β entrapped in liposomes could successfully inhibit liver fibrosis in an animal model. Besides, liposomes conjugated with retinol and loaded with siRNA against the rat protein gp46 (homologue of human heat shock protein 47), decreased collagen secretion and fibrosis (Du et al., 2007; Sato et al., 2008).

Cui et al. (2015) proposed liposomes for insulin delivery, aiming higher oral bioavailability. The authors explored herbal lipids as an alternative to conventional phospholipids in its composition, and simulated an environment for tests by mimicking gastrointestinal fluids. Liposomes formulated with ergosterol presented the best protective potential when compared to other herbal lipids such as β-sitosterol and stigmasterol. Ergosterol-based liposomes decreased blood glucose levels in approximately 50% after oral administration, and low toxicity was observed when liposomes were exposed to $CaCo_2$ cells.

Sadhukhan et al. (2015) developed micellar nanoparticles by synthesizing PEG-g-gellan copolymers to provide controlled release of simvastatin. The system was tested in hyperlipidemic rabbits. The oral administration of the drug in its free form resulted in reductions in total cholesterol, triglycerides and LDL levels by 13, 18 and 19%, respectively, and increased HDL levels in around 12%. On the other hand, the treatment with simvastatin-loaded nanoparticles decreased total cholesterol, triglycerides and LDL levels in 30, 20 and 44%, respectively.

Hu et al. (2015) functionalized poly(lactic-co-glycolic acid) nanoparticles with plasma membrane of human platelets, creating a biodegradable polymeric nanosystem of poor uptake by macrophages and poor complement activation. Antigens associated to immunoregulation and to cellular adhesion were used in the coating strategy, what provided platelet-like properties and protection from the body's immune responses. The authors used these nanoparticles in animal models of coronary restenosis, controlled with docetaxel, and of systemic bacterial infection, treated with vancomycin. These drugs, when bounded to the nanoparticles, showed superior efficacy when compared to non-bounded drugs. More than 90% of the nanoparticles were quickly distributed after intravenous injection, reaching primarily the liver and the spleen.

9.2.4 LIPOSOMES AND NANOPARTICLES IN INFECTIOUS DISEASES

In general, the main advantages of exploring liposomes and nanoparticles for the treatment of infectious diseases include reductions of minimum inhibitory concentration (MIC) and minimal bactericidal concentration

(MBC) (or minimal fungicidal concentrations) of synthetic or natural antimicrobial compounds. The antibiofilm potential of these nanodevices remains poorly described. The concentrations of antimicrobial compounds, due to controlled release properties of nanodevices, tend to be higher in tissues and biological fluids when compared to free compounds. Furthermore, these nanodevices have been described to be a promising strategy to overcome microbial resistance, a critical problem widely observed in microorganisms from different environments (water, soils, hospitals and human or veterinarian clinical isolates).

Solleti et al. (2014) investigated the antimicrobial activity of azithromycin entrapped in liposomes against planktonic cells and biofilms of *Pseudomonas aeruginosa* strains. The MIC and MBC values of liposome-entrapped azithromycin were considerably lower than the observed for free azithromycin. MIC and MBC values of free azythromycin ranged from 64 to 512 μg/mL and from 128 to 1024 μg/mL, respectively. Conversely, MIC values of liposome-entrapped azithromycin ranged from 8 to 128 μg/mL, and MBC values ranged from 16 to 256 μg/mL. Liposomal and free azithromycin were effective to eradicate biofilms at concentrations ranging from 1024 to 512 μg/mL, but liposome-entrapped azithromycin significantly reduced bacterial counts when compared to free azithromycin. Also, the production of virulence factors was decreased after exposure to liposome-entrapped azithromycin.

Gajra et al. (2015) developed polymeric lipid hybrid nanoparticles loaded with itraconazole and explored strategies to optimize the intestinal permeability of the hybrid. The nanoparticles were prepared with biodegradable polycaprolactone, soya lecithin and poly vinyl alcohol. No initial burst release was observed during *in vitro* assays, suggesting its safety. In an *ex vivo* tissue penetration model, the nanoparticles reached 30% of permeability after 4 h of exposure.

Ruiz-Rico et al. (2015) explored the antimicrobial potential of mesoporous silica nanoparticles MCM-41, which were prepared with tetraethylorthosilicate, N-cetyltrimethylammonium bromide and sodium hydroxide, and were loaded with caprylic acid, a food-grade GRAS (generally recognized as safe) ingredient that is also used in cosmetics and pharmaceutical formulations. Both free and MCM 41-loaded caprylic acid were effective against *Escherichia coli, Salmonella enterica, Staphylococcus aureus* and *Listeria monocytogenes*. However, curiously, the antimicrobial activity of free and loaded caprylic acid, assessed as MBC, fell within the same range of 18.5 to 20 mM for these pathogens. The authors proposed the use of MCM-41 in systems that might impair the antimicrobial activity of caprylic acid due to unwanted or unexpected chemical interactions.

Hang et al. (2015) evaluated the antiviral potential of cuprous oxide nanoparticles against hepatitis C virus in cell cultures. The nanoparticles inhibited infection at noncytotoxic concentrations and also inhibited the cellular entry of hepatitis C virus pseudoparticles. However, no effect on viral replication was seen.

Lopes et al. (2014) compared the *in vitro* and *in vivo* antileishmanial activity of liposomes and lipid nanoparticles loaded with oryzalin. The two formulations presented low cytotoxicity and poor hemolytic activity when compared to free oryzalin, and presented similar antileishmanial efficiency of glucantime. In in-vitro assays, the antileishmanial activity of liposomes and lipid nanoparticles loaded with oryzalin was inferior to the free drug (possibly due to problems associated to drug delivery on the culture media, according to the authors). However, in the murine model of visceral leishmaniasis, both liposome and nanoparticle loaded oryzalin were significantly more effective in reducing the parasitic burden when compared to the effects of the nonentrapped drug.

9.2.5 LIPOSOMES AND NANOPARTICLES IN NEUROLOGICAL TARGETS

The impermeability of the blood–brain barrier (BBB) to drugs and biomolecules, combined to the tight organization of endothelial cells in brain blood vessels, compose the most challenging technical problem in the pharmacological management of neurological diseases. The administration of drugs with the objective of reaching high doses in the bloodstream to treat neurological diseases has varied limitations. First, it is generally inefficient due to the poor distribution on brain tissues. Second, there are risks of side effects due to eventual interactions with targets not related with the disease. Third, it can be vastly expensive, because of the extensive administration of dosage forms. Thus, the use of nanotechnologies for in situ administration, or vectorized nanosystems, represents an important alternative to neural drug therapy.

The BBB may undergo varied changes in its integrity and functions when the brain is under neurological disorders. It has been proposed that the impaired BBB may provide strategic drug delivery opportunities into the brain. Thus, the use of nanodevices can optimize the entrance of drugs in this complex environment, due to the high biocompatibility with cell membranes. Efforts to overcome such technical difficulties have been proposed in different studies. Clark and Davis (2015) prepared gold nanoparticles loaded with transferrin, to bind to its receptors (TfRs). An acid-cleavable linkage between transferrin and the nanoparticle core was added to improve its access to the BBB. The authors demonstrated that the nanoparticles could cross BBB *in vitro* and also entered the brain parenchyma of mice in greater amounts *in vivo*, after systemic administration. This transferring-associated system is being proposed for both studying mechanisms of reaching the BBB and to deliver drugs more efficiently.

Moreover, Pedram et al. (2015) proposed a mathematical model of brain mapping technique using superparamagnetic nanoparticles, which was designed to be superior to electroencephalography, in order to define the boundary of epileptic neuronal networks more accurately. The system was developed to create super-paramagnetic aggregates, which could improve tissue contrast in magnetic resonance imaging.

Afergan et al. (2008) used negatively charged liposomes to deliver serotonin to the brain targeting monocytes. Endocytosis of liposomes by monocytes reached 60%, and 4 h after the administration through intravenous injection, liposome-entrapped serotonin concentration in the brain was of 0.138% of the dosage, and free serotonin concentration was 0.068% of the initial dosage.

Chhabra et al. (2015) developed zinc-loaded nanoparticles, which can cross the blood-brain barrier, for brain targeted rapid zinc delivery, in order to treat psychiatric and neurodegenerative disorders like schizophrenia, attention deficit hyperactivity disorder, depression, autism, Parkinson's and Alzheimer's diseases.

9.2.6 LIPOSOMES AND NANOPARTICLES IN THE DEVELOPMENT OF VACCINES

Jiao et al. (2003) developed a hepatitis C virus (HCV) DNA vaccine in cationic liposomes. The authors described that liposome-mediated DNA immunization induced greater immune responses than immunization with free DNA. Liposomal DNA provided strong antigen-specific Th1 type immune responses when composed of dimethyl-dioctadecylammonium bromide (DDAB) and egg yolk phosphatidylcholine (EPC). On the other hand, liposomes composed of DDAB and 1,2-dioleoyl-sn-glycerol-3-ethylphosp hocholine (DOEPC) induced Th2 type antigen-specific immune response.

Kamath et al. (2009) investigated the immunization with liposome-based formulations for *Mycobacterium bovis*. The antigens Ag85B-ESAT-6/CAF01 reduced the postchallenge bacterial growth of *M. bovis* BCG, being mediated by the induction of strong Th1 and Th17 responses in murine models. The protective effects remained adequate for at least 6 months, and this was attributed to a massive activation of dendritic cells.

Kaba et al. (2009) designed a prototypic malaria vaccine-based on a versatile self-assembling polypeptide nanoparticle, which can display antigenic epitopes repetitively, targeting a protein of *Plasmodium berghei*. The vaccine conferred a protective immune response to mice that lasted more than 6 months after the first challenge, and upto 15 months against a second challenge, and did not need a heterologous adjuvant to provide such effects.

Bal et al. (2011) co-encapsulated ovalbumin and a Toll-like receptor ligand (PAM3CSK4 (PAM) or CpG) in cationic liposomes in order to study their effects on the *in vitro* and *in vivo* (murine model) maturation steps of dendritic cells (DC). DC maturation induced by the liposomal formulation was superior when compared to free immunogens. Interestingly, the encapsulation of PAM and CpG did not influence IgG titers, but OVA/CpG liposomes provide stronger shift of IgG1 production to IgG2a, when compared to nonentrapped CpG.

9.3 LIPOSOME AND NANOPARTICLES IN FOOD SCIENCES

9.3.1 OVERVIEW

The application of nanotechnology in food sciences is considerably more recent when compared with other areas such as the production of pharmaceuticals. Within this framework, the use of nanotechnology has been focused on the development of products with functional features, nutraceuticals, development or optimization of processes in the food industry, improvement of biochemical characteristics of food, preservation and improvement of sensory characteristics of formulations and kinetics of nutrients, and in the development of packaging technology, improving shelf life of alimentary products (Assis et al., 2012).

The use of liposomes and nanoparticles in food formulations must meet the requirements such that they can be framed as GRAS (generally recognized as safe) products. This definition was provided by the Food and Drug Administration (FDA), which describes that substances added intentionally to food formulations are considered food additives, and these should be reviewed and approved – except in cases that qualified specialists have adequately demonstrated a safe employment of these in the intended formulations (FDA, 1997). Nanoparticles are obtained from food processing steps such as milling and homogenization. In addition to processing, nanoparticles can be associated with other food ingredients to form micelles or nanofibers, and may aid in the digestive process, as in the transport of lipids, due to micelle formation by fatty acids, monoglycerides, bile salts and phospholipids (Rogers, 2015).

The employ of nanotechnology in food sciences has become increasingly frequent, once nanomaterials characteristics have been improved regarding their safety for consumption. However, nanotechnologies in food sciences are still lacking specific legislation and technical regulations worldwide, in order to provide standardizations and monitoring of this growing market (Blasco and Pycó, 2011; Sozer and Kokini, 2009). Although the safety of nanosystems has been improved, it is important to note that due to increased contact surface, nanoscale particles can easily spread in tissues and increase toxicity risks, modify rheological properties of food and kinetics parameters of nutrients (Sozer and Kokini, 2009).

9.3.2 NANOSYSTEMS AND NUTRACEUTICAL PRODUCTS

Nutraceuticals are defined as dietary formulations-based in concentrated presentations of one or more pure bioactive compounds. They are originally isolated from plant foods, but nowadays, bioactive compounds of mineral and animal sources have been included. These formulations are prepared in different pharmaceutical dosage forms and are used in dosages many times superior to those detectable in their original sources to improve health in varied aspects (Dos Santos et al., 2015). Nutraceuticals can be isolated from its source of origin and have been increasingly incorporated in food matrices, since they possess properties of avoid chronic illnesses when employed in the diet (Dos Santos et al., 2015). However, the incorporation of these in food matrices is still challenging, as in addition to being required to be stable, they must resist processing steps. This is mostly due to the fact that, in some cases, they can only be added to the food matrix in the beginning of the industrial process, given that additions performed on the latest steps of the process can offer varied risks to consumers and are, thus, often forbidden by health regulatory agencies.

Oils, liposoluble vitamins, carotenoids and flavonoids, amidst other substances of nutraceutical interest, can be added to food matrices, what make these ingredients functional additives. For these situations, the use of colloidal systems is necessary, thus, improving characteristics such as stability, compatibility with other ingredients of the formulation, and release in the gastrointestinal tract. A food matrix is defined as the way in which it is organized, and its components or ingredients can be fresh, minimally processed, and/or proper for consumption. The number of advanced food formulations obtained with the use of nanomaterials has increased in the latest years, once properties such as bioavailability and sensorial aspects can be improved, as well as interesting industrial advantages such as prolonged conservation (Sozer and Kokini, 2009).

Such advances represent a set of challenges to the industrial sector, once several assessments must be performed concerning toxicity risks in human consumption of nanosystems, acceptance by consumers and eventual technical modifications of manufacturing processes by the addition of nanomaterials (Assis et al., 2012; Mcclements, 2015). Regarding the development of food with nutraceutical compounds in its formulations, for which the matrices were specifically developed with the aim of improving its bioavailability, nutraceuticals may be present within the food matrix, which can be unprocessed or processed, such as sauces, creams and pasta. Moreover, nutraceuticals can be embedded by simple mixture if they are soluble in the formulation, or associated to liposomes or nanoparticles (McClements, 2015; Nacka et al., 2001). Liposomes, nanoparticles and some nanoemulsions have been developed to protect and/or control the release of nutrients to the gastrointestinal tract. Still, critical research and development barriers in this context include *in situ* instability, tissue permeation and absorption efficiency of these nanosystems during the digestive process, given that bile salts are able to accelerate the rupture of these systems (Liu et al., 2012).

A general classification of substances into four groups regarding their permeability to biological membranes and the solubility in biological fluids is often considered in this context (Williams et al., 2013). Class I include substances of high solubility and high permeability; class II lists the substances with high permeability and low solubility. In turn, class III list substances with low permeability and high solubility, and finally, we can find in class IV substances with low permeability and low solubility. Once this classification is known, some inferences can be made. For instance, if a lipophilic substance is employed in a food matrix, and if this is considered a class II substance, it shall probably present a high permeability in intestinal tissues. However, it shall probably also present low solubility in water. Alternatively, the delivery in liposomes or nanoparticles can be explored in order to increase the solubility of these compounds in intestinal fluids, and consequently, their bioavailability (McClements, 2015; Williams et al., 2013).

The bioavailability of nutraceuticals can still be increased if they are consumed in combination with certain food. A good example would be piperine, naturally present in black pepper (*Piper nigrum*), which is able to increase the permeability of cell membranes, and also the intake of lipids concomitantly with vegetables of colors ranging from orange to red or yellow, increasing the bioavailability of carotenoids. Aiming to reach this increased membrane permeability, certain surface-active agents (of proven safety for human consumption) can be added as components of food matrices, which behave providing characteristics to the whole formulation that liposomes or nanoparticles could provide to a single compound (McClements, 2015).

9.3.3 USE OF NANOSYSTEMS TO INCREASE THE BIOAVAILABILITY OF NUTRIENTS

The entrapped compounds and surface properties of these nanosystems are the most influent issues regarding the digestion of nanosystems. A strategy to stabilize these nanosystems, protect them against deterioration, and reduce the release of contents during digestion, is their functionalization with different molecules such as polysaccharides, synthetic polymers and electrolytes; moreover, pH-sensitive strategies might be explored (Gregoriadis, 2007).

Liposomes formulated with negatively charged lipids tend to increase their diameter at first on the digestion in intestinal fluid, followed by a gradual decrease. Pancreatin and enzymes such as pancreatic lipase, cholesterol esterase and phospholipase A_2, are able to hydrolyze the liposomes and nanoparticles components, what can destabilize the nanosystems and make them loose their (Whitcomb and Lowe, 2007; Gallier et al., 2013).

Depending on the performance of various enzymes on the lipids that compose liposomes, modifications of vesicles structure, resulting in aggregation and fusion of particles, can be observed experimentally. Some researches provided evidence in this context by studying the process of liposome digestion, in which particles with larger diameter after certain time of digestion were found (Gallier et al., 2013; Kibat et al., 1986; Whitcomb and Lowe, 2007).

Liu et al. (2015) investigated the protection of bovine serum albumin in liposomes. The vesicles were formulated with negatively charged phospholipids by the addition of lactoferrin. The liposomes were tested in simulated gastrointestinal fluids. The authors showed that the average diameter of liposomes decreased after such exposure, reaching half of the initial diameter in the first 30 minutes and remaining constant thereafter; although the diameter remained unchanged in absence of pepsin and pancreatin, which are present in gastric and intestinal fluids, respectively, during the gastric digestion at pH 1.5. This fact can be partially explained due to hydrolysis of bovine serum albumin in aqueous medium, what shifts negatively charged liposomes to neutral charge vesicles (Liu et al., 2012, 2015; Nacka et al., 2001).

9.3.4 USE OF NANOSYSTEMS FOR KINETIC OPTIMIZATION OF ABSORPTION OF NUTRIENTS AND PROTECTION AGAINST OXIDATION

There are few studies about the kinetics of liposomes in food technology. When a liposome with an entrapped compound is resuspended in an aqueous medium without the presence of this compound, the system is not in thermodynamic equilibrium. This can trigger the release of the compound to the external environment until equilibrium is established, in a spontaneous fashion (Frézard et al., 2005; Liu et al., 2012). In cases that liposome entrapment of substances occurs in its structural framework and not in the aqueous core, the release will be dependent on the partition coefficient of the substance between the membrane and the aqueous phase (Frézard et al., 2005).

Nanoparticles have been explored to improve oral bioavailability due to mechanisms like increased stability in the gastrointestinal tract, improvement of solubility in intestinal fluids, improved absorption and decreased influence of first-pass hepatic metabolism (Yao et al., 2015). Nanoemulsions and nanoparticles systems have been used to encapsulate lipophilic nutraceutical compounds. After digestion, mixed micelles produced, carrying these nutraceuticals to the mucous layer making them available for absorption through the enterocytes. EDTA, chitosan and some surfactants may modulate the structure and integrity of intestinal cells, and if used in nanosystems, they tend to increase the absorption and bioavailability of entrapped compounds (Yao et al., 2015).

Amines, amino acids and proteins are able to trigger lipid peroxidation, what may alter liposome formulations and interfere with the characteristics of the entrapped compounds. If nanoparticles are prepared with lipid compounds, they are susceptible to these issues as well. These reactions are capable of producing a range of organic products. Aldehydes produced from lipid peroxidation can react with amine groups of proteins and these intermediaries continue to react with other compounds aldehydes, favoring aldolization (Korzac et al., 2004; Xiong, 2000). Aldolization reactions are capable of causing the formation of conjugated double bonds and the production of pigments. The products obtained by the oxidation of lipids and proteins are often precursors of fluorescence and color of these compounds, ranging from yellow, red and brown (Thanonkaew et al., 2007).

Marine species of mollusks belonging to the *Cephalopoda* class, known as squid or cuttlefish (*Sepia pharaonis*), present in their muscular layer a large amount of phospholipids. These membrane lipids, and a high content of polyunsaturated fatty acids, present a large surface area for being in the form of basecoat and as a result are more susceptible to oxidation (Thanonkaew et al., 2007). Studies with liposomes made from phospholipids obtained from *Sepia pharaonis* showed that lipid oxidation and the formation of yellow and brown pigments are increased at high temperatures. The loss of free amines is an alternative for monitoring interactions between phospholipids present in liposomes and lipid oxidation products. Similar results are described for liposomes of egg yolk lecithin (Thanonkaew et al., 2007). The most accepted mechanism for this phenomenon presents oxidative carbonyl compounds as ketones and aldehydes from lipid oxidation mainly forming products with amine groups characterized as products of Schiff base. The rearrangement of these products can occur in the form of polymerization, originating pigments ranging from yellow to brown. One of these compounds is phosphatidylethanolamine, which confers no enzymatic darkening in *in vitro* reactions (Thanonkaew et al., 2007).

Inorganic compounds including iron and ascorbic acid are able to make liposomes susceptible to oxidation if vesicles are dispersed in a system in which they are present. This phenomenon happens as iron in the Fe^{3+} state is reduced in the presence of ascorbic acid, resulting in Fe^{2+}, which accelerates the breakdown of lipids, resulting in reactive species such as hydrogen peroxide. On the other hand, compounds like sodium chloride are able to reduce lipid oxidation. It is possible that this is due to the availability of free chloride ions to form complexes with iron and other cations, thus decreasing their prooxidant activity (Mei et al., 1997; Thanonkaew et al., 2007).

To study the stability of liposomes and nanoparticles, one must consider the environment in which these come into contact and the characteristics of its components. Especially in the case of Liposomes, stability can be determined in part by the phospholipids composition, taking as an example those who present double layers and thus, are more unstable due to the possibility of peroxidation. An alternative to evaluate the stability of these liposomes would be the measure of turbidity or determining the size through the dynamic light scattering (Chorilli et al., 2007; Fidorra et al., 2006; Lima, 1998).

Soybean phosphatidylcholine liposomes and hydrogenated phosphatidylcholine liposomes with addition of cholesterol, when stored in buffers at 30°C, are able to increase the turbidity of the medium, indicating disruption of these liposomes. Soybean phosphatidylcholine liposomes are very stable if not subjected to extreme conditions. These liposomes are stable for more than six months at a temperature of 25°C pH 7.4, if oxygen and light-protected, preventing instability triggered by oxidation and photolysis, respectively (Chorilli et al., 2007).

9.3.5 NANOSYSTEMS IN DAIRY PRODUCTS

Studies on the protection of food against pathogenic and deteriorating microorganisms have been described, and bacteriocins have been explored in this context. Bacteriocins are ribosomally synthesized antimicrobial peptides produced by several bacteria, which are usually bioactive against phylogenetically related species. Due to factors like the digestion in gastrointestinal tract of animals and humans, bacteriocins have been shown to be safe and interesting for use at the food industry (Malheiros et al., 2012; Taylor et al., 2008).

Benech et al. (2002) developed a liposomal formulation to control the growth of *Listeria innocua* in cheddar cheese by entrapping nisin Z in proliposome H. Liposomes activity was compared to the effect of free nisin Z produced *in situ* by a strain of *Lactoccus lactis*. The nisin Z-producing strain and nisin Z-containing liposomes did not affect cheese production or chemical composition. After six months of monitoring, cheeses inoculated with *L. lactis* presented 10^4 CFU/g of *L. innocua* and 12% of the antimicrobial potential of the bacteriocin remained active. Cheeses made with encapsulated nisin Z, on the other hand, contained less than 10 CFU/g of *L. innocua,* and 90% of the initial nisin activity was preserved, suggesting that the nanosystem was 100 times more effective than the use of free nisin.

Laridi et al. (2003) also encapsulated nisin Z, exploring different proliposomes (Pro-lipo® H, Pro-lipo® S, Pro-lipo® C and Pro-lipo® DUO) and assessed their stability in milk under fermentation. Liposomes encapsulation efficiency ranged from 9.5 to 47%, and the best results were observed for proliposome H. Nisin Z in liposomes remained stable in upto 27 days at 4°C in different media including milk with different levels of fat content.

A nisin-like bacteriocin produced by a strain of *Bacillus* sp. (P34) isolated from aquatic environments of North region of Brazil showed antimicrobial activity against *Listeria monocytogenes* and *Bacillus cereus*, and was suggested to be used in the food industry as a preservative like nisin is currently employed (Malheiros et al., 2012). However, it has been proposed that bacteriocins may lose their antimicrobial activity due to interferences or cross reactions with food components like proteins and fats (Taylor et al., 2008; Vaucher et al., 2010). Liposome entrapment can overcome this limitation, as bacteriocins become more stable and protected against proteases and interactions with nutrients. Malheiros et al. (2012) showed that the antimicrobial efficiency of the nisin-like bacteriocin in food preservation was significantly increased when encapsulated into liposomes. As the liposomes were used in full and skimmed milk, it was possible to effectively inhibit the growth of *L. monocytogenes* (Malheiros et al., 2012).

Zohri et al. (2013) developed a hybrid nanoparticles made of alginate and chitosan, which was complexed with nisin for application in feta cheese conservation. The efficiency of the nanoparticles in inhibiting the growth of *L. monocytogenes* and *Staphylococcus aureus* was five times higher than free nisin, showing an increased preservation potential of this formulation. The sensorial acceptance and physicochemical characteristics of feta-type cheese have been improved after the application of nanotechnology, suggesting that these systems are promising for the dairy industry (Zohri et al., 2013).

As exposed, liposomes can be used to provide benefits to food technology. Milk usually presents low iron levels, and the deficiency of this mineral decreases hemoglobin levels and may evolve for anemia. The addition of minerals to milk can be an effective way of increasing their daily intake. Ferrous sulfate is a low-cost ingredient which provides iron with adequate bioavailability. However, it is able to reduce the shelf life of the products due to the oxidative potential, and to promote shifts of color and flavor (Jackson and Lee, 1991; Xia and Xu, 2005). Liposome entrapment of iron salts represents a way to effectively solve the dilemma of the fortification of this mineral, thereby preventing the previously mentioned problems. Ferrous sulfate liposomes prepared by Xia and Xu (2005) using lecithin showed the same bioavailability of free iron; however, these are thermally unstable. Thus, the authors explored the addition of cholesterol and polysorbate 80 and encapsulation by the reverse phase evaporation method. It was possible to reach encapsulation efficiency of 67% and a ferrous sulfate concentration of 15 mg/L, resisting sterilization process (Xia and Xu, 2005).

The encapsulation of enzyme in liposomes is also used in the manufacture of dairy products. The combination of bacterial protease, fungal protease, lipase (Palatase M) and Flavourzyme encapsulated in liposomes was explored by Kheadr et al. (2003). This liposome was added to milk for preparation of Cheedar cheese before the addition of rennet, in order to accelerate the proteolysis and lipolysis during the maturation phase. These enzymes allowed the obtaining of varied cheeses with different sensorial aspects. Samples in which liposomes containing bacterial protease and lipase were incorporated resulted in a more mature appearance and characteristic aroma, without appearance of eccentric flavors. Bitter flavors were detected in cheeses in which liposomes containing fungal protease and lipase were incorporated, making this formulation unsuitable, once the acceptance of this cheese may be impaired (Kheadr et al., 2003).

It is known that the clove oil has antimicrobial activity against *S. aureus* and *Escherichia coli*; nevertheless, its antimicrobial compounds are highly unstable. Thus, Cui et al. (2015) encapsulated this oil in liposomes in concentrations of 5.0 mg/mL. These liposomes impaired the production of pore-forming toxins of *S. aureus*, but lost the antimicrobial activity against *E. coli*. These liposomes exhibited similar activity against *S. aureus* in tofu cheese, being able to increase the shelf life of these products after opened for consumption (Cui et al., 2015).

9.3.6 LIPOSOMES IN THE PREPARATION OF FOOD SUPPLEMENTS

Fruits and vegetables and their preparations as juices are consumed worldwide, and contribute to provide the adequate daily intake of vitamins. However, during the process of industrialization and pasteurization, these compounds can be rapidly degraded (Couto and Canniatti-Brazaca, 2010). To increase the concentration of these vitamins in juices, it has been proposed to encapsulate them into phosphatidylcholine liposomes stabilized with stearic acid and calcium stearate. These liposomes, which were prepared by Marsanasco et al. (2011) using the dehydration-rehydration method, showed protective effect over the antioxidant activity of vitamins C and E, without altering sensory characteristics of the juices. The formulations showed microbiological stability after pasteurization when stored at ^{40} C over a period of 40 days (Marsanasco et al., 2011).

The use of nanosystems is still limited in food and nutritional supplements also due to the risk of liposomes leakage after long periods of storage. This could result in the loss of the protective effect over the entrapped compound. The sensitivity of liposomes the environmental conditions and electrostatic forces makes it required adding a component able to stabilize him, which must be approved by the organs of direction and food security. As an alternative for stabilization of these liposomes, a recent of Frenzel and Steffen-Heins (2015) proposed the use of isolated whey protein for liposome coating, was decreased the membrane permeability and the osmotic alterations caused on media with excess of salts or carbohydrates, however the coating appeared more sensitive during the gastric digestion. These liposomes coated with edible proteins show benefits for the stability and shelf life of products, expanding its applicability in food sciences (Frenzel and Steffen-Heins, 2015).

9.4 CONCLUSION

The use of liposomes and nanoparticles can be explored in several fields in pharmaceutical and food sciences, as described in this chapter. An intriguing question was made throughout this discussion: what is missing to these technologies finally reach the pharmaceutical and alimentary markets? Well, in view of the exposed, the production in large scale is still challenging. Also, there is lack of toxicological studies that offer enough safety for administrating liposomes or nanoparticles to other animal models than murine models. Several *in vitro* models, including *ex vivo* approaches have been proposed; however, their representativeness of the complexity that involves their processing in the organism is still very limited. Immune responses and metabolism reactions, for instance, are not easily simulated in the current models. Moreover, mathematical modeling of drug release by these nanodevices, and biophysical studies of their interaction with biological targets, may improve our understanding of the behavior of nanoparticles after administration (or addition, in case of food matrices), and thus, the pharmacological or biological effects may be somehow predicted. It is possible that more structured studies with humans exploring preliminary nanoformulations shall be a closer reality when the interactions of liposomes and nanoparticles with biological targets are better understood from a molecular biophyisicochemical point of view. Also, as the technologies become more popular, research costs shall decrease and make the investigations in this field available to more research groups.

ACKNOWLEDGEMENTS

The authors are thankful to Stephan Holzschuh (Bayer HealthCare Pharmaceuticals, Leverkusen), for kindly providing an important reference of this chapter. MVDS is head of GPqFAR and is currently supported by grants from Fundação de Amparo à Pesquisa do Estado de Minas Gerais (FAPEMIG). RMDS is currently recipient of a Master of Food Sciences fellowship from Conselho Nacional de Desenvolvimento Científico e Tecnológico (CNPq).

KEYWORDS

- **clinical use**
- **food preservation**
- **food technology**
- **liposomes**
- **nanoparticles**
- **treatments**

REFERENCES

Afergan, E.; Epstein, H.; Dahan, R.; Koroukhov, N.; Rohekar, K.; Danenberg, H. D.; Golomb, G. Delivery of serotonin to the brain by monocytes following phagocytosis of liposomes. *J. Control. Release*, **2008**, *132*, 84–90.

Assis, L. M.; Zavareze, E. R.; Prentice-Hernandez, C.; Souza-Soares, L. A. Review: characteristics of nanoparticles and their potential applications in foods. *Braz. J. Food Technol.*, **2012**, *15*(2), 99–109.

Bal, S. M.; Hortensius, S.; Ding, Z.; Jiskoot, W.; Bouwstra, J. A. Co-encapsulation of antigen and Toll-like receptor ligand in cationic liposomes affects the quality of the immune response in mice after intradermal vaccination. *Vaccine.*, **2011**, *29*(5), 1045–1052.

Benech, R. O.; Kheadr, E. E.; Laridi, R.; Lacroix, C.; Fliss, I. Inhibition of Listeria innocua in cheddar cheese by addition of Nisin Z in liposomes or by *in situ* production in mixed culture. *Appl Environ Microbiol*, **2002**, *68*, 3683–3690.

Bhartia, B.; Ward, J.; Guthrie, A.; Robinson, P. J. Hepatocellular carcinoma in cirrhotic livers: double contrast thin-section MR imaging with pathologic correlation of explanted tissue. *AJR Am. J. Grastroenterol.*, **2003**, *180*, 577–584.

Blanco, E.; Shen, H.; Ferrari, M. Principles of nanoparticle design for overcoming biological barriers to drug delivery. *Nat. Biotechnol.*, **2015,** *33*(9), 941–951.

Blasco, C.; Picó, Y. Determining nanomaterials in food. *Trends Analyt Chem.*, **2011**, *30*(1), 84–99.

Boks, MA.; Bruijns SC.; Ambrosini, M.; Kalay H.; van Bloois, L.; Storm, G.; Gruijl, T.; van Kooyk, Y. *In situ* Delivery of Tumor Antigen- and Adjuvant-Loaded Liposomes Boosts Antigen-Specific T-Cell Responses by Human Dermal Dendritic Cells. *J. Invest. Dermatol.,* **2015,** 135(11):2697-704.

Chatin, B.; Mével, M.; Devallière, J.; Dallet, L.; Haudebourg, T.; Peuziat, P.; Colombani, T.; Berchel, M.; Lambert, O.; Edelman, A.; Pitard, B. Liposome-based Formulation for Intracellular Delivery of Functional Proteins. *Mol. Ther. Nucleic. Acids.*, **2015**, *4*, e244, doi: 10.1038/mtna.2015.17.

Chhabra, R.; Ruozi, B.; Vilella, A.; Belletti, D.; Mangus, K.; Pfaender, S.; Sarowar, T.; Boeckers, T. M.; Zoli, M.; Forni, F.; Vandelli, M. A.; Tosi, G.; Grabrucker, A. M. Application of Polymeric Nanoparticles for CNS Targeted Zinc Delivery *In Vivo. CNS Neurol Disord Drug Targets.,* **2015**, *14*(8), 1041–1053.

Chorilli, M.; Rimério, T. C.; Oliveira, A. G.; Scarpa, M. V. Estudo da Estabilidade de Lipossomas Compostos de Fosfatidilcolina de Soja e Fosfatidilcolina de Soja Hidrogenada Adicionados ou Não de Colesterol por Método Turbidimétrico. *Latin Am J Pharmacy.*, **2007**, *26*(1), 31–37.

Clark, A. J.; Davis, M. E. Increased brain uptake of targeted nanoparticles by adding an acid-cleavable linkage between transferrin and the nanoparticle core. *Proc Natl Acad Sci U S A.*, **2015**, *112*(40), 12486–12491.

Couto, M. A. L.; Canniatti-Brazaca, S. G. Quantification of vitamin C and antioxidant capacity of citrus varieties. *Ciência Tecnol Alim.,* **2010**, *30*, 15–19.

Cui, H.; Zhao, C.; Lin, L. The specific antibacterial activity of liposome-encapsulated Clove oil and its application in tofu. *Food Control.*, **2015**, *56*, 128–134.

Du, S. L.; Pan, H.; Lu, W. Y.; Wang, J.; Wu, J.; Wang, J. Y. Cyclic Arg-Gly Asp peptide-labeled liposomes for targeting drug therapy of hepatic fibrosis in rats. *J. Pharmacol. Exp. Ther.*, **2007**, *322*, 560–568.

Dzamukova, M. R.; Naumenko, E. A.; Lvov, Y. M.; Fakhrullin, R. F. Enzyme-activated intracellular drug delivery with tubule clay nanoformulation. *Sci Rep.*, **2015**, *5*, 10560, doi: 10.1038/srep10560.

FDA. Generally recognized as safe (*GRAS*), 1997, available at: http://www.fda.gov/Food/IngredientsPackagingLabeling/GRAS/ Accessed Nov 04, 2015.

Fidorra, M.; L. Duelund; C. Leidy, A. C. Simonsen, L. A. *Bagatolli. Biophys.* **2006**, *9*, 4437–4451.

Floyd, J. A.; Galperin, A.; Ratner, B. D. Drug Encapsulated Aerosolized Microspheres as a Biodegradable, Intelligent Glioma Therapy. *J. Biomed. Mater. Res. A.*, **2016**, *104*(2), 544-52.

Frenzel, M.; Steffen-Heins, A. Whey protein coating increases bilayer rigidity and stability of liposomes in food-like matrices. *Food Chem.*, **2015**, *173*, 1090–1099.

Frézad, F.; Schettini, D. A.; Rocha, O. G. F.; Demicheli, C. Lipossomas: propriedades físicoquímicas e farmacológicas, aplicações na quimioterapia à base de antimônio. *Quim Nova.*, **2005**, *28*(3), 511–518.

Gajra, B.; Dalwadi, C.; Patel, R. Formulation and optimization of itraconazole polymeric lipid hybrid nanoparticles (Lipomer) using Box Behnken design. *Daru. J. Pharm. Sci.*, **2015**, *23*, 3. doi: 10.1186/s40199-014-0087-0

Gallier, S.; Shaw, E.; Cuthbert, J.; Gragson, D.; Singh, H.; Jiménez-Flores, R. Hydrolysis of milk phospholipid and phospholipid–protein monolayers by pancreatic phospholipase A2. *Food Res Internat.*, **2013**, *54*, 718–725.

Gharib, A.; Faezizadeh, Z; Mesbah-Namin, S. A.; Saravani, R. Preparation, characterization and *in vitro* efficacy of magnetic nanoliposomes containing the artemisinin and transferrin. *Daru J Pharm Sci.*, **2014**, *22*(44), doi: 10.1186/2008–2231–22–44.

Gregoriadis, G. Liposome Technology 3rd Edition. **2007,** CRC Press, Boca Raton, Volumes I–III.

Hang, X.; Peng, H.; Song, H.; Qi, Z.; Miao, X.; Xu, W. Antiviral activity of cuprous oxide nanoparticles against Hepatitis C Virus *in vitro*. *J. Virol. Methods.*, **2015,** *222*, 150–157.

Hatakeyama, H.; Akita, H.; Ishida, E.; Hashimoto, K.; Kobayashi, H.; Aoki, T.; Yasuda, J.; Obata, K.; Kikuchi, H.; Ishida, T.; Kiwada, H.; Harashima, H. Tumor targeting of doxorubicin by anti-MT1-MMP antibody-modified PEG liposomes. *Int. J. Pharm.*, **2007**, *342*, 194–200.

Holzschuh, S.; Kaeß, K.; Fahr, A.; Decker, C. Quantitative *In vitro* Assessment of Liposome Stability and Drug Transfer Employing Asymmetrical Flow Field-Flow Fractionation (AF4). *Pharm Res.*, **2016**, *33*(4), 842–855.

Hu, C. M. J.; Fang, R. H.; Wang, K. C.; Luk, B. T.; Thamphiwatana, S.; Dehaini, D.; Nguyen, P.; Angsantikul, P.; Wen, C. W.; Kroll, A. V.; Carpenter, C.; Ramesh, M.; Qu, V.; Patel, S. H.; Zhu, J.; Shi, W.; Hofman, F. M.; Chen, T. C.; Gao, W.; Zhang, K.; Chien, S.; Zhang, L. Nanoparticle biointerfacing by platelet membrane cloaking. *Nature* 526, **2015**, *526*(7571), 118–121.

Jackson, L. S.; Lee, K. Microencapsulated iron for food fortification. *J. Food Sci.*, **1991**, *56*(4), 1047–1050.

Jiao, X.; Wang, R. Y.; Feng, Z.; Alter, H. J.; Shih, J. W. Modulation of cellular immune response against hepatitis C virus nonstructural protein 3 by cationic liposome encapsulated DNA immunization. *Hepatology.*, **2003**, *37*(2), 452–460.

Kaba, S. A.; Brando, C.; Guo, Q.; Mittelholzer, C.; Raman, S.; Tropel, D.; Aebi, U.; Burkhard, P.; Lanar, D. E. A nonadjuvanted polypeptide nanoparticle vaccine confers long-lasting protection against rodent malaria. *J. Immunol.* **2009**, *183*(11), 7268–7277

Kamath, A. T.; Rochat, A. F.; Christensen, D., Agger, E. M., Andersen, P., Lambert, P.-H., & Siegrist, C. A. A Liposome-Based Mycobacterial Vaccine Induces Potent Adult and Neonatal Multifunctional T Cells through the Exquisite Targeting of Dendritic Cells. PLoS ONE, **2009**, *4*(6), e5771. http://doi.org/10.1371/journal.pone.0005771.

Kheadr, E. E.; Vuillemard, J. C.; El-Deeb, S. A. Impact of liposome-encapsulated enzyme cocktails on cheddar cheese ripening. *Food Res Internat.*, **2003**, 36, 241–252.

Kibat, P.; Stricker, H. Storage stability of dispersions of soybean lecithin liposomes. *Pharmaz Indus.* **1986**, *48*, 1184–1189.

Koçer, A. Functional liposomal membranes for triggered release. *Methods. Molec. Biol.*, **2010,** *605*, 243–255.

Kong, F.; Zhang, X.; Zhang, H.; Qu, X.; Chen, D.; Servos, M.; Mäkilä, E.; Salonen, J.; Santos, H. A.; Hai, M.; Weitz, D. A. Inhibition of Multidrug Resistance of Cancer Cells by Co-Delivery of DNA Nanostructures and Drugs Using Porous Silicon Nanoparticles Giant Liposomes. *Adv. Funct. Mater.*, **2015**, *25*, 3330–3340.

Korczak, J.; Hes, M.; Garmza, A.; Jedrusek-Golinska, A. Influence of fat oxidation on the stability of lysine and protein digestibility in frozen meat product. *Electr J Polish Agric Univers.*, **2004**, *7*, 1–13.

Laridi, R.; Kheadr, E. E.; Benech, R. O.; Vuillemard, J. C.; Lacroix, C.; Fliss, I. Liposome encapsulated nisin Z: optimization, stability and release during milk fermentation. *Intern. Dairy, J.*, **2003**, *13*, 325–336.

Liu, M.; Li, M.; Wang, G.; Liu, X.; Liu, D.; Peng, H.; Wang, Q. Heart-targeted nanoscale drug delivery systems. *J Biomed Nanotechnol.*, **2014,** *10*(9), 2038–2062.

Liu, W.; Ye A.; Liu, C.; Liu, W.; Han, J.; Singh, H. Behavior of liposomes loaded with bovine serum albumin during *in vitro* digestion. *Food Chem.*, **2015**, *175*, 16–24.

Liu, W.; Ye A.; Liu, C.; Liu, W.; Singh, H. Structure and integrity of liposomes prepared from milk- or soybean-derived phospholipids during *in vitro* digestion. *Food Res Internat.*, **2012**, *48*, 499–506.

Lopes, R. M.; Gaspar, M. M.; Pereira, J.; Eleutério, C. V.; Carvalheiro, M.; Almeida, A. J.; Cruz, M. E. Liposomes versus lipid nanoparticles: comparative study of lipid-based systems as oryzalin carriers for the treatment of leishmaniasis. *J. Biomed. Nanotechnol.*, **2014**, *10*(12), 3647–3657.

Malheiros, P. S.; Sant'Anna, V.; Utpott, M.; Brandelli, A. Antilisterial activity and stability of nanovesicle-encapsulated antimicrobial peptide P34 in milk. *Food Control.*, **2012**, *23*, 42–47.

Markoutsa, E.; Pampalakis, G.; Niarakis, A.; Romero, I. A.; Weksler, B.; Couraud, P. O.; Antimisiaris, S. G. Uptake and permeability studies of BBB-targeting immunoliposomes using the hCMEC/D3 cell line. Eur. *J. Pharm. Biopharm.*, **2011**, *77*, 265–274.

Marsanasco, M.; Márquez, A. L.; Wagner, J. R.; Alonso, S. V.; Chiaramoni, N. S. Liposomes as vehicles for vitamins E and C: An alternative to fortify orange juice and offer vitamin C protection after heat treatment. *Food Res Internat.*, **2011**, *44*, 3039–3046.

McClements, D. J. Enhancing nutraceutical bioavailability through food matrix design. *Innov Food Sci.*, **2015**, *4*, 1–6.

Mei, L.; McClements, D. J.; Wu, J.; Decker, E. A. Ironcatalyzed lipid oxidation in emulsion as affected by surfactant, pH and NaCl. *Food Chem.*, **1997**, *61*, 307–312.

Mura, S.; Nicolas, J.; Couvreur, P. Stimuli-responsive nanocarriers for drug delivery. *Nat Mater.*, **2013**, *12*(11), 991–1003.

Nacka, F.; Cansell, M.; Gouygou, J. P.; Gerbeaud, C.; Méléard, P.; Entressangles, B. Physical and chemical stability of marine lipid-based liposomes under acid conditions. *Coll Surf B: Biointerf.*, **2001**, *20*, 257–266.

Osawa, M.; Erickson, H. P. Liposome division by a simple bacterial division machinery. *Proc Natl Acad Sci USA*, **2013,** *110*(27), 11000–11004.

Pedram, M. Z.; Shamloo, A.; Alasty, A.; Ghafar-Zadeh, E. Toward Epileptic Brain Region Detection Based on Magnetic Nanoparticle Patterning. *Sensors (Basel).*, **2015**, *15*(9), 24409–24427.

Rashidinejad, A.; Birch, E. J.; Sun-Waterhouse, D.; Everett, D. W. Delivery of green tea catechin and epigallocatechin gallate in liposomes incorporated into low-fat hard cheese. *Food Chem.*, **2014**, *156*, 176–183.

Rogers, M. A. Naturally occurring nanoparticles in food. *Curr Opin Food Sci.*, **2015**, 7, 14–19.

Ruiz-Rico, M.; Fuentes, C.; Pérez-Esteve, E.; Jiménez-Belenguer, A. I.; Quiles, A.; Marcos, M. D.; Martínez-Máñez, R.; Barat, J. M. Bactericidal activity of caprylic acid entrapped in mesoporous silica nanoparticles. *Food Control*, **2015**, *56*, 77–85.

Sadhukhan, S., Bakshi, P., Datta, R. and Maiti, S. Poly(ethylene oxide)-g-gellan polysaccharide nanocarriers for controlled gastrointestinal delivery of simvastatin. *J. Appl. Polym. Sci.*, **2015**, *132*, doi: 10.1002/app.42399.

Sato, Y.; Murase, K.; Kato, J.; Kobune, M.; Sato, T.; Kawano, Y.; Takimoto, R.; Takada, K.; Miyanishi, K.; Matsunaga, T.; Takayama, T.; Niitsu, Y. Resolution of liver cirrhosis using vitamin A-coupled liposomes to deliver siRNA against a collagen-specific chaperone. *Nat. Biotechnol.*, **2008**, *26*, 431–442.

Solleti, V. S.; Alhariri, M.; Halwani, M.; Omri, A. Antimicrobial properties of liposomal azithromycin for *Pseudomonas* infections in cystic fibrosis patients. *J. Antimicrob. Chemother.*, **2015**, *70*(3), 784–796.

Sozer, N.; Kokini, J. L. Nanotechnology and its applications in the food sector. Trends in Biotechnology. **2009**, *27*(2), 82–89.

Tanimoto, A.; Kuribayashi, S. Application of superparamagnetic iron oxide to imaging of hepatocellular carcinoma. *Eur. J. Radiol.*, **2006**, *58*, 200–216.

Taylor, T. M.; Bruce, B. D.; Weiss, J.; Davidson, P. M. Listeria monocytogenes and Escherichia coli O157:H7 inhibition *in vitro* by liposome encapsulated nisin and ethylene diaminetetra acetic acid. *J. Food Safety*, **2008**, *28*, 183–197.

Thanonkaew, A.; Benjakul, S.; Visessanguan, W. Chemical composition and thermal property of cuttlefish (*Sepia pharaonis*) muscle. *J Food Compos Analysis.*, **2006**, *19*, 127–133.

Thanonkaew, A.; Benjakul, S.; Visessanguan, W.; Decker, E. A. Yellow discoloration of the liposome system of cuttlefish (*Sepia pharaonis*) as influenced by lipid oxidation. Food Chemistry. **2007**, *102*, 219–224.

Vaucher, R. A.; Motta, A. S.; Brandelli, A. Evaluation of the *in vitro* cytotoxicity of the antimicrobial peptide P34. *Cell Biol Internat.* **2010**, *32*, 317–323.

Wang, X.; Song, Y.; Su, Y.; Tian, Q.; Li, B.; Quan, J.; Deng, Y. Are PEGylated liposomes better than conventional liposomes? A special case for vincristine. *Drug Deliv.*, **2015,** *29*, 1–9.

Whitcomb, D.; Lowe, M. Human pancreatic digestive enzymes. *Digest Dis Sci.*, **2007**, *52*, 1–17.

Williams, H. D.; Trevaskis, N. L.; Charman, S. A.; Shanker, R. M.; Charman, W. N.; Pouton, C. W.; Porter, C. J. H. Strategies to address low drug solubility in discovery and development. *Pharmacol.* 2013, *65*, 315–499.

Xia, S.; Xu, S. Ferrous sulfate liposomes: preparation, stability and application in fluid milk. *Food Res Internat.*, **2005**, *38*, 289–296.

Xiao, L.; Gu, L.; Howell, S. B.; Sailor, M. J. Porous silicon nanoparticle photosensitizers for singlet oxygen and their phototoxicity against cancer cells. *ACS Nano*, **2011,** *5*, 3651–3659.

Xiong, Y. L. Protein oxidation and implication for muscle food quality. Antioxidants in muscle food. In: E. A. Decker, C. Faustman, & C. J. Lopez-Bote (Eds.). New York: John Wiley & Sons, Inc. **2000**, 85–111.

Yao, M.; McClements, D. J.; Xiao, H. Improving oral bioavailability of nutraceuticals by engineered nanoparticle-based delivery systems. *Curr Op Food Sci.*, **2015**, *2*, 14–19.

Zohri, M.; Alavidjeh, M. S.; Mirdamadi, S. S.; Behmadi, H.; Nasr, S. M. H.; Gonbaki, S. E.; Ardestani, M. S.; Arabzadeh, A. J. Nisin-loaded chitosan/alginate nanoparticles: a hopeful hidrid biopreservative. *J. Food Safety*, **2013**, *33*, 40–49.

CHAPTER 10

NANOTECHNOLOGY FOR COSMETIC HERBAL ACTIVES: IS IT A NEW BEAUTY REGIME?

RANJITA SHEGOKAR

Free University of Berlin, Kelchstr. 31, 12169 Berlin, Germany

CONTENTS

10.1 INTRODUCTION

Since ancient time till today, cosmetic products always attracted attention of man and woman. To state precisely, hair care and dermal care products are the everyday need of the modern woman. Men are not behind;

today the average percentage of male has become conscious of their looks and appearance. Natural products are among the favorite one and are traditionally used for beauty care. It is well known that Cleopatra used to apply donkey's milk to her skin. The science of cosmetology is believed to have originated in Egypt and India, but the earliest records of cosmetic substances and their application dates back to Circa 2500 and 1550 B.C.; to the Indus valley civilization.

The Federal Food, Drug, and Cosmetic Act (FD&C Act) has defined cosmetics, based on their intended use, as "articles intended to be rubbed, poured, sprinkled, or sprayed on, introduced into, or otherwise applied to the human body for cleansing, beautifying, promoting attractiveness, or altering the appearance" [FD&C Act, Sec. 201(i)]. The products category like skin moisturizers, perfumes, lipsticks, fingernail polishes, eye, and facial makeup preparations, cleansing shampoos, permanent waves, hair colors, and deodorants would be covered as cosmetic.

10.2 HERBAL COSMETIC TREND

Several thousands of cosmetic products are out in the market and heavily used by generation from all age groups. The synthetic cosmetics usually contain chemical ingredients like imidazolidinyl and diazolidinyl urea, parabens, petrolatum, propylene glycol, Polyvinyl pyrollidone/vinyl alcohol copolymer, surfactants like sodium lauryl sulfate, triethanolamine, synthetic colors, etc. However, the scientific community and published reports has shown growing concerns over the use of these synthetic components as they lead to adverse effectives. A few to include are blemishes, thinning of eyelashes, dermatitis, allergic reactions and carcinogenic.

The times of Cleopatra and the British queens has witnessed the use of natural cosmetics from organic ingredients like coconut, turmeric, neem, kokum butter, etc. However, with the surge of synthetic cosmetics the market trend changed due to the quick benefits shown by the synthetic components. Media is actively educating customers by comparing the safety of various personal care products. This contributed to the consumer awareness with an increased interest in much safer products. Being traditional in nature, the herbal cosmetics offer both aspects like functional advantages and safety over the synthetic compounds and lesser toxic effects. Scientifically "Aware" consumer is now turning back to the natural organic substances. This can be validated against steady growth in herbal cosmetic market. Beside awareness, fashion consciousness and purchasing power of the people has added to this growth momentum. As said earlier, the consumer of today is more knowledgeable about the ingredients that are used in the formulation of cosmetics and are opposed to synthetic beauty chemicals which can cause side-effects. This has led to the popularity of herbal cosmetics all over the world. Consumer is becoming suspicious of chemical ingredient and more aware about the composition of product they use. They believe in more long-term safe use of products based on organic-based or natural source-based products. The cosmetic industry is very much aware of this market shift and is trying to develop natural source and environmentally sustainable products based cosmetic products. There is a significant market value for natural base ingredient in this regard.

Herbal beauty products were first introduced to the world way back in the 70s and today there is a plethora of brands in the herbal cosmetics market. Yet, the market is still growing and there is scope for new entrants who can deliver quality natural products which can satisfy the consumer requirements. Following the financial crisis during 2008–2010, the overall global cosmetics market is recovering at a steady pace. The growth in economies largely encourages the growth of the market for global cosmetics. Among that the herbal cosmetics industry is experiencing a rising graph in terms of market and products which are being sold worldwide. Global cosmetic market is expected to reach $390.07 by 2020 according to Allied market research, 2015 report (Report Code: CO 15651). Women cosmetics products contribute almost double in generation of revenue by 2020 and most of the leads will be predicted by Europe market. The main area of attraction still remains skin and sun care products (https://www.alliedmarketresearch.com/cosmetics-market). Grand view recently published research which predicts $15.98 billion market by 2020 only for organic personal care products.

A shift of preference towards natural and organic beauty products, particularly in United States and European countries, has fostered the growth of the cosmetics market. The rising demand for natural, herbal and organic beauty products creates potential opportunities for manufacturers to innovate and develop new products in accordance to consumer preferences. L'Oréal Group, Avon Incorporation, Proctor and Gamble Corporation, Unilever, Oriflame Cosmetics, Revlon Incorporation, Kao Corporation, Estee Lauder Companies, Skinfood, Shiseido Incorporation, Procter & Gamble, Clinique from Lauder; Neutrogena, Johnson & Johnson; Avon; and the Estee Lauder brand are currently the top players in herbal/synthetic nanotechnology-based cosmetic market (Mu et al., 2010; Raj et al., 2012; Zippin and Friedman, 2009).

10.3 WHY HERBAL ACTIVES?

The herbal extracts based cosmetics contain bioactive ingredients or phyto-constituents which acts very effectively and can do magical wonders to the body. Photo-protective phyto- constituents have properties to protect and rejuvenate the skin from damage due to UVA and UVB radiations, environmental pollution, atmospheric temperature fluctuations, wrinkling and inflammation. These phyto-constituents act as natural sun blockers and as natural sources of antioxidants. The natural sun screeners in human skin are absorbing lipids, proteins and nucleotides. Plant or herbal extracts contain high concentration of these substances and provide easy and better protection to the peptide bonds of the skin's proteins as well as the sensitive lipids and nucleotides. Free radicals are one of the major destroyers of skin and to combat them are needed powerful antioxidants. Naturally occurring bioactive compounds like flavonoids, phenolic acids prevent the adverse effects of UV-radiation and stimulate the skin's blood circulation and repair.

10.4 CONTRIBUTION OF NANOPARTICLE IN DELIVERY OF HERBAL ACTIVES

Nanotechnology already proved its effectives in cosmetic, pharma and food market. Electronic and engineering markets use nanotechnology to high extent. Nanoparticles (NP) are single particles with a diameter by definition below 100 nm, although their agglomerates may be larger as per Scientific Committee on Consumer Products (SCCP, 18 December

2007, Safety of Nanomaterial in cosmetic products, p.10.). According to Mihranyan et al. (2011) nanotechnology-based products were first introduced in cosmetic industry followed by pharma and food industry and out of several products registered in 2009 under this "nano" group, almost 13% were for cosmetic use, and the turnover was around US$170 billion as per report published by Eurostaf, a French company. Pubmed search engine (Oct 2015), showed 120,000 published journal articles when searched using keyword "nanoparticle," compared to 6 article hit for customized period of 1970–1980s. Keyword "nanoparticle cosmetic" showed around 5000 hits while "nanoparticle herbal" resulted in 160 published articles. Recently published report by Vance et al. (2015) on use of nanotechnology in consumer products identified that the skin is primary route of exposure for nanomaterials. In 2005, the Woodrow Wilson International Center for Scholars and the Project on Emerging Nanotechnologies was formed to create the Inventory of Consumer Products-based on nanotechnology CPI in 2005. This database is revised in 2013 and in total 1814 consumer products are listed from 622 companies out of 32 countries. Silver is the most frequently used in nanoform (435 products, or 24%). Success of nanoparticles in cosmetic field also questions its safety, which a separate topic of discussion (Amenta et al., 2015; Nohynek et al., 2008; Takahashi et al., 2009).

Nanotechnology not only optimizes the manufacturing conditions for cosmetic formulations bearing multicomponent systems but also their effectiveness. Some of the applications are aimed to make fragrances last longer, sunscreens and antiaging creams to be more effective with sustained effect for longer hours, to stabilize the formulation, for color/appearance change of formulation, e.g.; curcumin, to increase solubility of actives, e.g.; Vitamin C, for better penetration through skin or hair cuticles. Nanocarriers used in cosmetic field in general do not act biologically except delivering the active like vitamins, antioxidants or chemical UV filters in skin or to scalp. Mainly they are used as carrier, to improve permeation, penetration, delivery of actives, solubility enhancement, enhancement of product appearance to provide stability to active, etc. The principals are quite similar to pharma nanoparticle concept. With the advent of nanotechnology in herbal cosmetics the highly sensitive active ingredients, such as vitamins and antioxidants can be encapsulated and used in treatments for antigraying and hair loss. Nanotechnology allows combined encapsulation of two or more actives and can be used to improve the UV protection in combination with organic sunscreens, such as 2-hydroxy-4-methoxy benzophenone. In cosmetic industry, nanoparticles are not new; they have already used for delivery of silver, gold, titanium, aluminum, silicon in the form of topical application (D'souza and Shegokar, 2017).

This chapter mainly discusses the use of nanotechnology in formulation development of herbal actives. It counts the success stories of nanoparticles which encapsulates herbal actives along with the recap on safety profile. Various types of drug delivery platforms which are used till date for cosmetic delivery of herbal actives are discussed in detail. At the end of chapter readers are guided on some formulation advise on type of delivery system selection and safety determination criteria. Application of nanotechnology for synthetic actives is out of scope of this chapter.

10.5 FORMULATION OF HERBAL COSMETICS NANOPARTICLES

Following sessions describe various drug delivery systems developed for formulating herbal actives for cosmetic use. Following sessions will touch on production process in short.

Readers are advised to refer other review papers describing manufacturing details of these nanoparticles.

10.5.1 LIPOSOMES

Liposomes are closed spherical colloidal systems of phospholipids together with chains of surfactants. Liposomes are the first to introduced and currently commonly used carrier in cosmetic market. This is one of the first "Some" based carrier developed and explored in cosmetic market (Figure 10.1) Liposomes are incorporated in cosmetics or dermal delivery for various purposes like, to increase active deposition in skin, to provide localized release of active, e.g.; hair cuticles and to improve occlusion effect on skin. In many cases, encapsulation of active in liposomes is used to protect the drug from light and heat, thereby maintain the stability.

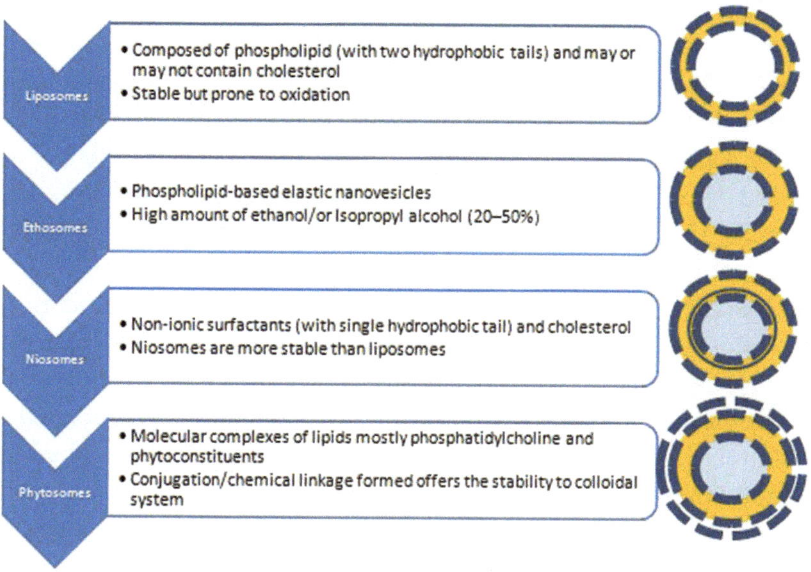

FIGURE 10.1 Differences between various 'somes.'

Takahashi et al. (2009) produced Aloe vera extract encapsulated soyabean lecithin-based liposomes, 200 nm. Liposomes showed improved penetration and higher proliferation rate when tested in human skin fibroblast and epidermal keratinocytes due to improved skin penetration.

Hibiscus sabdariffa calyx extract has powerful antioxidant activity was encapsulated in soyabean lecithin/Tween 80/deoxycholic acid (DA) liposomes to overcome skin permeation and dermal irritation problem associated with it. Liposome formulations showed good stability over 2 months and when tested for *in vitro* activities showed increased antioxidant activity, due increased dermal penetration and reduced dermal toxicity when tested on rabbit skin (Pinsuwan et al., 2010).

Polyphenols are well known and are intensively studied for their skin-whitening activity. *Artocarpus lakoocha* Roxb (Moraceae), heartwood extract has potent antioxidant activity as per DPPH assay. Commercially available Pro-lipo™ were used to produce liposomes of plant extract. *In vivo* efficacy study on lotions (0.1 mL/twice daily/forearm) containing pure plant extract and lotions containing liposomal (172 nm) extract in 10 female volunteers was performed. Skin whitening was determined by using Chromameter® CR 400 against L-glutathione as positive control. *Artocarpus lakoochaplant* extract showed higher skin whitening property for plant extract (tyrosinase inhibition – 77%) than L glutathione (tyrosinase inhibition – 50%). Liposomes not only enhanced the penetration but also improved the efficacy of active (Teeranachaideekul et al., 2013).

Lipoid Kosmetik AG, a Swiss-based company has developed four products for Herbasome line (apple, pear, carrot and coconut herbasomes), which has amazing skin hydration property.

Herbasomes, a next generation of liposome encapsulates plant-derived water. Company's information brochure displays effectiveness of herbasomes in maintaining skin hydration properties-based on the results from *in vivo* studies in 12 female volunteers. Herbasomes can be incorporated in other skin care formulations at level upto 10% (www.lipoid-kosmetik.com).

A study by Kaur and Saraf, compared the functionality of three colloidal carriers liposomes, ethosomes, and transfersomes in size range of 160 to 270 nm. The alcoholic *Curcuma longa* extract was used as active. The skin hydration potential (by Corneometer-CM 820) and sebum (by Sebumeter-SM 815) was used as responsive parameters for photo protective action. Extract loaded transfersomes creams showed highest hydration potential followed by ethosomes and liposomes in human volunteers (Kaur and Saraf, 2011).

Liposomes were studied as carriers for phenolic compounds extracted from Microalgae Spirulina Strain LEB-18 and Chlorella pyrenoidosa (de Assis et al., 2014). Schmid et al. (2004) have found that liposomes containing mycosporine-like amino acids from the red alga Porphyra umbilicalis (common name – Nori) can help in reducing UV-A-induced skin aging. Liposomes could improve the elasticity, hydration; reduce wrinkle depth, lipid perocidation and roughness in study conducted on 20 women (age 36–54) upon twice a day application under irradiation conditions (UV-A rays 10 J/cm^2). Same research group evaluated the rejuvenating effect of *Chlamydocapsa* species (snow algae). Anti-aging effects of liposomes encapsulated active were confirmed by studies conducted in cultures of skin cells and in clinical studies in human volunteer. The snow algae powder could interfere with intracellular pathways thereby repairing the skin functions. It promoted Klotho gene and AMPK activity in skin and induced calorie restriction mimetic effect thereby resulting in repair and improvement in cell metabolism (Schmid et al., 2014).

10.5.2 NANOEMULSIONS

The nanoemulsions are stable systems and easy to spread on body surfaces (nonsticky and nonoily, Figure 10.2). This system allows homogeneous distribution and deep penetration of the active into the substrate (skin/hair). The fluid nature of nanoemulsion gives a pleasant esthetic character and skin feel to users and hence can be applied for delivery of fragrances and as substitutes for liposomes and vesicles. Till dates various actives are incorporated in nanoemulsions and micro emulsion systems. Temperature stability could be only limiting factor for use of nanoemulsions systems.

Hydroalcoholic stem and leaf extract of *Vellozia squamata* Pohl, Velloziaceae, was incorporated in nanoemulsions by phase inversion temperature method containing Babaçu oil, sorbitan monoestearate and PEG-40 hydrogenated castor oil. The antioxidant activity of extract remained undisturbed even after conversion into nanoemulsions. These ready to use system can be further diluted in gel or cream base (Quintao et al., 2013).

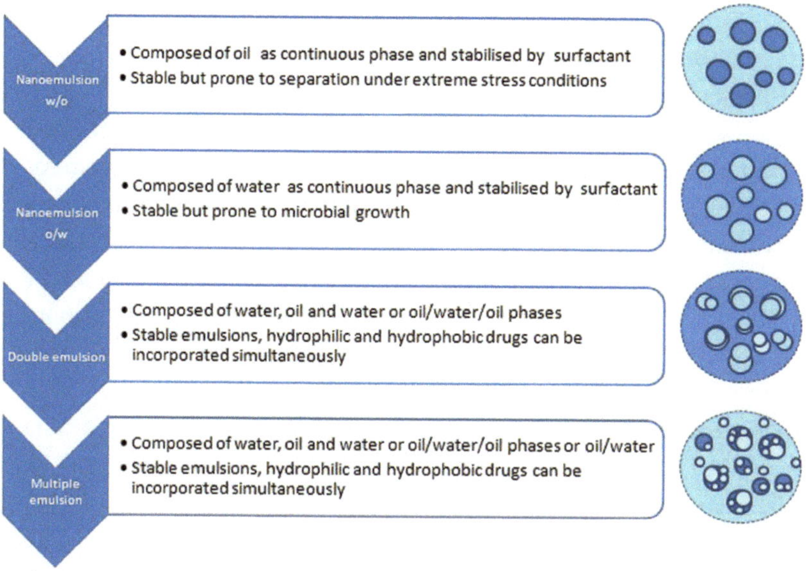

FIGURE 10.2 Types of emulsions for herbal actives.

Research group at Chiang Mai University have prepared nanoemulsion loaded with marigold flower extract (*tagetes erecta* linn) and incorporated then in gel preparation as antiwrinkles composition. Nanoemulsions were prepared by phase inversion temperature. Nanoemulsion based gels showed no toxicity to skin and were completely safe to apply. Skin irritation and antiwrinkle effect was determined in 30 healthy human confirmed efficacy of gel formulation. Improved hydration profile is could be due to better occlusion property offered by gel together with emulsion system (Leelapornpisid et al., 2014).

Glycyrrhizic acid is a primary constituent of licorice root (*Glycyrrhiza glabra* L.) and posses potent antioxidant and skin whitening properties. Glycyrrhizic acid nanoemulsion was prepared by HPH technique-based on jojoba oil, Plantacare® 2000 UP and Span® 85 systems. *In vivo* tape stripping test in 5 healthy human volunteers showed improved cumulative drug penetration and steady state flux compared to pure drug incorporated in gel (Mishra et al., 2011).

Nanbu and Hakata-Ku (2008, US 2009/0069253 A1), described the arbutin (hydroquinone glucoside in α, β form) based skin whitening cosmetic nanoparticle composition. Increased solubility of arbutin after conversion as nanoparticle offered great help to formulation scientist. The emulsified nanoparticles in presence of lecithin produced particle size of around 10–150 nm. This emulsified nanodispersion can be added to gel or cream-based cosmetic preparation to alleviate skin problems like wrinkles, dullness, and pigmentation. In another work, a herbal nanoemulsion containing lemon juice and essential oils like tea tree oil, rosemary oil, tulsi oil, lavender oil, and mentha oil were developed for various possible uses (Patent CA 2746566 A1). Hair products are using nanoemulsions to encapsulate active ingredients and carry them deeper into hair shafts. RBC Life Science's Nanoceuticals Citrus Mint Shampoo and Conditioner are made with Nano Clusters™, to give hair a healthy shine.

Many other actives like ethanolic extract of *Phyllanthus urinaria* (*P. urinaria*) (Mahdi et al., 2011), Sea buckthorn pulp oil and Vitamin E (Mibelle Biochemistry), Coenzyme Q10, Vitamin E acetate (Mibelle Biochemistry), Vitamins A, E, and C (Mibelle Biochemistry), lycopene [Butnariu et al., 2011; Lopes et al., 2010), capsaicin (Kim et al., 2014), astaxanthin (Affandi et al., 2011) and lutein (Mitri et al., 2011)] are used in the form of nanoemulsion-based skin antiaging products.

10.5.3 SOLID LIPID NANOPARTICLES AND NANOSTRUCTURED LIPID CARRIERS

At the beginning of 90s, first and second generation of lipid nanoparticles (40 to 1000 nm) called solid lipid nanoparticles (SLNs) and nanostructured lipid carriers (NLCs) offered combined effect of emulsion, liposomes and polymeric nanoparticles (Müller et al., 2000). Till date several pharmaceutical, cosmeceutical and nutraceutical actives are loaded on these lipid nanoparticles. Actives are mainly loaded in fatty acid chains or in between lipid layers and imperfections of lipid matrix. Lipid nanoparticles offer, improvement in chemical stability of actives, controlled/sustained delivery of active, occlusion and hydration properties to skin (Figure 10.3). These lipid carriers can encapsulate flavors, perfumes, colors and various UV blockers. Several methods like high-pressure homogenization (HPH), microemulsion technique, emulsification-solvent evaporation, solvent displacement method, phase inversion method, ultrasonication and membrane contractor technique offer production of lipid nanoparticles. Various

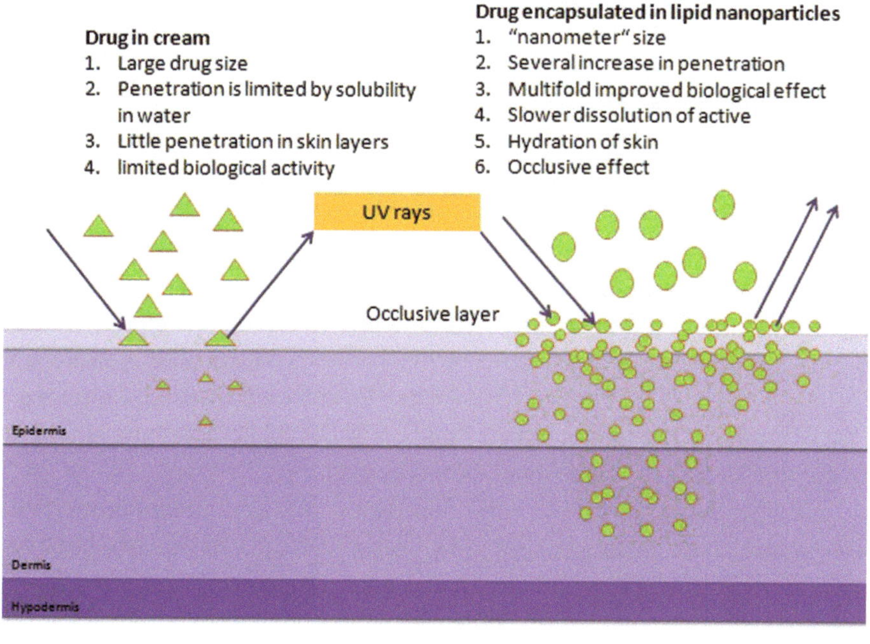

FIGURE 10.3 Dermal mechanism of action of lipid nanoparticles.

oils can be used as composition units or for encapsulation of actives in NLCs. Some e.g. are listed in Table 10.1.

Lutein was incorporated in various nanocarriers (150–300 nm) like SLNs, NLCs and nanoemulsions which were prepared by high pressure homogenization. Permeation studies on fresh pig ear skin showed, that no (for SLNs, NLCs) or very little lutein (0.4% after 24 h) permeated through skin. Thus the active remained in the skin and is not systemically absorbed. SLN, showed 0.06% degradation after irradiation with 10 MED (Minimal Erythema Dose), in NLC 6–8%, compared to 14% in the NE, and to 50% as lutein powder suspended in corn oil, this confirmed UV blocking effects of formulations (Mitri et al., 2011, 2012). Several other actives are incorporated in lipid nanoparticles like CoQ10, lutein, resveratrol, lycopene, sacha-inchi oil, curcumin, quercetin, etc. (Hommoss et al., 2007, Hommoss, 2008).

Marine algae possess potent antioxidative activity beside rich in polyunsaturated fatty acids and proteins (Mercurio et al., 2015; Wang, 2015). Shegokar et al., developed algae extract from species *Nannochloropsis oculata* and *Chaetocerous diatom* loaded NLC dispersions (200–300 nm) for topical treatment. Feasibility of production by homogenization and stability of dispersion was studied (unpublished lab data). Several companies like Oceanwell, L'Oreal's, La Mer Jason Natural Cosmetics, Beauty au Naturel, body-shop and blue lagoon Iceland stores offer seaweed-based cosmetic products. NLC technology-based product called NanoLipid Restore CLR® (produced by Chemisches Laboratorium Dr. Kurt Richter, Germany/distributed by Pharmacos, India). It uses black current seed oil which is rich in ω-3 and ω-6 fatty acids. This skin care product is designed for regenerative skin especially the dry and aged skin. It acts by restoring the skin barrier and reducing transepidermal water loss. The same NLC product is used in leading cosmetic line IOPE® from Amore Pacific, South Korea. Additionally, other NLC based products like Nanorepair Q10® (cream and serum) and Nanovital Q10® (cream) from Cutanova® (Dr. Rimpler, Germany) and Surmer® from Isabelle Lancray (France) (Kaur and Saraf, 2011) are available in market.

Zingiber zerumbet oil has antiinflammatory and antioxidant properties. To explore these properties, Zingiber zerumbet oil was encapsulated in lipid nanoparticles. The NLCs of mean particle size ~ 97 nm and of zeta potential around – 40 mV zeta potential were prepared by ultrasonication technique for transdermal delivery (Rosli et al., 2015).

TABLE 10.1 NLC Based Formulations on Different Types of Oils (Either Composed of Or Encapsulated)

Oil	Source	Reference
Fish oil	Aquatic animal/ tissues	Pharmazie, **2009**, *64*(8), 499–504
Argan oil	Plant/nut	Pharmazie, **2011**, *66*(3), 187–191
Coconut oil	Plant/nut	http://www.jurnalteknologi.utm.my/index.php/jurnalteknologi/article/view/1248
Neem oil	Plant/fruit	http://www.jurnalteknologi.utm.my/index.php/jurnalteknologi/article/view/1248
Rice oil	Plant/fruit	http://www.ist.cmu.ac.th/researchunit/pcrnc/paper/seminar/proc2.pdf
Black Currant seed Oil	Plant/fruit	http://www.diss.fu-berlin.de/diss/servlets/MCRFileNodeServlet/FUDISS_derivate_000000005075/thesis_Aiman_Hommoss.pdf
Eugenol + oleic acid	Plant/fruit	Int J Pharm Pharm Sci, **2011**, *3*(4), 138–143
Olive oil	Plant/fruit	10.1109/TNB.2012.2232937
Hydrogenated Palm Oil	Plant/fruit	10.1109/TNB.2012.2232937
Sunflower oil	Plant/seed	Braz. Arch. Biol. Technol. *56*(4)
Soyabean oil	Plant/seed	Int J Res Pharm Biomed Sci, **2013**, *4*(2), 683–691
Zingiber	Plant/rhizomes	J Adv Res in Applied Mechanics, **2015**, *11*(1)
zerumbet oil		16–23
Turmeric oil encapsulated Miglyol-based NLC	Plant/rhizomes	Drug Dev Ind Pharm, **2010**, *36*(7), 773–80
Algae oil	Algae	J Food Sci, **2014**, *79*(2), E169–E177
Polyoxyl castor oil	Plant/fruits	J Nanobiotechnol **2015**, *13*, 47
Green coffee oil	Plant/fruits	http://www.cifarp.com.br/historico/2013/painel/_inc/trabalhos/resumo1304.pdf

Lipid-based nanosystems were studied for transdermal drug delivery. In one of the study, NLCs encapsulated lappacontine and ranaconitine isolated from *Aconitum sinomontanum*. Cellular uptake of fluorescence-labeled nanoparticles using laser scanning confocal microscopy and fluorescence-activated cell sorting confirmed faster and significant penetration of active in the form of NLC (Guo et al., 2015). Green coffee oil is another active used in cosmetic application for its emollient property. To improve its regenerative, moisturizing, UV protection and cellulite reducer in cosmetic applications Nosari and Freitas prepared lipid nanoparticles (mean particle size 250 to 770 nm). The production feasibility of green coffee oil encapsuled lipid nanoparticles was assessed by hot melt nano-emulsion method (Nosari and Freitas, 2013).

10.5.4 NANOCRYSTALS

Nanocrystals (100 to 1000 nm) comprises surfactant stabilized drug particles without any polymer or lipid matrix. Nanocrystal technology is increasingly growing area in cosmetic market; it offers not only solubility enhancement but also penetration and stability improvement. Nanocrystals (nanosuspension) can be produced by variety of top down and bottom up approaches, like homogenization, milling, solvent precipitation, microfluidization and combination of them. This industrially feasible technology offers formulator ease of production and formulation of poorly soluble active (Shegokar and Müller, 2010). In pharma already several poorly soluble compounds are processed as nanocrystals. Cosmetic actives like flavonoids (antioxidants), lutein, beta carotene, coenzyme Q10 and apigenin are produced as nanocrystals and exhibited improved features of compound (Mitri et al., 2011; Shegokar, 2014). Figure 10.4 shows mechanism of action and benefits of nanocrystals when applied dermally.

Nanocrystals of lutein, a potent antioxidant active was produced by Mitri et al. (2011) to improve solubility after dermal application. Nanosuspensions were processed by two techniques HPH to produce particle size below <500 nm. Type of production method and production settings significantly affected the particle size distribution of final product. The increased saturation solubility (by ~26 fold) of lutein after application dermally resulted in 18 times higher skin penetration compared to coarse drug (Mishra et al., 2012).

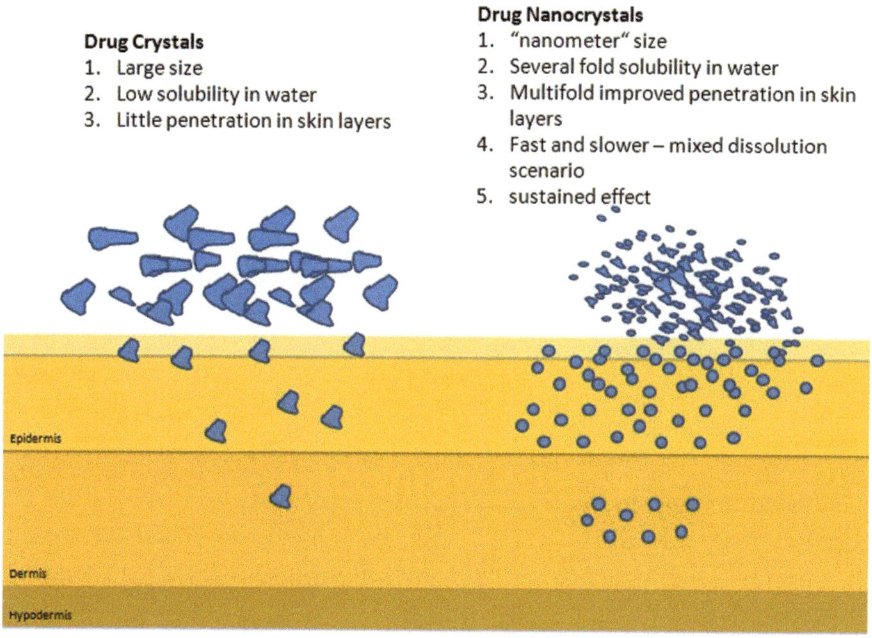

FIGURE 10.4 Dermal mechanism of action of nanocrystals.

Several attempts to improve solubility and dissolution rate of curcumin which is another potent antioxidant were done. Wei et al., employed wet ball milling for production of oily nanosuspension (Wei et al., 2013), homogenization and milling by Rachmawati et al. (2013, nanoprecipitation by Moorthi et al. (2013), Spray dried curcumin nanocrystals prepared by high-pressure homogenization technique (Ravichandran, 2013), by solvent evaporation technique (Chidambaram et al., 2013; http://sphinxsai.com/2014/PTVOL6/PT=56(842–849)AJ14.pdf) for various administration routes of curcumin.

Nanocrystals of glycyrrhetinic and glycyrrhizic acids were developed using three different processes (homogenization, milling, smartCrystal® technology) to exert improved skin whitening effects. smartCrystal process resulted in smallest particle size of 158 nm followed by milling 270 nm and homogenization 325 nm. Glycyrrhetinic acid (5% w/w) was dispersed in aqueous plantacare 2000 UP solution 1% (w/w). Combination technique resulted in sharp reduction in particle size up to 158 nm compared to 269 nm by milling and 325 by homogenization. Overall solubility was increased by 20 fold and penetration in human skin by 11 fold compared to coarse active.

In another study, apigenin nanosuspensions of mean size 300–400 nm was produced by smartCrystal technology at industrial batch size. The effect of various preservatives like ethanol, alkyl polyglyceride, TPGS, cetyl pyridium chloride, 1,3 pentandiol (hydrolite-5), triclosan, Euxyl 9010 and multiEx on stability was evacuated. Two-fold increase in antioxidant activity and 3.4 fold enhancement in skin penetration were observed (Al Shaal et al., 2011)

Zhang et al. (2014) found that baicalin, a flavonoid can reduce UVB-induced epidermal thickening in mouse model upon topical application. Baicalin processed as nanocrystals (248 nm) by high-pressure homogenization showed improved bioavailability in Wister rat over coarse drug, this could be due to increased dissolution kinetics (Jin et al., 2013). Similar effect was also observed when baicaline nanocrystals produced by ultrasonic-homogenization-fluid bed drying technology (Shi-Ying et al., 2014).

Several other cosmetic actives like resveratrol (Kobierski et al., 2011; http://abstracts.aaps.org/Verify/aaps2013/postersubmissions/W4097.pdf), hesperetin (Chen et al., 2013, http://www.conference.net.au/chemeca2011/papers/401.pdf, http://abstracts.aaps.org/Verify/AAPS2014/PosterSubmissions/M1191.pdf), ursorlic acid (Song et al., 2014), oleanolic acid (Jun, 2005), quercetin (Corrias, 2014; Kakran et al., 2012; Sahoo et al., 2011), Coenzyme Q10 (ubiquinone) (Lai et al., 2013; Mauludin, 2008; http://mt.china-papers.com/2/?p=244461), magnolol ((Lin 2013) and tretenoin (Lai et al., 2013) are produced as nanocrystals and exhibited increased solubility. Juvena, Switzerland, introduced four cosmetic products-based on rutin nanocrystal.

10.5.5 OTHER NANOPARTICLES

A lecithin-capsule nanoparticulate systems was developed for delivery of actives from *Zanthoxylum piperitum*, *Torilis japonica* fruit, *Salvia miltiorrhiza*, *Safflower*, *Canidium officinale makino*, Green tea leaf, Pomegranate, Pine tree leaf, Red ginseng, Ginseng, Angelicae, vitamin mixture of nicotinamide and tocopherol acetate by Kim (2010) for alopecia treatment. Lecithin-based nanoparticles can stimulate and activate hair follicles thereby increasing metabolism and inhibited depilation to promote hair growth. Nanoparticles helped to increase penetration through these follicles (Hommoss, 2008). In another study, mesoporous microparticulate systems were formed-based on silicon and were loaded with extracts from Aloe vera, Asian ginseng, Capsicum species, Cascara sagrada, Garlic, Ginger by Canham et al. (Canham et al., 2011) to treat skin problems like acne, oily skins, wrinkles (Corrias, 2014). Micellar dispersion was prepared by hydrating mixture of various oils (almond oil, coconut oil, corn oil, cottonseed oil, linseed oil, olive oil, soybean oil, peanut oil, mineral oils) and stabilized it using tween 60, tween 80, and nonylphenol polyethylene glycol ethers, and mixture of them. The dispersion produced has particle distribution between 10–1000 nm. The micellar dispersion can improve skin penetration due to small size.

10.6 VARIOUS "SOMES"

10.6.1 NIOSOMES

Niosomes are structurally similar to liposomes, however the composition is different (Figure 10.1). Niosomes are prepared mainly using nonionic surfactants (with single hydrophobic tail) and cholesterol. Liposomes are mainly composed of phospholipid (with two hydrophobic tails) and may or may not contain cholesterol. Niosomes are more stable than liposomes, as the phospholipids are more susceptible to oxidation. Niosomes can encapsulate lipophilic and amphiphilic actives. Various techniques like thin film hydration, microfluidization, reverse phase evaporation technique, transmembrane pH ingredient, bubble methods, etc. are used for preparation of niosomes.

To explore the bioactivity of curcuminoids transdermally niosomes (~12 μm) based on sorbitan monooleate, cholesterol, and Solulan C-24 was prepared by Rungphanichkul et al. (2011). *In vitro* permeability through shed snake skin showed improved penetration while no penetration effect was observed for methanolic solution. High flux was observed for curcumin niosomes followed by desmethoxy curcumin, and bis-desmethoxycurcumin niosomes.

Sucontphunt et al. (2013) encapsulated *Piper nigrum* extract which has potential antielastase and antioxidant activities in nano-niosomes. Stable nano-niosomes were composed on sorbitan monooleate and PEG 400 and showed particle size around 200 nm. Beside *Piper nigrum*, author evaluated various other thai plants for similar properties.

In another separate study, skin antiaging properties of gallic acid sourced from *Terminalia chebula* Retz. (Combretaceae) extract loaded niosomes in gel were studied. Two types of niosomes were prepared; elastic and nonelastic. Author determined skin antiaging of loaded niosomes in human volunteers by measuring the skin elasticity and roughness. Skin irritation test using the closed patch test in rabbit showed that niosomes loaded formulations are nonirritant while pure gallic acid fraction (without encapsulation) showed irritation to skin. Skin elastic recovery was faster for elastic niosomes loaded gels compared (32.57%) due to easy penetration compared to nonelastic niosomes gels (28.73%). Similar improvement pattern was observed in skin elastic extension for elastic (~23%) and nonelastic gels (~21%). A significant decrease in skin roughness was observed for gels loaded with elastic and nonelastic gels. Niosomes offered chemical stability, biosafety and improved antiaging activity to gallic acid due to improved penetration in targeted skin layers (Manosroi et al., 2011).

In other work, Bhramaramba et al. (2015), prepared uniformly spreadable gels-based on *Terminalia chebula* leave extract to exert antibacterial and antifungal, analgesic, antiinflammatory, sun burns and wound healing were prepared. Skin irritation degree was determined in rats after continuous application of gel on skin once a day for upto 7 days showed no sign of skin irritation. Gels can be loaded up to 5% of plant extract.

Naturalis Life Technologies, an Italian-based cosmetic company offers several niosomes and phytosome-based compositions for improved performance (http://www.naturalislife.com/nio-active.html).

10.6.2 ETHOSOMES

Ethosomes are phospholipid-based elastic nanovesicles. They contain a high amount of ethanol/or isopropyl alcohol (20–50%), which aid elastic behavior to ethosomes. This elastic behavior or deformability allows them to permeate intact through the human skin. Ethosomes can be prepared by cold or hot method and is used various transdermal and cosmetic application. It is a carrier of choice for pilosebaceous targeting.

Topical application of ethosomes encapsulated with tetrandrine was explored against liposomes. The drug flux after in terms of *ex vivo* permeation and deposition was higher for ethosomes compared to liposomes. Furthermore, results were confirmed by confocal scanning microscopy observations on skin. The study outcome verifies the potential of ethosomes for delivery of actives, synthetic or herbal (Fan et al., 2013). Curcuma longa extract was incorporated in ethosomes to exert antiwrinkle effect on skin by penetrating into deeper layers. Ethosomes of mean particle size around 216 nm (zeta potential −30 mV) were prepared-based on soya phosphatidylcholine and ethanol. Increase in viscoelasticity property on human skin was evaluated by Cutometer and showed that cream incorporated with ethosomes could improve skin condition by 10–50% in terms of elasticity and sagginess. This could be due to improved penetration of encapsulated active (Jeswani and Saraf, 2014). Builders et al. (2014), explored hair growth promoting activity of *Moringa oleifera* leaf extract in ethosomal composition. Skin irritation study lasting 1 month in rat ($n = 5$) showed that ethosomal extract exhibited concentration dependent activity and improved hair growth indicated by increasing number of follicles in anagen phase compared to positive control minoxidil. No skin irritation sign in terms of edema or odema or irritation was observed. Table 10.2 lists herbal actives, which are incorporated in various "SOMES" vesicles for, improve active delivery to skin.

Blue-green Klamath algae constituent C-phycocyanin, a high molecular weight protein phycocyanin was isolated and encapsulated in various phospholipid vesicles of 100–200 nm in size like liposomes, ethosomes and penetration enhancer (like propylene glycol or Transcutol P) containing vesicles. Rhodamine labeled vesicles showed enhanced penetration and distribution throughout skin layers confirmed by confocal laser scanning microscopy. Penetration enhancer containing vesicles showed superior performance in terms of penetration and hydration compared to the other lipid vesicles (Caddeo et al., 2013). Similar type of vesicles were prepared for delivery of quercetin (80–200 nm) to target deeper skin layers and results were confirmed by release and tape striping experiments in one-day-old Goland–Pietrain hybrid pigs. In this study, vesicles based on four-penetration enhancer like propylene glycol, polyethylene glycol

TABLE 10.2 List of Herbal Actives Encapsulated in Various "SOMES"

Name of plant	Active	Reference
Ethosomes		
Glycerrhiza glabra	Ammonium glycyrrhizinate	World J Pharm Pharmaceutical Sci, *4*(6), 972–989
Cannabis sativa	tetrahydrocannabi-diol	J Control Release, **2003**, *93*(3), 377–387
Tripterygium wilfordii	Triptolide	African J Pharm and Pharmacol, **2012**, *6*(13), 998–1004
Sophora alopecuerides	Alkaloid extract	AAPS Pharm Sci Tech, **2010**, *11*(3), 1350–1358
Curcuma longa	Curcumin	J Cosmet Dermatol, **2011**, *10*(4), 260–265
Liposomes		
	Magnolol	https://mibellebiochemistry.com/app/uploads/2015/03/MAXnolia_Magnolia_Derived_Honokiol_and_Magnolol_Fight_Against_Skin_InflammAging_CosmeticsDesign_Feb_2010.pdf
Nux vomica		*Acta Pharmacologica Sinica* **2007**, *28*, 1851–1858
	Quercetin	J Photochem and Photobiol B: Biol, **2013**, *127*, 8–17
	Diospyrin	Int. Res J Pharm. App Sci, **2013**, *3*(4), 40–50
Myrtus communis		Microbiol Res, **2014**, *169*(4), 240–254
Artemisia arborescens		Int J Nanomedicine, **2007**, *2*(3), 419–425
	Puerarin	Int J Cosmetic Science, **2008**, *30*(4), 285–295
Nanoparticles		
	Berberine	Int J Nanomedicine, **2009**, *4*, 227–232
	Quercitrin	Int J Nanomedicine, **2011**, *6*, 1621–1630
	Hypocrellin	US 20140170229 A1
	Silybin	Alternative Medicine Review, *16*(3), 239–249
	Ginseng	EP 2218447 A1
Radix Salvia miltiorrhiza		Int J Nanomedicine, **2014**, *9*, 1–15

TABLE 10.2 (Continued)

Name of plant	Active	Reference
Lipid nanoparticles		
	Curcumin	Res J of Pharmaceutical, Biological and Chem Sci, **2013**, *4*(1), 784–801
	Curcumoids	Int J Pharm. **2012**, 28, *423*(2), 440–451
	Beta carotene	Braz. Arch. Biol. Technol, **2013**, *56*(4), 663–671
	Lycopene	Pharmazie, **2013**, *68*(9), 723–31
Phytosomes		
	Quercetin	Pharmaceutical Sciences, **2014**, *20*, 96–101
	Oxymatrine	Res J of Pharm Dosage Forms and Technol, **2014**, *6*(1), 44–49
Ginkgo biloba		Phytother Res. **2006**, *20*(11), 1013–6.
	Marsupium	AAPS Pharm Sci Tech. **2008**, *9*(1), 129–137.
	Embelin	J. Incl. Phenom. Macrocycl. Chem. **2011**, *69*, 139–147.
	Naringenin	Phytosome: Drug Delivery System for Polyphenolic Phytoconstituents, **2011**, *7*(4), 209–219.
Silibium marianum	Silybin	World J Gastroenterol, **2011**, 17(18), 2288–2301.
Vitis vinifera	Catechin, epicatechin	Int Res J Pharm, **2011**, *2*(6), 28–33 http://leucoselect.phytosome.info/
Curcuma longa	Curcumin	http://www.nowfoods.com/Curcumin-Phytosome-60-Veg-Capsules.htm
Thea sinensis	Epigallocatechin, epicatechin	http://greenselect.phytosome.info/ Altern Med Rev, **2009**, *14*(2), 154–60
Panax ginseng	Ginseng	http://www.indena.com/products/personal-care/ginselect-phytosome-ginseng-idb-2/
Ginko biloba	Ginkgoic acid	http://naturalfactors.com/product/ginkgo-biloba-phytosome/
Transfersome		
Capsicum annum	Capsaicin	China Journal of Chinese Materia Medica, **2006**, *31*(12)
Curcuma longa	Curcumin	African Journal of Pharmacy and Pharmacology, **2011**, *5*, 1054–1062
Catharanthus roseus	Vincristine	Chinese Journal of Experimental Traditional Medical Formulae, **2012**

400, labrasol and Transcutol showed potential in formation of vesicles (Chessa, 2009).

10.6.3 PHYTOSOMES

Phytosomes are molecular complexes of lipids mostly phosphatidylcholine and phytoconstituents (Gupta et al., 2007). Conjugation/chemical linkage in complexes offers the stability to colloidal system due to which drug loading is quite high compared to other carriers. Quercetin phytosomes were investigated by Maiti et al. (2005) for its hepatoprotective effects and bioavailability enhancement. Same group also reported use phytosomes for naringenin and curcumin to exert antioxidant effects (Maiti et al., 2007). Cao et al., investigated various ways of topical delivery of Oxymatrine, a quinolizidine alkaloid compounds (*Sophora flavescens*). Oxymatrine-phospholipid complex and microemulsion were chosen to improve skin permeation and occlusion (Fa-Hao et al., 2010).

Various other phytosomes-based studies evaluated effectiveness of phytosomes in delivery of herbal actives. Some of the examples include phytosomes of *Gingko biloba* terpenes for topical application (Loggia et al., 1996), silymarin phytosomes with biological effects and activity toward skin phospholipid (Yanyu et al., 2006), improved transdermic action of ginseng saponin phytosomes due to improved penetration and hydration of cutaneous layer (Kidd, 2009). Gupta and Dixit evaluated liposomes, niosomes, and phytovesicles for encapsulation of curcumin. They observed that phytovesicles composed of phosphatidylcholine could exert improved antioxidant and anti aging properties are due to the amphiphilic nature of the complex compared to liposomes and niosomes (Gupta and Dixit, 2011). Similar results were obtained by Zaveri et al. (2011) for curcumin-phospholipid complex which showed 60% improved permeation through rat skin. *Curcuma longa* extract (0.5 to 2%) loaded liposomes, ethosomes, and transfersomes in size range of 167 to 262 nm are prepared by Saraf et al. (2011) and Kaur and Saraf (2011). The photoprotective effect of vesicles incorporated in cream was determined by assessing skin hydration (Cutometer) and sebum content (Sebumeter) up to 6 weeks. All the vesicles excreted improved skin hydration effect, the effect was much more noteworthy for transfersomes loaded creams flowed by ethosomes and liposome-based creams.

10.6.4 PHARMACOSOMES

Pharmacosomes are either micelles, vesicles or form of hexagonal assembly of colloidal drug attached covalently to the phospholipid. They can be loaded with active or attached covalently to the active (Sharma, 2013). Microparticulate lipoidal carriers or lipid-based supramolecular vesicular systems can serve as other interesting drug delivery platforms for topical applications (Kumar et al., 2012). These particles are currently evolving in cosmetic research.

10.7 LIMITATIONS OF NANOTECHNOLOGY IN HERBAL COSMETICS

Application of nanotechnology to herbal cosmetics is growing very fast and can be seen by increasing number of publications and products in the market. Nanoparticles due to smaller particle size, large surface area and lipophilic nature could cross skin barriers and exert multifold increase in beneficial effects. Various carriers like liposomes, SLNs/NLCs, nanoemulsions, ethosomes, niosomes, micelles and many others showed promising results in delivering herbal actives to different skin layers.

With the advent of novel drug delivery systems and new methods of preparation several patents in cosmetics have been granted in the recent years. Some examples to state are like compositions and methods for administering collagen to humans have been developed by Pinksky et al. (EP1951762 A1/2008). The collagen-containing lipid vesicles could deliver the human collagen which eliminates problems of chemical and physical instability of the collagen as well as immune responses to nonhuman collagen. Recently, Kim et al., has patented a cosmetic composition (WO 2010093065 A1/2010), which contains xanthan gum up to 5 parts of the herbal powder for dead cell removal and skin regeneration with DDS mechanism, in which irritation relief, efficiency, effects, and usability are improved. Xanthan gum, along with a hydrophilic moisturizer, a skin regenerative/functional ingredient or a natural antiinflammatory ingredient (*Portulaca Oleracea* extract), and natural herb oil (kalrip deuce, tea tree) are mixed in an herbal powder at predetermined ratio, which is a mixture of nine kinds of herbal powders. A multi component controlled release system for anhydrous cosmetic compositions patented by Shefer & Shefer of Salvona LLC (WO 2003088894/A2/2003) which describes an improved controlled release system that can be incorporated into anhydrous cosmetic formulations and can encapsulate different types of fragrances, flavors, active ingredients, or combinations of them. The controlled delivery system is a free-flowing powder of solid hydrophobic nano-spheres that are encapsulated in a moisture sensitive microspheres composition. The fragrances, flavors, and active ingredients encapsulated in the nano-spheres can be the same or different from those encapsulated in the microsphere. A Chinese patent (CN 103230355 B/2015) on cosmetic composition discloses a sunscreen cosmetic composition comprising nano-crystalline cellulose, including nano-crystalline cellulose, and titanium dioxide. The composition of the present invention includes glyceryl monostearate, petrolatum, xanthan gum, aloe gel, glycerol, Span-60, Tween-60 and paraben. The formulation protects from 320 ~ 400 nm UVA and UVB region area of 280 ~ 320 nm UV irradiation, its sun protection factor (SPF) values > 15 showing high sunscreen efficiency. Another Chinese patent (CN 104116651/2014), discloses a herbal essence skin care product composed of a nano-liposome and water-soluble active matters, that comprise, by weight, 4–6 parts of *Rhodiola rosea*, 3–5 parts of Herba Leonuri, 2–4 parts of Lonicera japonica, 1–3 parts of Salvia miltiorrhiza and 1–2 parts of *Centella asiatica*. The nano-liposome carrier has good physiological compatible safety, and can enter hypodermal cells in a physical form. The transmission and the efficacies of the active substances can be well and fast performed by adopting a nano-liposome loading technology. The water-soluble part comprised *Rhodiola rosea*, *Herba Leonuri*, *Lonicera japonica*, *Salvia miltiorrhiza* and *Centella asiatica* are wrapped by a nano-liposome wrapping technology, so the above five-kinds herbal essence has the characteristics of no irritation to skins, fine, smooth, nature comfortable hand feeling, and increase of the skin residence time and the absorption utilization rate by times. These patents are a testimony of the advances in the field of herbal nanocometics showing promising future and changes in the global cosmetics research.

Beside ongoing discussion on toxicity related nanoparticle, scalability is another issue. Till date, only few papers describe about production scalability of herbal nanoparticles. PharmaSol GmbH, Germany and Dr. Rimpler offers scaling up for lipid-based formulations and scaled up till date CoQ10, argan oil loaded SLNs beside several flavonoid and ascorbic acid nanocrystals for dermal applications. In exploring nanoparticles for cosmetic delivery, excipient compatibility of herbal active or extract especially with synthetic excipients should be evaluated to avoid formation of any harmful complex or product formation during stability. Long-term stability studies on formulation are equally important which are studied limitedly. The threshold toxicological concern and dermal sensitization threshold can be considered as important parameters for the assessment of nanoparticle safety at long-term. Beside local skin irritation test, it is highly recommended to be aware of allergy issue for particular herb constituent or plant. An optimized nanotoxicity study *in vivo* testing protocol needs to be designed to get complete idea on irritation, toxicity and allergy of nanoparticle by selecting optimum study methodology, cell line or animal model (Shegokar et al., 2015). In addition to that, testing protocol for both *in vitro* and *in vivo* evaluation of herbal cosmetics needs to be properly set up to get trustworthy results and complete information of product performance. Table 10.3 lists various tests that are currently

TABLE 10.3 List of Test Used in Evaluation of Skin Care Products

Test	Use
Transepidermal water loss (TEWL)	To measure transdermal water loss
Occlusion effect	To Determine occlusion effect by various tests like hydration *Or* by Microdialysis methodology which measures dermal interstitial fluid or by skin conductance, electrical impedance
Strip testing- penetration	To determine in vivo penetration of active in skin layers
Franz diffusion release	To check release kinetics of formulation either using artificial membrane or pig, human cadever skin, etc.
Skin irritation	It is a *vivo* test to determine irritation index of formulation when applied topically on shaved rat, rabbit, mouse and human skin

TABLE 10.3 (Continued)

Test	Use
Rabbit eye test	It is a *in vivo* test to determine irritation index of nanomaterials for ocular care formulations
Skin whitening effect	It is a *vivo* test to determine skin whitening by determining malanine content and or biomarker effect or by using Chromameter® CR 400
Anti Wrinkle effect	To determine the wrinkle reduction before and after application of formulation using techniques like skin autofluorescence (SAF) or high frequency ultrasound measurement of digital dermal thickness
Hair loss or thinning analysis	To determine hair growth by epiluminescence microscopy (ELM) and other techniques
Skin pH determination	To determine skin pH using pH meter
Skin firmness	To determine antiwrinkle effect and occlusive effect indirectly by Twistometer or Dermal Torque Meter
Skin firmness	To determine viscoelasticity property on human skin by Cutometer

used in cosmetic evaluation. However, available literature data on safety studies shows that herbal nanoparticles are safe for topical application, based on the excipients chosen. Green chemistry is another eye catching research area which is used to synthesize nanoparticles mainly mediated by microbes, plants, and chemical routes via sonication, microwaving, and many others methods could be another safe way to produce metals nanoparticles and natural vesicles.

10.8 CONCLUSION

Modern world is fascinated by herbal cosmetic products and they are well accepted by consumer for their application effects. Variety of drug delivery system offers delivery of active on skin surface or in deeper layers of skin besides offering physical chemical stability to active. Herbal active associated challenges like physical and chemical instability, solubility can be overcome by encapsulation in vesicles or lipid nanoparticles. Till date different carriers like nanocrystals, liposomes, lipid nanoparticles and niosomes have been explored for dermal delivery and found to be safe and effective carriers. Skin surface or stratum corneum localization is relatively nontoxic and stable. However, Cosmetic formulations based on herbal actives needs to follow standard set of testing and strict protocol to avoid any complications. Till date limited research has been performed on dose dependent and repeated exposures of herbal or synthetic actives. Nonetheless, future of herbal natural cosmetic actives in form of nanoparticles is bright and most awaited by consumer worldwide as next beauty regime.

KEYWORDS

- **beauty**
- **cosmetics**
- **dermal**
- **drug delivery systems**
- **ethosomes**
- **herbal actives**
- **liposomes**
- **nanocrystals**
- **skin**
- **solid lipid nanoparticle**

REFERENCES

Affandi, M.; Julianto, T.; Majeed, A.; Development and Stability Evaluation of Astaxanthin Nanoemulsion, *Asian Journal of Pharmaceutical and Clinical Research*, **2011**, *4*.

Al Shaal, L.; Müller, R. H.; Shegokar, R.; smartCrystal combination technology–scale up from lab to pilot scale and long-term stability, *Pharmazie* **2010**, *65*, 877–884.

Al Shaal, L.; Shegokar, R. et al. "Production and characterization of antioxidant apigenin nanocrystals as a novel UV skin protective formulation." International Journal of Pharmaceutics **2011**, *420*(1), 133–140.

Al Shaal, L.; Shegokar, R.; Muller, R. H.; Production and characterization of antioxidant apigenin nanocrystals as a novel UV skin protective formulation, *International Journal of Pharmaceutics* **2011**, *420*, 133–140.

Amenta, V.; Aschberger, K.; Arena, M.; Bouwmeester, H.; F. Botelho Moniz, Brandhoff, P.; Gottardo, S.; Marvin, H. J.; Mech, A.; Quiros Pesudo, L.; Rauscher, H.; Schoonjans, R.; Vettori, M. V.; Weigel, S.; Peters, R. J.; Regulatory aspects of nanotechnology in the agri/feed/food sector in EU and non-EU countries, *Regulatory toxicology and pharmacology: RTP* **2015**, *73*, 463–476.

Bhramaramba R. et al., Formulation and Evaluation of Herbal Gel Containing Terminalia chebula Retz.; Leaves Extract, *Sch. Acad. J. Pharm.* **2015**, *4*, 172–176.

Builders, P. F.; Iwu, I. W.; Mbah, C. C.; Iwu, I. W.; Builders, M. I.; Audu, M. M.; Moringa Oleifera Ethosomes a Potential Hair Growth Activator: Effect on Rats. *J Pharm Biomed Sci* **2014**, *4*, 611–618.

Butnariu, M.; Giuchici, C.; The use of some nanoemulsion-based on the aqueous propolis and lycopene extract in the skins protective mechanisms against UVA radiation, *Journal of Nanoscience and Nanotechnology*. **2011**, *9*, 1–9.

Caddeo, C.; Chessa, M.; Vassallo, A.; Pons, R.; Diez-Sales, O.; Fadda, A. M.; Manconi, M.; Extraction, purification and nanoformulation of natural phycocyanin (from Klamath algae) for dermal and deeper soft tissue delivery, *Journal of Biomedical Nanotechnology* **2013**, *9*, 1929–1938.

Canham, L.; Qurrat, S. U. A.; et al. "Composition containing loaded and capped porous silica particles" US 9243144 B2, **2011**.

Cao Fa-Hao, Ou Yang, W.; Wang, Y.; Yue, P.; Li, S.; A combination of a microemulsion and a phospholipid complex for topical delivery of Oxymatrine, *Archieves of Pharmcal Research* **2010**, *34*, 551–562.

Chen, R.; Tailor-made antioxidative nanocrystals: production and in vitro efficacy, Ph.D. Thesis Freie Universität Berlin, Germany (2013).

Chessa, M.; Innovative Liposomes for overcoming biological barriers, University of Cagliari, PhD thesis, http://veprints.unica.it/840/1/Chessa_PhD_Thesis.pdf (2009).

Chidambaram, M.; Chinnamaruthu, S.; Sellappan, M.; Krishnasamy, K.; Anionic Surfactant Based Topical Curcumin Nanosuspension: Fabrication, Characterization and Evaluation, *Nano Biomed Eng* **2013**, *5*, 86–89.

Corrias, F.; Nanocarriers for drug targeting and improved bioavailability, PhD Thesis, *Universita' degli Studi di Cagliari* http://veprints.unica.it/942/ (2014).

D'souza, A.; Shegokar, R. "Applications of Metal Nanoparticles in Pharma." In: Metal nanoparticles in Pharma, Ed: Rai M.; Shegokar R.; Springer Verlag, 2017.

Fan C. et al., Enhanced Topical Delivery of Tetrandrine by Ethosomes for Treatment of Arthritis, *BioMed Research International Article ID 161943*, **2013**, 13.

Guo, T.; Zhang, Y.; Zhao, J.; Zhu, C.; Feng, N.; Nanostructured lipid carriers for percutaneous administration of alkaloids isolated from Aconitum sinomontanum *Journal of Nanobiotechnology* **2015**, *13*, 47.

Gupta, A.; Ashawat, M. S.; Saraf, S.; Saraf, S.; Phytosome: A Novel Approach Towards Functional Cosmetics, *Journal of Plant Sciences* **2007**, *2*, 644–649.

Gupta, N. K.; Dixit, V. K.; Development and evaluation of vesicular system for curcumin delivery, *Achieves of Dermatological Research* **2011**, *303*, 89–101.

Hommoss, A.; Al-Samman, M.; Müller, R. H.; UV radiation blocking activity booster depending on nanostructured lipid carriers (NLC), *Annual Meeting of the American Association of Pharmaceutical Scientists (AAPS)*, San Diego, USA, **2007**, pp. 2269.

Hommoss, A.; Nanostructured lipid carriers (NLC) in dermal and personal care formulations, PhD thesis, Free University of Berlin, Berlin, 2008.

Hommoss, A.; Peter, M.; Müller, R. H.; Sun protection factor (SPF) increase using nanostructured lipid carriers (NLC), Annual Meeting of the Controlled Release Society (CRS), Long Beach, USA, **2007**, pp. 760.

http://sphinxsai.com/2014/PTVOL6/PT=56(842–849)AJ14.pdf.

Jeswani, G.; Saraf, S.; Topical Delivery of Curcuma longa Extract Loaded Nanosized Ethosomes to Combat Facial Wrinkles, *J Pharm Drug Deliv Res* **2014**, *3*, doi: 10.4172/2325-9604.1000118.

Jin, S. Y.; Yuan, H. L.; Jin, S. X.; Lv, Q. Y.; Bai, J. X.; Han, J.; Preparation of baicalin nanocrystal pellets and preliminary study on its pharmacokinetics, *China Journal of Chinese materia medica* **2013**, *38*, 1156–1159.

Jun, C. Y.; Studies on Oleanolic Acid Nanosuspensions, Huazhong University of Science and Technology, **2005**, http://www.dissertationtopic.net/doc/1597424.

Kakran, M.; Shegokar, R.; Sahoo, N. G.; L. Al Shaal, Lin, L.; Müller, R. H.; Fabrication of Quercetin Nanocrystals: Comparison of Different Methods, *European Journal of Pharmaceutics and Biopharmaceutics*, **2012**, *80*, 113–121.

Katz, L. M.; Dewan, K.; Bronaugh, R. L.; Nanotechnology in cosmetics, Food and chemical Toxicology: An International Journal Published for the *British Industrial Biological Research Association*, **2015**.

Kaur, C. D.; Saraf, S.; Topical vesicular formulations of Curcuma longa extract on recuperating the ultraviolet radiation-damaged skin, *Journal of Cosmetic Dermatology* **2011**, *10*, 260–265.

Kidd, M.; Bioavailability and activity of phytosome complexes from botanical polyphenols: The silymarin, curcumin, Green tea and Grape seed extracts, *Alternative Medicines Review* **2009**, *14*, 226–246.

Kim, J. H. (2010). "Nanoparticle composition for prevention of hair loss and promotion of hair growth " US 20100104646 A1.

Kim, J. H.; Ko, J. A.; Kim, J. T.; Cha, D. S.; Cho, J. H.; Park, H. J.; Shin, G. H.; Preparation of a capsaicin-loaded nanoemulsion for improving skin penetration, *Journal of Agricultural and Food Chemistry* **2014**, *62*, 725–732.

Kobierski, S.; Ofori-Kwakye, K.; Muller, R. H.; Keck, C. M.; Resveratrol nanosuspensions: interaction of preservatives with nanocrystal production, *Die Pharmazie* **2011**, *66*, 942–947.

Kumar, D.; Sharma, D.; Singh, G.; Singh, M.; Rathore, M. S.; Lipoidal soft hybrid biocarriers of supramolecular construction for drug delivery, *ISRN Pharmaceutics* 2012 (2012) 474830.

Lai, F.; Pireddu, R.; Corrias, F.; Fadda, A. M.; Valenti, D.; Pini, E.; Sinico, C.; Nanosuspension improves tretinoin photostability and delivery to the skin, *International Journal of Pharmaceutics* **2013**, *458*, 104–109.

Leelapornpisid, P. et al., Nanoemulsion loaded with marigold flower extract (*Tagetes Erecta* Linn) in gel preparation as antiwrinkle cosmeceutical International Journal of Pharmacy and Pharmaceutical Sciences **2014**, *6*, 231–236.

Lin, S. (2013). "Magnolol nanosuspension increased the oral bioavailabilityof magnolol and its metabolites in rats, 19–20 Oct." Annual International Symposium on traditional Chinese medicine and the 28th Symposium on natural medicine, Ministry of health and welfare, National Institute of traditional Chinese medicine.

Loggia, R. D.; Sosa, A. T.; Morazzoni, P.; Bombardelli, E.; Anti-Inflammatory activity of some Ginkgo biloba constituents and their Phospholipid complexes, *Fitoterapia* **1996**, *3*, 257–273.

Lopes, L. B.; Van DeWall, H.; Li, H. T.; Venugopal, V.; Li, H. K.; Naydin, S.; Hosmer, J.; Levendusky, M.; Zheng, H.; Bentley, M. V.; Levin, R.; Hass, M. A.; Topical delivery of lycopene using microemulsions: enhanced skin penetration and tissue antioxidant activity, *J Pharm Sci* **2010**, *99*, 1346–1357.

Mahdi, E. S.; Noor, A. M.; Sakeena, M. H.; Abdullah, G. Z.; Abdulkarim, M. F.; Sattar, M. A.; Formulation and in vitro release evaluation of newly synthesized palm kernel oil esters-based nanoemulsion delivery system for 30% ethanolic dried extract derived from local *Phyllanthus urinaria* for skin antiaging, *International Journal of Nanomedicine* **2011**, *6*, 2499–2512.

Maiti, K.; Mukherjee, K.; Gantait, A.; Ahamed, H. N.; Saha, B. P.; Mukherjee, P. K.; Enhanced therapeutic benefit of quercetin-phospholipid complex in carbon tetrachloride induced acute liver injury in rats: a comparative study, *Iranian Journal of Pharmacology & Therapeutics* **2005**, *4*, 84–90.

Maiti, K.; Mukherjee, K.; Gantait, A.; Saha, B. P.; Mukherjee, P. K.; Curcumin-phospholipid complex: preparation, therapeutic evaluation and pharmacokinetic study in rats, *International Journal of Pharmaceutics* **2007**, *330*, 155–163.

Manosroi, A.; Jantrawut, P.; Akihisa, T.; Manosroi, W.; Manosroi, J.; In vitro and in vivo skin antiaging evaluation of gel containing niosomes loaded with a semipurified fraction containing gallic acid from Terminalia chebula galls. *Pharm Biol* **2011**, *49*, 1190–1203.

Marques de Assis, A. et al., Development and Characterization of Nanovesicles Containing Phenolic Compounds of Microalgae Spirulina Strain LEB-18 and Chlorella pyrenoidosa, *Advances in Materials Physics and Chemistry* **2014**, *4*, 7.

Mauludin, R.; Nanosuspensions of poorly soluble drugs for oral administration, Freie Universität Berlin, Germany, 2008.

Mercurio, D. G.; Wagemaker, T. A.; Alves, V. M.; Benevenuto, C. G.; Gaspar, L. R.; P.M. Maia Campos, In vivo photoprotective effects of cosmetic formulations containing UV filters, vitamins, Ginkgo biloba and red algae extracts, *Journal of Photochemistry and Photobiology. B, Biology* **2015**, *153*, 121–126.

Mihranyan, A. E. A.; Current status and future prospects of nanotechnology in cosmetics, *Progress in Materials Science* **2011**, *57*, 875–910.

Mishra, M.; Shegokar, R.; Müller, R. H.; Gohla, S.; Glycyrrhetinic acid SmartCrystal® with improved Skin Penetration, American Association of Pharmaceutical Scientist (AAPS) Annual Meeting and Exposition, Washington DC, 2011.

Mishra, M.; Shegokar, R.; Müller, R. H.; Mitri, K.; Gohla, S.; Lutein nanosuspension with improved skin penetration in human, July 15–18, 39th Annual Meeting & Exposition of the Controlled Release Society, Canada, 2012.

Mitri, K.; Shegokar, R.; Gohla, S.; Anselmi, C.; Muller, R. H.; Lipid nanocarriers for dermal delivery of lutein: preparation, characterization, stability and performance, *International Journal of Pharmaceutics* **2011**, *414*, 267–275.

Mitri, K.; Shegokar, R.; Gohla, S.; Anselmi, C.; Muller, R. H.; Lutein nanocrystals as antioxidant formulation for oral and dermal delivery, *International Journal of Pharmaceutics* **2011**, *420*, 141–146.

Moorthi, C.; Kathiresan, K.; Fabrication of highly stable sonication assisted curcumin nanocrystals by nanoprecipitation method, *Drug Invention Today* **2013**, *5*, 66–69.

Mu, L.; Sprando, R. L.; Application of nanotechnology in cosmetics, *Pharmaceutical Research* **2010**, *27*, 1746–1749.

Müller, R. H.; Dingler, A.; Schneppe, T.; Gohla, S.; Large scale production of solid lipid nanoparticles (SLN™) and nanosuspensions (DissoCubes™), in: D. Wise (Ed.), Handbook of Pharmaceutical Controlled Release Technology, Marcel Dekker Inc.; New York, 2000, pp. 359–376.

Nohynek, G. J.; Dufour, E. K.; Roberts, M. S.; Nanotechnology, cosmetics and the skin: is there a health risk? *Skin Pharmacology and Physiology* **2008**, *21*, 136–149.

Nosari, L.; Freitas, P.; http://www.cifarp.com.br/historico/2013/painel/_inc/trabalhos/resumo1304.pdf, (2013).

Patent CA 2746566 A1, Topical herbal formulation for treatment of acne and skin disorders

Pinsuwan, S.; Amnuaikit, T.; Ungphaiboon, S.; Itharat, A.; Liposome-containing Hibiscus sabdariffa calyx extract formulations with increased antioxidant activity, improved dermal penetration and reduced dermal toxicity, *Journal of the Medical Association of Thailand (= Chotmaihet thangphaet) 93 Suppl* **2010**, *7*, S216–226.

Quintao, F. et al., Hydroalcoholic extracts of Vellozia squamata: study of its nanoemulsions for pharmaceutical or cosmetic applications, *Rev. Bras. Farmacogn.* **2013**, *23*, 101–107.

Rachmawati, H.; Al Shaal, L.; Muller, R. H.; Keck, C. M.; Development of curcumin nanocrystal: physical aspects, *Journal of Pharmaceutical Sciences* **2013**, *102*, 204–214.

Raj, S.; Jose, S.; Sumod, U. S.; Sabitha, M.; Nanotechnology in cosmetics: Opportunities and challenges, *Journal of Pharmacy and Bioallied Sciences* **2012**, *4*, 186–193.

Ravichandran, R.; Studies on Dissolution Behavior of Nanoparticulate Curcumin Formulation, *Advances in Nanoparticles* 2 (2013).

Rosli, N. A.; Hasham, R.; Abdul Aziz, A.; Aziz, R.; Formulation and characterization of nanostructured lipid carrier encapsulated *Zingiber zerumbet* oil using ultrasonication technique, *Journal of Advanced Research in Applied Mechanics* **2015**, *11*, 16–23.

Rungphanichkul, N.; Nimmannit, U.; Muangsiri, W.; Rojsitthisak, P.; Preparation of curcuminoid niosomes for enhancement of skin permeation, *Die Pharmazie* **2011**, *66*, 570–575.

Sahoo, N. G.; Kakran, M.; Shaal, L. A.; Li, L.; Muller, R. H.; Pal, M.; Tan, L. P.; Preparation and characterization of quercetin nanocrystals, *Journal of Pharmaceutical Sciences* **2011**, *100*, 2379–2390.

Saraf, S.; Jeswani, G.; Kaur, C. D.; Saraf, S.; Development of novel herbal cosmetic cream with Curcuma longa extract loaded transfersomes for antiwrinkle effect, African *Journal of Pharmacy and Pharmacology* **2011**, *5*, 1054–1062.

Schmid, D.; Belser, E.; Zülli, F.; Rejuvenating effect of snow algae analyzed, PERSONAL CARE https://mibellebiochemistry.com/app/uploads/2015/03/Snow-Algae-Powder_Rejuvenating_Effect_of_Snow_Algae_Analyzed_Personal_Care_Magazine_Europe_04_2014.pdf (2014).

Schmid, D.; Schürch, C.; Fred, Z.; Cosmetics and Toiletries Manufacture Worldwide, UV-A sunscreen from red algae for protection against premature skin aging https://mibellebiochemistry.com/app/uploads/2015/03/Helioguard-365_UV-A-sunscreen-from-red-algae-for-protection-against-premature-skin-aging_CT-2004.pdf.

Sharma, P. A. P.; Pharmacosomes: An Emerging Novel Vesicular Drug Delivery System for Poorly Soluble Synthetic and Herbal Drugs, ISRN Pharmaceutics (2013).

Shegokar, R.; Mitri, K.; Carotenoid lutein: a promising candidate for pharmaceutical and nutraceutical applications, *Journal of Dietary Supplements* **2012**, *9*, 183–210.

Shegokar, R.; Müller, R. H.; Nanocrystals: Industrially feasible multifunctional formulation technology for poorly soluble actives, *International Journal of Pharmaceutics* **2010**, *399*, 129–139.

Shegokar, R.; Nanotechnology Based Apigenin Drug Delivery Systems, in: N.M. Stacks (Ed.), Natural Sources, Pharmacology and Role in Cancer Prevention: Handbook on Flavonoids: Dietary Sources, Properties and Health Benefits, Nova Science Publishers, NY, US, 2014.

Shegokar, R.; Nanotoxicity: must consider aspect of nanoparticle development, in: M. Lungu, Neculae, A.; Bunoiu, M.; Biris, C. (Ed.), Nanoparticles' Promises and Risks – Characterization, Manipulation, and Potential Hazards to Humanity and the Environment, Springer International Publishing, 2015, pp. 87–102.

Shi-Ying, J.; Jin, H.; Shi-Xiao, J.; Qing-Yuan, L.; Jin-Xia, B.; Chen, H. G.; Rui-Sheng, L.; Wei, W.; Hai-Long, Y.; Characterization and evaluation in vivo of baicalin-nanocrystals prepared by an ultrasonic-homogenization-fluid bed drying method, *Chinese Journal of Natural Medicines* **2014**, *12*, 71–80.

Song, J.; Wang, Y.; Song, Y.; Chan, H.; Bi, C.; Yang, X.; Yan, R.; Zheng, Y.; Development and characterization of ursolic acid nanocrystals without stabilizer having improved dissolution rate and in vitro anticancer activity, *AAPS Pharm. Sci. Tech.* **2014**, *15*, 11–19.

Sucontphunt, A.; Eaknai, W.; Ondee, T.; Pongpunyayuen, S.; Chansriniyom, C.; Bunwatcharaphansakun, P.; Sirival, N.; Tangkananond, W.; Nimmannit, U.; Abstract, Anti-oxidant, Anti-elastase activity and nano formulation of herbal extracts, *AAPS Conference, Abstract W4126* (2013).

Takahashi, M.; Kitamoto, D.; Asikin, Y.; Takara, K.; Wada, K.; Liposomes encapsulating Aloe vera leaf gel extract significantly enhance proliferation and collagen synthesis in human skin cell lines, *Journal of Oleo Science* **2009**, *58*, 643–650.

Teeranachaideekul V. et al., Liposomes: A novel carrier system for Artocarpus lakoocha extract to improve skin whitening, *JAASP* **2013**, *2*, 243–253.

Vance, M.; Kuiken, T.; Vejerano, E.; McGinnis, S.; Hochella, Jr. M.; Rejeski, D.; Hull, M.; Nanotechnology in the real world: Redeveloping the nanomaterial consumer products inventory, *Beilstein J. Nanotechnol.* **2015**, *6*, 1769–1780.

Wang, D.; Exploring the potential of algae in cosmetics, *Bioresource Technology* **2015**, *184*, 355–362.

Weblink 1, http://www.conference.net.au/chemeca2011/papers/401.pdf (2011).

Weblink 1a, https://www.alliedmarketresearch.com/cosmetics-market, assessed on 30th October 2015.

Weblink 2, http://abstracts.aaps.org/Verify/aaps2013/postersubmissions/W4097.pdf (2013).

Weblink 2a, www.lipoid-kosmetik.com.

Weblink 3, http://abstracts.aaps.org/Verify/AAPS2014/PosterSubmissions/M1191.pdf (2014).

Weblink 5, http://www.naturalislife.com/nio-active.html (2015).

Weblink 6, http://mt.china-papers.com/2/?p=244461 (2011).

Wei, X. L.; Han, Y. R.; Quan, L. H.; Liu, C. Y.; Liao, Y. H.; Oily nanosuspension for long-acting intramuscular delivery of curcumin didecanoate prodrug: preparation, characterization and in vivo evaluation, *European Journal of Pharmaceutical Sciences: Journal of the European Federation For Pharmaceutical Sciences* **2013**, *49*, 286–293.

Yanyu, X.; Yunmei, S.; Zhipeng, C.; Qineng, P.; The preparation of silybin-phospholipid complex and the study on its pharmacokinetics in rats, *International Journal of Pharmaceutics*, **2006**, *307*, 77–82.

Zaveri, M.; Gajjar, H.; Kanaki, N.; Patel, S.; Preparation and evaluation of drug phospholipid complex for increasing transdermal penetration of phytoconstituents, International *Journal of Institutional Pharmacy and Life Sciences* **2011**, *1*, 80–93.

Zhang, J. A.; Yin, Z.; Ma, L. W.; Yin, Z. Q.; Hu, Y. Y.; Xu, Y.; Wu, D.; Permatasari, F.; Luo, D.; Zhou, B. R.; The protective effect of baicalin against UVB irradiation induced photoaging: an in vitro and in vivo study, *PloS one* **2014**, *9*, e99703.

Zippin, J. H.; Friedman, A.; Nanotechnology in cosmetics and sunscreens: an update, *Journal of Drugs in Dermatology: JDD* **2009**, *8*, 955–958.

CHAPTER 11

ANTIMICROBIAL ACTIVITY OF NANOTECHNOLOGICAL PRODUCTS

LEONARDO QUINTANA SOARES LOPES,[1]
MÁRCIA EBLING DE SOUZA,[1]
RODRIGO DE ALMEIDA VAUCHER,[1] and
ROBERTO CHRIST VIANNA SANTOS[2,*]

[1]*Microbiology Laboratory Research, Centro Universitário Franciscano, Santa Maria-RS, 97010-032, Brazil*

[2]*Microbiology and Parasitology Department, Federal University of Santa Maria, Santa Maria-RS, 97105-900, Brazil,*
**E-mail: robertochrist@gmail.com*

CONTENTS

11.1 INTRODUCTION

The nanoscience is the basic scientific study between one and hundred nanometers, aiming acquire and create theories, concepts, fundamentals and technical knowledge which help in understanding, creation and manipulation of nanostructures. Also exhibit fundamental control on the physical and chemical properties of structures on a molecular scale (Ratner and Ratner, 2003). The prefix "nano" is related to a measurement scale where one nanometer represents 1 billionth of a meter or 10^{-9} of meter.

The nanotechnology is a new term and refers to the ability to manipulate individual atoms and molecules to produce nanostructured materials with different chemical and physical-chemical properties and different behavioral those shown in bulk form. The matter on nanoescale, change their properties creating materials have greater surface area, higher reactivity, increased conductivity (Safari and Zarnegar, 2014).

11.2 HISTORY AND APPLICATIONS

The year 1959 is considered the starting point of nanotechnology, when Richard Feynman held a lecture entitled "There's Plenty of Room at Bottom" at the annual meeting of the American Physical Society. Feynman discussed the possibilities to condense on a pin head, all pages of the Encyclopedia Britannica, suggesting a technology able to build nano-objects atom by atom, molecule by molecule extrapolating the known physical laws. Feynman also referred to different objects and scientific areas that could be optimized with the development of technology at the nanoscale, such as faster computers and advances in the biological sciences (Toumey, 2008).

Years later, in 1981, Gerd Binnig and Heinrich Rohrer developed the scanning tunneling microscope in laboratories of International Business Machines (IBM) in Switzerland. The microscope was very advanced for its time, as well as produce images at the atomic scale, allowing individual movement of atoms, as suggested by Feynman in 1959, the microscope was awarded the Nobel Prize in 1986. Thus nanotechnology found the path of development using these new tools (Binnig and Rohrer, 1982).

Nanotechnology is one of emerging technologies of the last decades, show an inter and multidisciplinary field, it covers areas such as Chemistry, Physics, Biology, Computer Science, Engineering and Medicine, which act simultaneously in the search for new materials and products contributing to the rapid development of this science (Basavaraj, 2012).

Nanotechnology and nanoscience are increasingly present in the media because they are attracting interest not only scientific but also industrial and financial.

The use of this technology also comes revolutionizing the field of pharmaceutical sciences, pharmaceutical nanotechnology is involved in the development, characterization and application of therapeutic systems in nanometric scales, aimed to direct and control the release of drugs became the therapy more effective (Kumar, 2010). There are many materials used in the production of nanostructures and these determine the type, the properties and characteristics of incorporated drug release (Vauthier et al., 2003). The main reason for the differences in behavior between the composite materials and nanocomposites is related to the large surface area of the material, resulting in intense interaction between the matrix and nanoparticles (Jena et al., 2013).

The main types of nanostructures used by the pharmaceutical industry for the encapsulation of active are: liposomes, polymeric nanoparticles, lipid nanoparticles, cyclodextrins and the nanoparticles which the drugs can be associated with metals, fullerenes, dendrímers and carbon nanotubes (Domingo and Saurina, 2012). The nanostructures have a promising future in the pharmaceutical industry, especially in the drug vectorization, which may be synthesized to improve the pharmacological and therapeutic properties of the drug, cause they have a sustained and controlled release, a higher selectivity (thus increasing the therapeutic index), a decrease of side effects and protection of degradation in the gastrointestinal tract, thus increasing bioavailability, the possibility of incorporating hydrophilic and lipophilic substances in the devices, reducing the therapeutic dose, number of doses and increase acceptance of therapy by the patient (Jena et al., 2013).

11.2.1 ANTIMICROBIAL ACTIVITY

The discovery of antimicrobial drugs provided a great advance in medicine reducing significantly the mortality rates, revolutionizing the treatment of infectious diseases. However, these drugs have limitations as increased bacterial resistance, high production costs, limited accessibility and adherence issues of patient (Alvan et al., 2011).

Use of nanoparticles is one of the strategies to overcome resistance to antimicrobial drugs in patients with cystic fibrosis (CF). It represents a promising approach to overcome the mucus barrier and to extend the antibiotic retention in the lung. The mucus buildup results in chronic lung infections caused by pathogenic bacteria, among which the most prevalent are the *Pseudomonas aeruginosa* (Lyczak et al., 2002).

The powder Terramycin is an important antibiotic used for the treatment of pulmonary infections caused by Gram-negative bacteria by inhalation. However, in CF patients its effectiveness is limited due to the low ability to achieve sufficient concentrations at the site infection, the rapid elimination of free drug and difficulty of penetration on mucus (Hadinoto and Cheow, 2014; Tseng et al., 2013). The study developed nanoparticles-based alginate and chitosan containing tobramycin (NP) and evaluated the antimicrobial activity in vitro and in vivo. The antimicrobial activity of NP against *P. aeruginosa* PA01 was equivalent to non-encapsulated tobramycin (minimal inhibitory concentration of 0.625 mg/mL). However, the in vivo model performed with *Galleria mellonella,* the NP showed a protective effect, providing 80% survival while free drug showed only 40% of survival after 96 h. The NPs were then functionalized with dornase alpha (recombinant human deoxyribonuclease I, DNase), which reduces the viscoelasticity of mucus by DNA cleavage demonstrating improve the penetration of NP in the sputum. The NPs tobramycin, both with and without functionalized DNase, exhibited anti-*Pseudomonas* effects. Demonstrating that the polymeric NP has the potential to provide a prolonged action on the sites of infection, a sustained release and therapeutic effect on the specific target (Deacon et al., 2015).

This technology has been used to improve the stability of natural products such as curcumin which is a yellow-orange powder extracted from the root of *Curcuma* or turmeric and has antimicrobial (Gunes et al., 2015) and healing (Kulac et al., 2013) activities. However, healing potential curcumin for therapeutic use has been difficult by poor oral bioavailability, poor water solubility and rapid degradation, limiting the clinical applicability. Then they were produced and characterized curcumin nanoparticles (CURC-NP) which inhibited in vitro growth of methicillin resistant *Staphylococcus aureus* (MRSA) and *Pseudomonas aeruginosa* in dose-dependent manner, in vivo inhibited MRSA growth and improved healing in a model of wound in mice. This may be related to the reduced size and increased surface area of the nanoparticles promoting passage through biological barriers, and also with the increase of interaction with host cells and microbial agents (Krausz et al., 2015).

In a research performed by Low et al. (2013), the authors produced two different types of liposomes, one with the essential oil of *Melaleuca alternifolia* and another with the same oil associated to silver. All liposomes demonstrated a controlled release improving their antimicrobial efficacy as well as reduced the required effective concentration. These findings may have an impact on the toxicity caused by the need for high doses of free drugs, or the microbial resistance, where long-term application is necessary.

Other research in the production of solid lipid nanoparticles containing *Melaleuca alternifolia* essential oil showed that these nanostructures enhance the stability of the essential oil while avoiding or reducing volatilization. Furthermore it keeps antimicrobial activity against 6 species of *Mycobacterium* and 11 fungal, indicating that the use of nanotechnology may represent an alternative for the treatment of mycobacterial and fungal diseases (de Souza et al., 2014).

Nanotechnology is also being used to produce biodegradable films derived from quinoa seeds. The use of these films is applied primarily on cover and food packaging, in order to maintain quality. Current trends include the development of packaging which interact with foods, may incorporate antimicrobials and antioxidants in these films, or additives that tend to retard the deterioration of product (Kechichian et al., 2010). The development of a biodegradable film of quinoa starch with the active packaging with gold nanoparticles (AuNPs) showed a significant increase in tensile strength comparing to standard film, an increase in UV absorption and a decreased solubility, which provides improved protection for food packaging increasing the possibilities of application. Also the thermal stability of these films was impressive, around 270°C. Simultaneously, the films containing gold nanoparticles showed positive results for antimicrobial activity, showing the effectiveness in inhibiting the growth of pathogenic organisms, particularly *S. aureus* (Pagno et al., 2015).

The copper-based nanoparticles are increasingly recognized due to its antimicrobial activity, the mechanism of action is not fully known. The bactericidal activity of these nanoparticles depends on the size, morphology, stability and concentration (Misra et al., 2014). Laha et al. (2014) developed a spherical copper oxide nanoparticles and on shaped sheets. The spherical nanoparticles showed higher bacterial activity against Gram-positive bacteria, while the nanoparticles in sheet form showed greater antimicrobial property front Gram-negative bacteria. It is suggested that a large release of ions, followed by a production of reactive oxygen species, causing DNA damage and breaks the membrane is the main cause of the bactericidal action of both forms of particles. However, nanoparticles may join with bacteria via various types of interactions, such as Van der Waals forces, hydrophobic and electrostatic interactions that help to damage the cell membrane (Singh et al., 2012).

Many antimicrobial agents of high molecular weight, shows difficulty in penetrating the cell membrane, it occurs in some species of microorganisms that carry this inherent strength. The fusogenic liposomes consist of lipid which becomes the bilayer more fluid and can promote destabilization of biological membranes releasing the active inside the cell. Experimental results of fusogenic liposomes loaded with fusidic acid demonstrated that the liposomes maintained the in vitro effect of free drug, becoming active against Gram-negative bacteria which the free drug is inactive. It is suggested that these vesicles are capable to interact with the cell membrane through a fusion mechanism, promoting diffusion of the drug into the cytoplasm (Nicolosi et al., 2015).

Inorganic nanomaterials exhibit physical properties which demonstrate advantages in biological applications. The multi-walled carbon nanotubes (MWNTs), for example, have a low antimicrobial activity, although limited, this can be increased with the use of drugs or nanoparticles (Yuan et al., 2004). A research group developed MWNTs with Cephalexin immobilized on its surface using PEG as a linker. The NTCM-Cephalexin improved significantly the antimicrobial activity also act by inhibiting bacterial adherence preventing the biofilm formation (Qi et al., 2012).

11.2.2 ANTIBIOFILM ACTIVITY

Research covering the use of nanodrugs for the prevention and treatment of biofilms has been growing steadily in recent years. The biofilm is a community of bacterial cells adhered to a surface enclosed by a extracellular self-secreted matrix of polymeric substances (Dunne et al., 2002). The structure of the biofilm, as well as certain communication mechanisms between microorganisms and their phenotypic changes are factors which inhibit the penetration of antimicrobials (Lewis, 2001). The biofilm formation involves a coordinated process of molecular events, starts with microbial adhesion to surface (reversible and irreversible), the slightly adhered microorganisms materialize the accession process with the exopolysaccharides production starting the maturation process (micro colonies and macro colonies) and dispersion (Monds and O'Toole, 2009).

The vast majority of infections caused by biofilms are associated with the use of invasive medical implants. The bacterial infection begins with the membership, so it is important to inhibit the adhesion of bacteria on the surface of the implant, producing implants with antimicrobial surfaces (Guo et al., 2014). Faced with this, one of the strategies that have been studied to overcome these limitations is orthopedic implants titanium coated with titanium nanotubes (NT-Ti). In this study the implant was coated with two layers of NT-Ti, a loaded with an antimicrobial (Gentamicin) and the other with a coating-based on biopolymers (PLGA and chitosan) to improve adhesion of osteoblasts. The coated device showed a simultaneous capacity, providing release of gentamicin over a prolonged period (i.e., upto 3–4 weeks), excellent osteoblast adhesion, and effective antibacterial properties, becoming potential candidates for future biomedical applications (Kumeria et al., 2015).

Using silver nanoparticles (AGNPS), researchers evaluated the antimicrobial activity in different maturation stages of biofilm formed by *Pseudomonas putida*. The phases were identified from ATP activity data, number of cells, expression of genes associated with biofilm and EPS amount. Biofilms presenting the stages 2 and 3 showed little or no reduction in ATP activity after exposure to silver nanoparticles. However, the same treatment reduced the ATP activity on more than 90% in less mature biofilms phase (Thuptimdang et al., 2015). In contrast studies in *Candida* biofilms intermediate stages of biofilm maturation and development did not affect the susceptibility of *C. albicans* and *C. glabrata* the AGNPS (Monteiro et al., 2015).

With the increasing number of products containing silver nanoparticles in recent years is unavoidable that domestic and industrial leavings reach to the effluent, may endanger the biological wastewater treatment process that is performed by the microbial community (Chen et al., 2013). A relatively large proportion of current biological wastewater treatment systems, such as biological disks (RBCs) and biological filters use the biofilms. Thus an analysis was performed on biofilms of industrial water treatment, which were exposed to the AGPNS. The tests provided the same pH, ionic strength, and natural organic matter present in the plant. It was observed that biofilms were not significantly damaged in the presence of silver nanoparticles. A large part of the nanoparticles suffer aggregation and sulfidization in the EPS matrix, not reaching the microbial cells. Furthermore, the structural stability of the wastewater biofilm probably contributed to the AGNPS tolerance. However, some bacteria were killed, and there may be a reduction of the compositional diversity biofilms making them more sensitive and potentially reduce the stability of the system (Sheng and Liu, 2015).

Alhajlan et al. (2013) developed liposomes of Clarithromycin with negative charge, reducing toxicity, improving the activity of clarithromycin against *P. aeruginosa* highly resistant when compared with the antimicrobial agent in free form. Also eradicated completely the biofilm, meanwhile clarithromycin only reduced the formation. The improved efficacy of liposomal formulations probably occurs due to the constant contact of the liposomes with the target and the slow release of clarithromycin, accelerating the penetration of the antimicrobial in the biofilm.

The metal nanoparticles have antimicrobial activity against a wide variety of species (Vargas-Reus et al., 2012), are able to support high temperatures and have low toxicity to mammalian cells when stabilized at surfaces (Manusadžianas et al., 2012). Front of these properties a study was performed using zinc oxide nanoparticles and hydroxyapatite as nanoscale coating on the titanium dental implants. Dental implants are prone to failure, favoring the colonization by pathogenic oral bacteria that can lead to biofilm formation, bone destruction and ultimately failure of the implant (Lindhe and Meyle, 2008). The results demonstrated several benefits in using composites which consist of multiple nanoparticles, promoting prevention of infection opportunistic pathogens, due to the synergistic effect of nanoparticles and providing greater osseous integration (Abdulkareem et al., 2015).

Another study conducted a porous coating aggregation of titanium surfaces with nanoparticles and chlorhexidine hexametaphosphate. This coating maintained a sustained release of chlorhexidine throughout the duration of the study, reducing the growth of *Streptococcus gordonii* on the surface of titanium coated with titanium as compared to exposed to an aqueous solution of chlorhexidine (Wood et al., 2015).

The reduced size allows for manipulation of magnetic properties of materials, making them a potential for diagnostic, noninvasive treatments also may be used as any pharmaceutical delivery vehicle. The magnetic nanoparticles have a greater biological stability. With this magnetic nanoparticles of iron oxide were coated with Cy5.5 azide, an infrared dye, showing through a super resolution microscopy (Leica SR GSD 3D Super Resolution 3D Ground State Depletion Microscope) they can be used to observe the fate of nanoparticles inside the biofilm, allowing direct observation of individual nanoparticles in 3D with individual bacteria within the biofilm (Stone et al., 2015). Several studies show that nanoparticles of various compositions are able to disrupt the bacterial biofilms, but the mechanisms are still unknown (Raftery et al., 2013). This study will assist future investigations using labeled magnetic nanoparticles and high-resolution microscopy helping to clarify the form of association between the nanoparticles and bacteria inside biofilms.

11.2.3 ANTIVIRAL STRATEGIES

The constant appearance of new viruses with the general absence of availability of specific antiviral therapeutics for a variety of viruses has made the search for antiviral drugs and therapeutics a challenging research task. Viral infections cause a considerable challenge to the body's immune system as they hide inside cells, making it difficult for antibodies to reach them. In contrast to bacterial infections, which are mostly treated using antibiotics, immunization against viral infections is not always possible. Moreover, some viruses are capable of mutating from one person to the next, making vaccination a difficult task as the viruses have already changed their format by the time vaccines are available (Milroy and Featherstone, 2002).

Efforts in the search of new alternatives strategies are crucial for the development and improved antiviral therapy. A well-known strategy is to change the physic-chemical properties of antiviral drugs such as acyclovir or vidarabine by chemical modifications (Shen et al., 2009; Zhang et al., 2014). Another alternative for the delivery of antiviral drugs is the use of controlled-release delivery vehicles in the form of tablets and patches. Such formulations reduce the administered dose and aim to overcome problems of noncompliance and loss of drug activity. The ideal delivery platform would release the antiviral drug at a constant dose over a long time. Acyclovir delivery vehicles-based on silicone polymer were proposed as implants for viable and long-time suppressive therapy of HSV-1 (Johnson et al., 2007).

The design of nanomaterials-based delivery systems has several advantages. Nanomaterials have the characteristics of high surface-to-volume ratios, enabling the packaging of multiple antiviral agents onto the same nanoparticles. Using these nanomaterials, it might be possible to overcome problems associated with the use of high doses of antiviral drugs.

The dendrimers has shown primarily as complexing carrier molecules. They have been shown to possess an inherent biological activity, which could be seen to have a significant contribution in the antiviral field (Bourne et al., 2000; Marx, 2008; McCarthy et al., 2005). The antiviral activity of dendrimers is attributed to the synthetic modification of dendrimer molecules such that they include functional groups in the surface layer, which are capable of forming complexes with cell or viral receptors, pivotally resulting in the disruption of normal virus–cell interactions, including the initial virus–cell binding. The primary antiviral mechanism of dendrimers thus occurs early in the infection process through blockade of the virus attachment to the cell, or interference with adsorption. However, certain compounds may possess secondary mechanisms of action (Bourne et al., 2000).

Dendrimers has been demonstrated in vitro activity of against the influenza virus, respiratory syncytial virus (RSV), measles virus, and HIV (Marx, 2008; McCarthy et al., 2005). The stark resemblance between the three-dimensional architecture of dendrimers and natural bio-macromolecules, particularly proteins, has instigated 'bio-inspired' applications. One such dendrimer-based application that is significant in the field of topical microbicides is VivaGel™ (Starpharma, Melbourne, Australia), a topical microbicide, which has been granted 'Fast Track' status as an Investigational New Drug (IND) for the prevention of HIV transmission and other STDs (McCarthy et al., 2005).

After dendrimers, fullerenes are the most investigated as potential nano-microbicides. Marcorin et al. (2000) synthesized fullerene derivatives and exploited their potential as inhibitors of HIV aspartic protease enzyme for the elucidation of a novel anti-HIV system. They defined the active region of HIV-protease as a cylindrical hydrophobic cavity (diameter ~10 Å), containing two amino acid residues, aspartate 25 and aspartate 125. Binding at these sites caused suppression of protein slicing and inhibited viral replication. The side chains (containing NH2 or NH3 + groups) of the water-soluble fullerene derivatives which were synthesized underwent electrostatic and/or hydrogen bond interactions with Asp 25 and Asp 125 (Bakry et al., 2007; Marcorin et al., 2000).

The results appear positive in terms of their large surface area. However, as carbon derivatives, the toxicological definition for fullerene is still quite controversial. Recently, fullerenes have been suggested to be carcinogenic (Burlaka et al., 2004; Sayes et al., 2004) and genotoxic, but only upon photosensitization (Sayes et al., 2004).

The anti-HIV activity of noble-metal nanoparticles has been showed (Elechiguerra et al., 2005). Bowman et al. (2008) have documented the first application of small-molecule coated gold nanoparticles as effective inhibitors of HIV fusion. Their concept arose from the premise of Mammen et al. (1998) that 'biological systems exploit multivalency in the synthesis of high-affinity ligands, because they allow an organism to take advantage of an existing set of monovalent ligands without the need for evolving completely new molecules for every required function.' The concept of 'multivalent therapeutics' is well conceived by the described system. The gold nanoparticles employed as a platform (2.0 nm diameter, mercaptobenzoic acid modified gold particles) transformed a weakly binding and biologically inactive small molecule into a multivalent conjugate that effectively inhibited HIV-1 fusion to human T-cells. Of significance is the similarity of this class of gold particles to proteins and dendrimers in terms of their atomic precision and mono-disperse nano-size (Bowman et al., 2008).

The concept of surface chemistry of silver nanoparticles predictability of interactions with external systems has been challenged in their investigation via the development and testing of silver nanoparticles with three different surface chemistries, namely, foamy carbon, poly (N-vinyl-2-pyrrolidone) (PVP), and bovine serum albumin (BSA). Contrary to expectations, they established congruency among all the formulations in that only nanoparticles below ~10 nm attached to the viral envelope and this occurred independently of the formulations' surface chemistry. Additionally, a regular spatial arrangement with equivalent center-to-center distances between the nanoparticles bound to the virus–cell was found. They ameliorated their findings with regard to both the spatial arrangement of nanoparticles and the size dependence of interaction in terms of the HIV-1 viral envelope. These investigations ultimately provided a deeper understanding into the mode of interaction between the virus and nanoparticles. Silver nanoparticles proposed undergo specific interaction with HIV-1 via preferential binding with the gp120 subunit of the viral envelope glycoprotein, through interaction of the silver nanoparticle with the exposed disulfide bonds of the gp120. This indicated the aforementioned probability that other noble-metal nanoparticles may also exhibit similar activity (Elechiguerra et al., 2005).

However, the toxicity and inhibition results differed, despite the congruency among the surface modified nanoparticles with reference to their interaction with HIV-1 (Elechiguerra et al., 2005). The differential behavior was attributed to the capping agents employed for each nanoparticle preparation. BSA- and PVP-protected nanoparticles displayed slightly lower inhibition because the nanoparticle surface was directly bound to and encapsulated by the capping agent. Contrarily, the carbon-coated nanoparticles exhibited a greater inhibitory effect due to their essentially free surface area. The free surface nanoparticles demonstrate higher cytotoxicity because of their surface chemistry, and the carbon-coated form shows comparatively free surfaces, which is able to interact strongly with the host cells, thus increasing their toxicity (Elechiguerra et al., 2005).

Liposomes composed of phospholipid bilayer structures were investigated as antiviral agents due to discovery that certain lipid compositions can bind to HIV particles and modulate their infectivity. Moreover, lipid-based nanoparticles have been designed for vaginal delivery of drugs or small interfering RNA to silence HIV gene expression and prevent the establishment of primary infection (Malavia et al., 2011; Wang et al., 2011).

Finally, the nanoparticles are under development as useful and powerful tools in the field of antiviral microbicides, they display a series of limitations that should be addressed because they can affect the transfer of these platforms to clinical trials. These limitations include toxicity, the appearance of nondesirable biological interactions, bioaccumulation, the enzymatic degradation of these nanosystems, their penetrance and absorption features into different tissues and their high manufacturing cost; this last limitation is especially a problem for scale-up production (Mamo et al., 2010).

11.2.4 ANTIFUNGAL STRATEGIES

Antifungal drugs are considerably fewer in number because of emergence of newer pathogenic fungi causing deep-seated mycosis. Clinically used major groups of antifungal agents are polyene antibiotics, azole derivatives, allylamines-thiocarbamates, morpholines and miscellaneous compounds such as 5-fluorocytosine and griseofulvin. Polyenes and azoles are most commonly used. Polyene antifungal agents used for the treatment of human diseases are amphotericin B (AmB) nystatin and natamycin. The only parenteral preparation with broad range of antifungal activity is AmB. Over the past several years, augmented efforts in both basic and clinical antifungal pharmacology have resulted in a number of exclusively new, reengineered or reconsidered compounds, which are at various stages of preclinical and early clinical development (Georgopapadakou and Walsh, 1996; Hay, 1994; Maesaki, 2002).

New et al. (1981) examined for the first time, the effects of liposome of AmB (L-AmB), using leishmania model and reported that L-AmB had a lower toxicity than AmB itself and the treatment with a higher dose of L-AmB could be feasible. Afterward, the validity of the L-AmB for mice histoplasmosis (Taylor et al., 1982), cryptococcosis (Graybill et al., 1982), and candidiasis (Lopez-Berestein et al., 1983) was assessed. In all cases, the L-AmB showed a lower toxicity than AmB to the host animals and thus could be administered at higher doses. Drugs incorporated in liposome were also shown to distribute mainly to reticule endothelial tissues including liver, spleen, and lung (Abra and Hunt, 1981). Later, a clinical trial performed in cancer patients who co-developed fungal infection confirmed that the L-AmB showed a higher tolerance than AmB even in human (Lopez-Berestein et al., 1985).

The development of new formulations of liposomes with specific properties has been encouraged. Immunoliposomes is one of these, contains fungus-specific antibodies on their surface which target them directly to the fungal cells. AmB coated with immunoliposomes abridged mortality appreciably in mice with invasive pulmonary aspergillosis as compared conventional liposomes of AmB(L-AmB) (100% vs. 16.7% survival rate). AmB coated with immunoliposomes was also more effective than AmB integrated with long-circulating liposomes (100% vs. 83.3% survival rate) (Otsubo et al., 1998). Likewise, treatment of murine candidiasis and cryptococcosis with AmB integrated with immunoliposomes proved enhanced activity compared to that with conventional L-AmB (Belay et al., 1993; Dromer et al., 1990).

The econazole (ECN; commonly used as nitrate salt) is an imidazole antifungal structurally related to another imidazole derivative, miconazole (Thienpont et al., 1975) and used to treat fungal infections such as *tinea pedis* and *cruris*, *pityriasis versicolor*. Several nano-formulation strategies have been studied for delivering econazole through targeted skin sites (Bachhav et al., 2011; Kumar et al., 2014; Passerini et al., 2009; Sanna et al., 2007; Yordanov, 2012). The first nano-carrier-econazole product for antimycotic therapy (Epi-Pevaryl C1 Lipogel) marketed in 1988 was a liposome preparation produced by Cilag company in Switzerland (Müller et al., 2002; Naeff, 1996).

The antifungal activity of chitosan is also augmented when incorporated into nanomaterials. Previous investigations suggest that these nanomaterials exert their antifungal activity in much the same way as chitosan itself. It has been proposed that chitosan nanoparticles have tighter membrane binding with fungal cells, and diffuse into fungal cells at a higher rate-based on in vitro studies (Ma and Lim, 2003; Qi et al., 2004). These findings have been attributed to their higher surface charge density and small size. Ing et al. (2012) showed that the MIC_{90} (minimum concentration needed to inhibit 90% of growth) of chitosan nanoparticles was significantly less than chitosan alone when used against *C. albicans* and *F. solani*. Interestingly, they showed that *C. albicans* was more easily inhibited by nanoparticle preparations made from lower molecular weight chitosan, while *F. Solani* was more susceptible to nanoparticles made with higher molecular weight chitosan. This relationship was not seen in treatments made from chitosan alone. It was also found that nanoparticles of smaller sizes more easily inhibited the growth of both species, supporting the hypothesis that nanoparticle size itself imparts more antifungal activity. However, the same study showed resistance of *A. niger* to chitosan nanoparticles. This resistance is explained by the fact that *A. niger* contains 10% of chitin in its cell wall, a factor that has been associated with resistance to chitosan by other fungi (Allan and Hadwiger, 1979; Klis et al., 2007). Chitosan nanoparticles have also been shown to inhibit mycelial growth and germination of *Alternaria alternata*, *Macrophomina phaseolina* and *Rhizoctonia solani* in vitro (Saharan et al., 2013). These results taken together suggest that the versatility of chitosan and the ease of its manipulation allow for specific formulations to address different pathogens. However, these preparations do not show universal antifungal activity. Chitosan nanoparticles can also be used as a vehicle to optimize other antifungal treatments. Modi et al. have developed a ketoconazole-loaded chitosan nanoparticle that overcomes poor gastric absorption of the drug (Modi et al., 2013). They showed high binding of the nanoparticles to pig mucin. An *ex vivo* mucosal diffusion assay using mouse stomachs showed sustained diffusion of the nanoparticle drug after gastric emptying. However, when treated with ketoconazole alone, diffusion was halted after gastric emptying. The nanoparticle formulation also showed sustained release over time. A similar study done by Zhou et al. (2013) studied amphotericin-B loaded poly(lactic acid)-grafted-chitosan nanoparticles and their ocular use. They similarly found high interaction between their nanoparticles and mucin, showing a decrease in mucin's negative charge when exposed to nanoparticles. This supports the notion that the positive surface charge of the nanoparticles is a primary mechanism of their mucosal adherence.

Nanosilver is an effective antifungal agent against a broad spectrum of common fungi. Kim et al. (2009) investigated nanosilver particles (NSP) antifungal properties on a total of 44 strains of six fungal species, and found that NSPs can inhibit the growth of *Candida albicans*, *Candida glabrata*, *Candida parapsilosis*, *Candida krusei*, and *Trichophyton mentagrophytes* effectively (Kim et al., 2008). Nasrollahi et al. (2009) and Kim et al. (2009) observed that NSPs can disrupt cellular membrane and inhibit the normal budding process; however, the exact mechanisms of action of nanosilver against fungi are still not clear.

A study with Ag Nanoparticles (AgNPbio) produced by *Fusarium oxysporum* was shown to have fungicidal effect against Fluconazole resistant *C. albicans*, with no cytotoxicity to mammalian cells. This inhibitory effect of AgNPbio on FLC-susceptible *C. albicans* had been previously described (Kim et al., 2009; Musarrat et al., 2010). Similar to our results, Ishida et al. (2014) reported a fungicidal effect of AgNPbio produced by *F. oxysporum* on *C. albicans*, with MIC and MFC values of 1.68 and 3.40 μg/ml, respectively. Cytotoxic concentration of AgNPbio found for HEp-2 cells was similar to Marcato et al. (2013). Besides, Lima et al. also showed non-cytotoxicity up to a concentration of 10 μg/ml of silver biogenic nanoparticles on 3T3 cells (Lima et al., 2013). Interestingly, our results highlight the decrease of FLC MIC of FLC-resistant *C. albicans* when combined to AgNPbio. The resistance to FLC in *C. albicans* is strongly associated with overexpression of genes encoding efflux pumps or lanosterol 14α-demethylase (Odds et al., 2003). On the other hand, the mechanisms of C. albicans death induced by Ag-NPs are not completely understood. AgNP-treated *C. albicans* exhibits a disrupted cell wall and cytoplasmic membrane (Kim et al., 2008). In addition, AgNPs cause an increase in reactive oxygen species and hydroxyl radical production, which can also contribute to cell membrane damage (Hwang et al., 2012). Since AgNPbio can alter the permeability of the cell membrane, we can hypothesize that they may facilitate the entry of FLC, which interferes with ergosterol biosynthesis. Regarding the effect of AgNPbio effect against biofilm, previous studies have shown an antifungal effect of chemically produced AgNPs in initial stages of biofilm formation of *C. albicans* (Monteiro et al., 2011). Importantly, silver nanoparticles produced by *F. oxysporum* are stable for several months due to protein capping, which occurs in the biogenic process, as observed by transmission electron microscopy (Marcato et al., 2012). Therefore, these results indicate the potential of AgNPbio for the development of new strategies for the treatment of FLC resistant *C. albicans* infections. Further studies are needed to establish the mechanism of yeast death and the usefulness of AgNPbio in medicine.

11.3 PROSPECTS

Against the growth number of researches which has been developed in last years, is possible visualize that the nanotechnology study associated with the science of life has been opening doors to many applications, leading a great impact in biomedical area, improving many troubles related to diagnosis of diseases, therapies, drug delivery system offering interest solutions to many disorders which threaten the life. This technology has been developing and changing a wide range of products, improving the life quality of people, clinical practice and public health.

The investment evolving the nanotechnology in the entire world is funded by billions of dollars (Gao et al., 2016). Therefore, there no have doubts that this science tends further advance. It is possible that soon has to be a better understanding of the physical concepts, chemical and biological related phenomena manifested on the nanoscale. However, there are challenges involving ethics, safety standards, cost-effective, and toxicological tests must be overcome before these new materials reach the market.

KEYWORDS

- **antimicrobial activity**
- **biofilm**
- **gram negative**
- **gram positive**
- **nanoparticles**
- **yeast**

REFERENCES

Abdulkareem, E. H.; Memarzadeh, K.; Allaker, R. P.; Huang, J.; Pratten, J.; Spratt, D. Elsevier Ltd, **2015**.

Abra, R. M.; Hunt, C. A. *Biochimica et biophysica acta*. **1981**, pp. 493–503.

Alhajlan, M.; Alhariri, M.; Omri, A. *Antimicrob. Agents Chemother.* **2013**, *57*(6), 2694–2704.
Allan, C. R.; Hadwiger, L. A. *Exp. Mycol.* **1979**, 3(3), 285–287.
Alvan, G.; Edlund, C.; Heddini, A. *Drug Resist. Updat.* **2011**, *14*(2), 70–76.
Bachhav, Y. G.; Mondon, K.; Kalia, Y. N.; Gurny, R.; Möller, M. *J. Control. Release* **2011**, *153*(2), 126–132.
Bakry, R.; Vallant, R. M.; Najam-ul-Haq, M.; Rainer, M.; Szabo, Z.; Huck, C. W.; Bonn, G. K. *Int. J. Nanomedicine* **2007**, *2*(4), 639–649.
Basavaraj, K. H. *Indian journal of dermatology*. **2012**, pp. 169–174.
Belay, T.; Hospenthal, D. R.; Rogers, A. L.; Patterson, M. J. *Mycopathologya* **1993**, *123*(1), 9–17.
Binnig, G.; Rohrer, H. *Surf. Sci.* **1982**, *126*(126), 236–244.
Bourne, N.; Stanberry, L. R.; Kern, E. R.; Holan, G.; Matthews, B.; Bernstein, D. I. *Antimicrob. Agents Chemother.* **2000**, *44*(9), 2471–2474.
Bowman, M. C.; Ballard, T. E.; Ackerson, C. J.; Feldheim, D. L.; Margolis, D. M.; Melander, C. *J. Am. Chem. Soc.* **2008**, *130*(22), 6896–6897.
Burlaka, A. P.; Sidorik, Y. P.; Prylutska, S. V; Matyshevska, O. P.; Golub, O. A.; Prylutskyy, Y. I.; Scharff, P. *Exp. Oncol.* **2004**, *26*(4), 326–327.
Chen, H.; Zheng, X.; Chen, Y.; Mu, H. *RSC Adv.* **2013**, *3*, 9835–9842.
Deacon, J.; Abdelghany, S. M.; Quinn, D. J.; Schmid, D.; Megaw, J.; Donnelly, R. F.; Jones, D. S.; Kissenpfennig, A.; Elborn, J. S.; Gilmore, B. F.; Taggart, C. C.; Scott, C. J. *J. Control. Release* **2015**, *198*, 55–61.
Domingo, C.; Saurina, J. *Anal. Chim. Acta* **2012**, *744*, 8–22.
Dromer, F.; Barbet, J.; Bolard, J.; Charreire, J.; Yeni, P. *Antimicrob. Agents Chemother.* **1990**, *34*(11), 2055–2060.
Dunne, W. M. *Clin. Microbiol. Rev.* **2002**, *15*(2), 155–166.
Ebling de Souza, M.; Quintana Soares Lopes, L.; de Almeida Vaucher, R.; Nunes Mário, D.; Hartz Alves, S.; Albertina Agertt, V.; Vendruscolo Bianchini, B.; Iensen Felicidade, S.; Matiko Anrako de Campus, M.; Augusti Boligon, A.; Linde Athayde, M.; Genro Santos, C.; Platchek Raffin, R.; Gomes, P.; Christ Vianna Santos, R. *J. Drug Deliv. Sci. Technol.* **2014**, *24*(5), 559–560.
Elechiguerra, J. L.; Burt, J. L.; Morones, J. R.; Camacho-Bragado, A.; Gao, X.; Lara, H. H.; Yacaman, M. J. *J. Nanobiotechnology* **2005**, *3*, 6.
Gao, Y.; Jin, B.; Shen, W.; Sinko, P. J.; Xie, X.; Zhang, H.; Jia, L. *Nanomedicine* **2016**, *12*(1), 13–19.
Georgopapadakou, N. H.; Walsh, T. J. *Antimicrob. Agents Chemother.* **1996**, *40*(2), 279–291.
Graybill, J. R.; Craven, P. C.; Taylor, R. L.; Williams, D. M.; Magee, W. E. *J. Infect. Dis.* **1982**, *145*(5), 748–752.
Gunes, H.; Gulen, D.; Mutlu, R.; Gumus, A.; Tas, T.; Eren Topkaya, A. *Toxicol. Ind. Health* **2013**, October.
Guo, Z.; Chen, C.; Gao, Q.; Li, Y.; Zhang, L. *Mater. Lett.* **2014**, *137*, 464–467.
Hadinoto, K.; Cheow, W. S. *Colloids Surf. B. Biointerfaces* **2014**, *116*, 772–785.
Hay, R. J. *J. Am. Acad. Dermatol.* **1994**, *31*(3), S82–S86.
Hwang, I. S.; Lee, J.; Hwang, J. H.; Kim, K. J.; Lee, D. G. *FEBS J.* **2012**, *279*(7), 1327–1338.
Ing, L. Y.; Zin, N. M.; Sarwar, A.; Katas, H. *Int. J. Biomater.* **2012**, *2012*, 632698.

Ishida, K.; Cipriano, T. F.; Rocha, G. M.; Weissmüller, G.; Gomes, F.; Miranda, K.; Rozental, S. *Mem. Inst. Oswaldo Cruz* **2014**, *109*(2), 220–228.

Jena, M.; Mishra, S.; Jena, S.; Mishra, S. *Int. J. Basic Clin. Pharmacol.* **2013**, 2(4), 353.

Johnson, T. P.; Frey, R.; Modugno, M.; Brennan, T. P.; Margulies, B. J. *Int. J. Antimicrob. Agents* **2007**, *30*, 428–435.

Kechichian, V.; Ditchfield, C.; Veiga-Santos, P.; Tadini, C. C. *LWT – Food Sci. Technol.* **2010**, *43*(7), 1088–1094.

Kim, K. J.; Sung, W. S.; Moon, S. K.; Choi, J. S.; Kim, J. G.; Lee, D. G. *J. Microbiol. Biotechnol.* **2008**, *18*, 1482–1484.

Kim, K.-J.; Sung, W. S.; Suh, B. K.; Moon, S.-K.; Choi, J.-S.; Kim, J. G.; Lee, D. G. *Biometals* **2009**, *22*(2), 235–242.

Klis, F.; Ram, a.; Groot, P. *Biol. Fungal Cell* **2007**, 97–120.

Krausz, A. E.; Adler, B. L.; Cabral, V.; Navati, M.; Doerner, J.; Charafeddine, R. A.; Chandra, D.; Liang, H.; Gunther, L.; Clendaniel, A.; Harper, S.; Friedman, J. M.; Nosanchuk, J. D.; Friedman, A. J. *Nanomedicine* **2015**, *11*(1), 195–206.

Kulac, M.; Aktas, C.; Tulubas, F.; Uygur, R.; Kanter, M.; Erboga, M.; Ceber, M.; Topcu, B.; Ozen, O. A. *J Mol Histol* **2013**, *44*(1), 83–90.

Kumar, C. S. S. R. *Mater. Today* **2010**, *12*, 24–30.

Kumar, J. Raja; Muralidharan, S.; Parasuraman, S. *J. Pharm. Sci. Res.* **2014**, *6*(5), 229–235.

Kumeria, T.; Mon, H.; Aw, M. S.; Gulati, K.; Santos, A.; Griesser, H. J.; Losic, D. *Colloids Surf. B. Biointerfaces* **2015**, *130*, 255–263.

Laha, D.; Pramanik, A.; Laskar, A.; Jana, M.; Pramanik, P.; Karmakar, P. *Mater. Res. Bull.* **2014**, *59*, 185–191.

Lewis, K. *Antimicrobial Agents and Chemotherapy*. 2001, pp. 999–1007.

Lima, R.; Feitosa, L. O.; Ballottin, D.; Marcato, P. D.; Tasic, L.; Durán, N. *J. Phys. Conf. Ser.* **2013**, *429*(1), 012020.

Lindhe, J.; Meyle, J. *J. Clin. Periodontol.* **2008**, *35*(Suppl. 8), 282–285.

Lopez-Berestein, G.; Fainstein, V.; Hopfer, R. *J. Infect. Dis.* **1985**, *151*(4), 704–710.

Lopez-Berestein, G.; Mehta, R.; Hopfer, R. L.; Mills, K.; Kasi, L.; Mehta, K.; Fainstein, V.; Luna, M.; Hersh, E. M.; Juliano, R. *J. Infect. Dis.* **1983**, *147*(5), 939–945.

Low, W. L.; Martin, C.; Hill, D. J.; Kenward, M. a. *Lett. Appl. Microbiol.* **2013**, *57*(1), 33–39.

Lyczak, J. B.; Cannon, C. L.; Pier, G. B. *Clin. Microbiol. Rev.* **2002**, *15*(2), 194–222.

Ma, Z.; Lim, L. Y. *Pharm. Res.* **2003**, *20*(11), 1812–1819.

Maesaki, S. *Curr. Pharm. Des.* **2002**, *8*(6), 433–440.

Malavia, N. K.; Zurakowski, D.; Schroeder, A.; Princiotto, A. M.; Laury, A. R.; Barash, H. E.; Sodroski, J.; Langer, R.; Madani, N.; Kohane, D. S. *Biomaterials* **2011**, *32*(33), 8663–8668.

Mammen, M.; Choi, S. K.; Whitesides, G. M. *Angewandte Chemie – International Edition.* **1998**, pp. 2754–2794.

Mamo, T.; Moseman, E. A.; Kolishetti, N.; Salvador-Morales, C.; Shi, J.; Kuritzkes, D. R.; Langer, R.; von Andrian, U.; Farokhzad, O. C. *Nanomedicine (Lond).* **2010**, *5*(2), 269–285.

Manusadžianas, L.; Caillet, C.; Fachetti, L.; Gylytė, B.; Grigutytė, R.; Jurkonienė, S.; Karitonas, R.; Sadauskas, K.; Thomas, F.; Vitkus, R.; Férard, J.-F. *Environ. Toxicol. Chem.* **2012**, *31*(1), 108–114.

Marcato, P. D.; Nakasato, G.; Brocchi, M.; Melo, P. S.; Huber, S. C.; Ferreira, I. R.; Alves, O. L.; Duran, N. *J. Nano Res.* **2012**, *20*, 69–76.
Marcato, P. D.; Parizotto, N. V.; Martinez, D. S. T.; Paula, A. J.; Ferreira, I. R.; Melo, P. S.; Durán, N.; Alves, O. L. *J. Braz. Chem. Soc.* **2013**, *24*(2), 266–272.
Marcorin, G. L.; Da Ros, T.; Castellano, S.; Stefancich, G.; Bonin, I.; Miertus, S.; Prato, M. *Org. Lett.* **2000**, *2*(25), 3955–3958.
Marx, V. *Nat Biotech* **2008**, *26*(7), 729–732.
McCarthy, T. D.; Karellas, P.; Henderson, S. A.; Giannis, M.; O'Keefe, D. F.; Heery, G.; Paull, J. R. A; Matthews, B. R.; Holan, G. *Mol. Pharm.* **2005**, 2(4), 312–318.
Milroy, D.; Featherstone, J. *Nat. Rev. Drug Discov.* **2002**, *1*(1), 11–12.
Misra, S. K.; Nuseibeh, S.; Dybowska, A.; Berhanu, D.; Tetley, T. D.; Valsami-Jones, E. *Nanotoxicology* **2014**, *8*(4), 422–432.
Modi, J.; Joshi, G.; Sawant, K. *Drug Dev. Ind. Pharm.* **2013**, *39*(4), 540–547.
Monds, R. D.; O'Toole, G. A. *Trends Microbiol.* **2009**, *17* (January), 73–87.
Monteiro, D. R.; Gorup, L. F.; Silva, S.; Negri, M.; de Camargo, E. R.; Oliveira, R.; Barbosa, D. B.; Henriques, M. *Biofouling* **2011**, *27*(7), 711–719.
Monteiro, D. R.; Takamiya, A. S.; Feresin, L. P.; Gorup, L. F.; de Camargo, E. R.; Delbem, A. C. B.; Henriques, M.; Barbosa, D. B. *J. Prosthodont. Res.* **2015**, *59*(1), 42–48.
Müller, R. H.; Radtke, M.; Wissing, S. A. *Adv. Drug Deliv. Rev.* **2002**, *54* (Suppl.), 131–155.
Musarrat, J.; Dwivedi, S.; Singh, B. R.; Al-Khedhairy, A. A; Azam, A.; Naqvi, A. *Bioresour. Technol.* **2010**, *101*, 8772–8776.
Naeff, R. *Adv. Drug Deliv. Rev.* **1996**, *18*(3), 343–347.
Nasrollahi, Y.-K.; Kim, B. H.; Jung, G. *Plant Dis.* **2009**, *93*(10), 1037–1043.
New, R. R.; Chance, M. L.; Heath, S. *J. Antimicrob. Chemother.* **1981**, *8*(5), 371–381.
Nicolosi, D.; Cupri, S.; Genovese, C.; Tempera, G.; Mattina, R.; Pignatello, R. *Int. J. Antimicrob. Agents* **2015**, *45*(6), 622–626.
Odds, F. C.; Brown, A. J. P.; Gow, N. A. R. *Trends Microbiol.* **2003**, *11*(6), 272–279.
Otsubo, T.; Maruyama, K.; Maesaki, S.; Miyazaki, Y.; Tanaka, E.; Takizawa, T.; Moribe, K.; Tomono, K.; Tashiro, T.; Kohno, S. *Antimicrob. Agents Chemother.* **1998**, *42*(1), 40–44.
Pagno, C. H.; Costa, T. M. H.; de Menezes, E. W.; Benvenutti, E. V; Hertz, P. F.; Matte, C. R.; Tosati, J. V.; Monteiro, A. R.; Rios, A. O.; Flôres, S. H. *Food Chem.* **2015**, *173*, 755–762.
Passerini, N.; Gavini, E.; Albertini, B.; Rassu, G.; Di Sabatino, M.; Sanna, V.; Giunchedi, P.; Rodriguez, L. *J. Pharm. Pharmacol.* **2009**, *61*, 559–567.
Qi, L.; Xu, Z.; Jiang, X.; Hu, C.; Zou, X. *Carbohydr. Res.* **2004**, *339*(16), 2693–2700.
Qi, X.; Gunawan, P.; Xu, R.; Chang, M. W. *Chem. Eng. Sci.* **2012**, *84*, 552–556.
Raftery, T. D.; Lindler, H.; McNealy, T. L. *Microb. Ecol.* **2013**, *65*(2), 496–503.
Ratner, B. M.; Ratner, D. *Nanotechnology: A Gentle Introduction to the Next Big Idea*, **2003**, Vol. 6.
Safari, J.; Zarnegar, Z. *J. Saudi Chem. Soc.* **2014**, *18*(2), 85–99.
Saharan, V.; Mehrotra, A.; Khatik, R.; Rawal, P.; Sharma, S. S.; Pal, A. *Int. J. Biol. Macromol.* **2013**, *62*, 677–683.
Sanna, V.; Gavini, E.; Cossu, M.; Rassu, G.; Giunchedi, P. *J. Pharm. Pharmacol.* **2007**, *59*(8), 1057–1064.

Sayes, C. M.; Fortner, J. D.; Guo, W.; Lyon, D.; Boyd, A. M.; Ausman, K. D.; Tao, Y. J.; Sitharaman, B.; Wilson, L. J.; Hughes, J. B.; West, J. L.; Colvin, V. L. *Nano Lett.* **2004**, *4*(10), 1881–1887.

Shen, W.; Kim, J. S.; Mitchell, S.; Kish, P.; Kijek, P.; Hilfinger, J. *Nucleosides Nucleotides Nucleic Acids* **2009**, *28*(1), 43–55.

Sheng, Z.; Liu, Y. *Water Res.* **2015**, *45*(18), 6039–6050.

Singh, J.; Kamboj, S. S.; Bakshi, M. S.; Khullar, P.; Singh, V.; Mahal, A.; Dave, P. N.; Thakur, S.; Kaur, G.; Singh Kamboj, S.; Singh Bakshi, M. *J. Phys. Chem. C* **2012**, *116*(15), 8834–8843.

Stone, R. C.; Fellows, B. D.; Qi, B.; Trebatoski, D.; Jenkins, B.; Raval, Y.; Tzeng, T. R.; Bruce, T. F.; Mcnealy, T.; Austin, M. J.; Monson, T. C.; Huber, D. L.; Mefford, O. T. **2015**, *459*, 175–182.

Taylor, R. L.; Williams, D. M.; Craven, P. C.; Graybill, J. R.; Drutz, D. J.; Magee, W. E. *Am Rev Respir Dis* **1982**, *125*(5), 610–611.

Thienpont, D.; Van, C. J.; Van Nueten, J. M.; Niemegeers, C. J.; Marsboom, R. *Arzneimittelforschung.* **1975**, *25*(0004–4172 (Print)), 224–230.

Thuptimdang, P.; Limpiyakorn, T.; McEvoy, J.; Prüß, B. M.; Khan, E. *J. Hazard. Mater.* **2015**, *290*, 127–133.

Toumey, C. *Techné* **2008**, *12*(3), 133–168.

Tseng, B. S.; Zhang, W.; Harrison, J. J.; Quach, T. P.; Song, J. L.; Penterman, J.; Singh, P. K.; Chopp, D. L.; Packman, A. I.; Parsek, M. R. *Environ. Microbiol.* **2013**, *15*(10), 2865–2878.

Vargas-Reus, M. a.; Memarzadeh, K.; Huang, J.; Ren, G. G.; Allaker, R. P. *Int. J. Antimicrob. Agents* **2012**, *40*(2), 135–139.

Vauthier, C.; Dubernet, C.; Fattal, E.; Pinto-Alphandary, H.; Couvreur, P. *Adv. Drug Deliv. Rev.* **2003**, *55*(4), 519–548.

Wang, L.; Sassi, A. B.; Patton, D.; Isaacs, C.; Moncla, B. J.; Gupta, P.; Rohan, L. C. *Drug Dev. Ind. Pharm.* **2011**, February, 1–13.

Wood, N. J.; Jenkinson, H. F.; Davis, S. A.; Mann, S.; O'Sullivan, D. J.; Barbor, M. E. *J. Mater. Sci. Mater. Med.* **2015**, *26*(6), 201.

Yordanov, G. *Colloids Surfaces A Physicochem. Eng. Asp.* **2012**, *413*, 260–265.

Yuan, J.; Chen, L.; Jiang, X.; Shen, J.; Lin, S. *Colloids Surfaces B Biointerfaces* **2004**, *39*(1–2), 87–94.

Zhang, Y.; Gao, Y.; Wen, X.; Ma, H. *Asian J. Pharm. Sci.* **2014**, *9*(2), 65–74.

Zhou, W.; Wang, Y.; Jian, J.; Song, S. *Int. J. Nanomedicine* **2013**, *8*, 3715–3728.

CHAPTER 12

DRUG TARGETING: PRINCIPLES AND APPLICATIONS

RUSLAN G. TUGUNTAEV,[1] AHMED SHAKER ELTAHAN,[1] SATYAJEET S. SALVI,[2] and XING-JIE LIANG[1,*]

[1]*Laboratory of Controllable Nanopharmaceuticals, CAS Key Laboratory for Biomedical Effects of Nanomaterials and Nanosafety, National Center for Nanoscience and Technology of China, Beijing, 100190, China, *E-mail: liangxj@nanoctr.cn*

[2]*Hampshire College, Amherst, MA 01002, USA*

CONTENTS

ABSTRACT

Drug targeting is a promising strategy for efficient treatment of various serious diseases such as cancer. The primary goal of this concept is to provide precise delivery of therapeutic agents into pathological area, while avoiding negative impact onto healthy tissues. Through this approach, high therapeutic efficacy of the drug can be attained while experiencing minimum side effects. Recently, nano-sized carriers have received great attention as delivery systems which can load, bring, and release the drugs in a localized area. Accurate drug delivery can be achieved either via passive targeting, which is-based on enhanced vascular permeability in affected zone, or active targeting, which can be achieved by decoration of nanocarriers surface with ligands that have high affinity toward the targeted area. In this chapter, general methods and means of drug targeting-based on nanomedicine have been discussed.

12.1 INTRODUCTION

Use of conventional drugs raises many problems, which are mainly associated with their biodistribution throughout the body. Thus, in order to reach the pathological site, pharmaceutical drug has to overcome many biological barriers, such as organs, cells, and intracellular compartments which lead to its inactivation and reducing therapeutic efficacy. Therefore, in order to achieve required therapeutic result, the concentration of the drug should be enhanced, which in turn, will lead to the drug's increased interaction with normal tissues and as a result, the risk of undesirable side effects will be increased. Altogether, these factors heavily influence, the overall therapeutic effect largely determined by precise drug accumulation.

Drug targeting can offer a new way to overcome the challenges, associated with the lack of drug specificity. Using this approach in the therapy of certain diseases has following advantages: (a) a higher concentration of the drug at the site of action, while the total dosage of administrated drug is low; (b) reduced drug's toxicity toward healthy tissues and, as a result, minimized side effects; (c) simplified drug administration protocols; (d) decreased cost of therapy (Torchilin, 2000). Nowadays, drug targeting is-based on applying pharmaceutical nanocarriers, whose function includes: (1) improved delivery of poorly water-soluble drugs; (2) targeted delivery of drugs specifically to their site of action at a high dose; (3) protection of a drug from the environment (e.g., acidic environment in the stomach or high levels of enzymes in the bloodstream);

(4) controlled drug release; (5) co-delivery of multiple agents for combination therapy and diagnostics. Drug delivery systems include liposomes, micelles, nanoparticles, dendrimers, polymer-drug conjugates, carbon nanotubes and others. The physicochemical properties of nanomaterials can be tuned to achieve required conditions for successful delivery of therapeutic agents. For example, size, structure, shape, morphology, surface and stimuli-responsive properties can be controlled when fabricating nanotherapeutics (Sun et al., 2014).

Drug targeting delivery, i.e., drug accumulation at site of action with a sufficient dose, can be accomplished in two ways: passive and active targeting (Figure 12.1). Passive targeting – drug accumulation in the areas with leaky vasculature. This phenomenon has been called enhanced

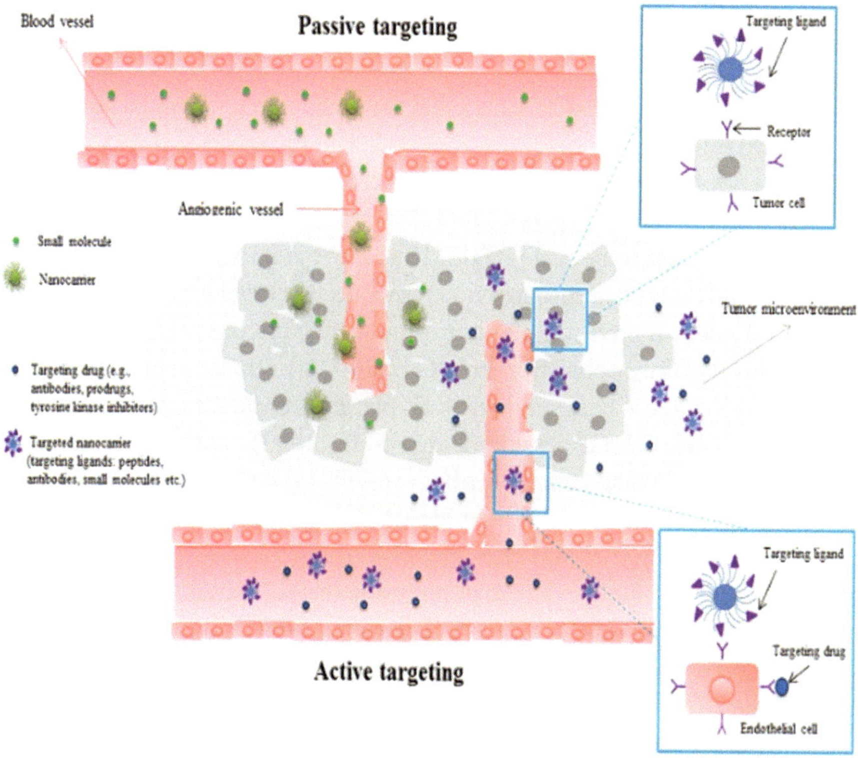

FIGURE 12.1 Schematic illustration of passive and active targeting strategies.

permeability and retention (EPR) effect (Matsumura et al., 1986; Maeda et al., 2001; Maeda et al., 2009). Active targeting involves the use of different ligands, which have high affinity toward the targeted area. The following substances can be used as targeting ligands: small molecules, antibodies, and their fragments, peptides, aptamers, etc. (Allen, 2002; Blanco et al., 2012; Zhong et al., 2014). In addition, active targeting can be associated with either internal physical parameters, such as, altered pH or temperature in pathological area, or with external magnetic field, which can influence on transport of carriers possessing ferromagnetic properties (Morachis et al., 2012).

Tumor targeting, as an example, can fully demonstrate the potential of drug targeting strategies.

12.2 PASSIVE TARGETING

In order to accumulate in certain area, drugs need to cross through vascular walls into the target site. However, the structure of normal blood vessels allows only small molecules to extravasate through the tight junctions between endothelial cells. Under certain conditions, blood vessels can be formed with defective structure, and this is due to increased permeability which causes them to be leaky (Jain, 1987). Irregular vasculature is a typical feature of neoplastic progression. Once tumor reached the certain size (3 mm^3) it needs a growing number of nutrients and oxygen to supply the rapidly proliferating cells (Bergers et al., 2003). In order to initiate angiogenesis, neoplastic tissue secretes multiple growth factors that trigger the formation of new blood vessels (Bates et al., 2002). Thus, tumor creates favorable conditions to sustain further growth and development. However, newly formed blood vessels have highly disorganized structure which is characterized by lack of pericytes and absent or defective basement membrane (Jain et al., 2010). These abnormalities lead to enhanced vascular permeability due to presence of fenestrations between endothelial cells. When blood components, including macromolecules and relatively large nanocarriers (e.g., liposomes and micelles), reach such areas they can extravasate and accumulate inside the interstitial space (Jain et al., 2010). Moreover, unlike normal tissues in which constant drainage renews interstitial fluid, tumor tissues have a dysfunctional lymphatic system, resulting in inefficient drainage from the tumor tissue (Noguchi et al., 1998; Padera et al., 2004; Swartz et al., 2007). Therefore, large molecules and nanocarriers that entered into the tumor tissue cannot be removed efficiently and hence are retained in the tumor. This spontaneous drug accumulation in leaky areas has been widely known as EPR effect, which was first discovered by Maeda and Matsumura (Maeda et al., 2001, 2009; Matsumura et al., 1986). The EPR effect is the basis of passive targeting.

Unlike normal tissues, tumor tissue has high interstitial fluid pressure that hinders diffusion-mediated transport of drugs toward tumor cells. Increased interstitial pressure, especially at the center of the tumor, reduces extravasation of drugs and decreases their uptake into solid tumor. In general, high oncotic pressure pulls the fluid back from the tumor center towards the periphery, thus preventing the interaction between therapeutics with tumor cells (Heldin et al., 2004; Jain, 1987). As a result, drugs, including nanocarriers, should overcome enhanced pressure and migrate through interstitial space and reach neoplastic cells in sufficient concentration in order to be considered as successful passive targeting agents. Due to their great size, nanocarriers bearing the pharmaceutical drug are less affected by enhanced interstitial pressure (Noguchi et al., 1998; Padera et al., 2004; Swartz et al., 2007). In this regard, nanotherapeutics' size can be concluded as one of the most important parameters in passive targeting. Indeed, the particle size has a crucial importance in both the interstitial and vasculature transport and extravasation. For successful extravasation, the size of the material should be smaller than the usual cut-off of the fenestrations in the neo-vasculature, which size usually ranges from 100 nm to 2000 nm and depends on the type and stage of the cancer (Haley et al., 2008; Hashizume et al., 2000). Moreover, large particles can be eliminated from the blood stream by spleen and liver more rapidly. On the other hand, small nanoparticles can be inactivated by kidneys (Sun et al., 2014). In addition, small agents will be able to rapidly diffuse back from interstitial space into the vascular compartment due to enhanced interstitial pressure (Dreher et al., 2006). Overall, it was found that the most optimal particle size in passive targeting ranges between 30 and 200 nm (Jain et al., 2010).

Among physicochemical parameters, surface charge of nanocarriers plays a significant role in EPR-based drug targeting. Surface charge can influence circulation at the time of intravenously administered drug, and therefore this can change the intra tumoral profile. Tumor cells surfaces are generally charged negatively due to presence of inner layer anionic components such as phosphatidylserine, glycoproteins and proteoglycans, anionic phospholipids. Nano-carriers with positive surface charge are preferentially seized by the tumors and retained for a longer time in comparison to neutral or negatively charged nanoparticles. Moreover, the extracellular pH is lower in neoplastic cells than in normal cells. Since the pH gradient becomes acidic, the electrostatic potential of nanoparticles becomes positive, which in turn, facilitates electrostatic interaction between nanoparticles and cancer cells (Honary et al., 2013, 2013). Since the surface charge of endothelial cells is negative, this is consistent with the fact that specific tumor vasculature targeting is conducted by primarily positively charged nanoparticles loaded with antiangiogenic drugs (Bertrand et al., 2014; Honary et al., 2013). Thus, the positive charged nanocarriers appears to be effective for tumor-specific passive targeting approach. On the other hand, negatively charged plasma proteins and cells in the organs of mononuclear phagocyte system (MPS) can increase blood clearance toward positively charged particles, thus reducing their therapeutic efficacy. In contrast to positively charged nanocarriers, negatively charged particles have prolonged blood circulation time due to less formation the aggregates with blood proteins and are recognized by cells of MPS (Bertrand et al., 2014). In order to achieve stability of nanoparticles in suspensions and prevent their aggregation, the average electrostatic potential (zeta potential) should be above 30 mV (positive or negative), hence enabling the particles to repel one another (Sun et al., 2014). In addition, it was demonstrated that transport of charged agents (either positive or negative) in the interstitium is limited due to interaction with extracellular matrix, while neutral compounds do not linger in the interstitial space and hence, can diffuse faster through the tissue (Lieleg et al., 2009).

Another parameter which was demonstrated to have an ability to affect biodistribution is the nanocarriers shape. Some studies suggested that the extravasation and penetration into tumor tissue of elongated agents is more efficient than that of spherical one (Chauhan et al., 2011; Decuzzi et al., 2009; Shukla et al., 2012). However, rod-like nanoparticles are hydro-dynamically more unstable while traveling in the blood stream (Sun et al., 2014). Taking this into account, it is clear that the geometrical parameters of nanocarriers also should be considered in order to improve their ability to cross the gaps in the vascular wall and penetrate deeply into the tumor tissues.

The strategy of passive targeting is based on EPR effect, which is one the most significant features of tumor angiogenesis. In order to achieve maximum therapeutic effect using passive targeting approach, nano-therapeutics should have appropriate parameters, such as size, surface charge, and shape. Moreover, to improve EPR effect the circulation time of drug-carrier should be prolonged by coating drug carriers mostly with polyethylene glycol (PEG) (Gref et al., 2000). The approach of passive targeting facilitates solving the problem associated with effective delivery of pharmaceutical drugs to the site of the tumor. Transport through vascular walls in areas with increased vascular permeability can help to avoid non-specific drug accumulation. Retained amount of drugs in the interstitial space can maintain sufficient therapeutic concentration, and therefore can leads to enhanced therapeutic index. Despite many advantages of this strategy, passive targeting still has its limitations. In particular, uneven distribution of blood vessels in solid tumors results in inhomogeneous drug permeability throughout the tumor, which in turn, leads to incomplete therapeutic effect. Also, in cases of small tumors or metastatic lesions, where strong angiogenesis is not exhibited, passive targeting EPR effect is ineffectual to the lack of numerous blood vessels.

12.3 ACTIVE TARGETING

The concept of active targeting is rather simple: conjugate the drug with another molecule, capable to recognize and bind to the specific site, and as the result, the affected drug will be selectively directed to the area of interest. This method increases the therapeutic index of the nanocarrier, and is widely regarded as the most promising strategy, since minimal side effects are manifested, and this method compensates for the lack of drug specificity.

In case of cancer, conjugation with targeting moiety provides preferential accumulation of nanotherapeutics in tumor microenvironment or tumor cells, which expresses an appropriate receptor. Targeting ligand will enhance binding affinity and facilitate receptor-mediated endocytosis, and thereby, will increase intracellular drug concentration (Nori et al.,

2003). In addition, the specificity towards certain factors, located in tumor microenvironment, also can be achieved, resulting in reduced support of tumor development.

There are several methods of targeting ligand conjugation. One the most common approaches is to attach targeting moiety to the surface of nanocarrier, which bears the cytotoxic drug. In recent years many works, regarding antitumor therapy by using targeted nanotherapeutics, have been reported. However, just few of them reached clinical trials (e.g., doxorubicin-loaded immuno-liposomes, modified with human monoclonal antibody or transferrin-targeted oxaliplatin liposomes) (Matsumura et al., 2004; Suzuki et al., 2008). Another approach relies on direct coupling of a therapeutic molecule to a targeting moiety. For example, radio-immuno-therapeutic agent ibritumomab tiuxetan (Zevalin®), which is used for treating of lymphoma, is a conjugation between monoclonal antibody ibritumomab with the chelator tiuxetan, that is functionalized with radioactive isotope (either yttrium-90 or indium-111). Some other monoclonal antibodies, such as trastuzumab (Herceptin®) or bevacizumab (Avastin®), can play the role of both targeting ligand and drug (Allen, 2002; Danhier et al., 2010).

12.3.1 ACTIVE TUMOR CELLS TARGETING

Active tumor cell targeting can be achieved by modifying the nanocarriers with specific targeting ligands such as peptides, aptamers, antibodies, small molecules (e.g., folic acid), and polysaccharides (Gu et al., 2007). Some of the ligands such as antibodies or aptamers are highly specific, while other ligands, such as folic acid, are less target specific. The aim of targeting moieties is to recognize the cell-surface receptors, overexpressed by cancer cells and bind these moieties which have high affinity to these receptors, to trigger receptor endocytosis. As a result, cellular uptake and intracellular accumulation occurs through the use of these moieties. Thus, by using nanocarriers, which are decorated with targeting ligands, these nanocarriers are considered as an attractive medium for specific delivery of anticancer agents. The ability of nanotherapeutics to enhance intracellular accumulation leads to higher antitumor efficacy of the drugs with less impact on normal cells. Usually selection of the targeting ligand depends on the type of cellular receptors. Targeting nanosized carriers can be prepared by grafting ligands onto their outer surface via diverse linkers. In the case of prodrugs, targeting moiety can be attached to the carrier backbone.

The most established cell proliferation receptors which are widely used for active targeting schemes include folate receptor, human epidermal receptors, transferring receptor and glycoproteins (Table 12.1).

12.3.1.1 Folate Receptor

The folate receptor is one of the most extensively investigated tumor markers that bind to the vitamin folic acid (FA), which is required for essential cell function- synthesis of purines and pyrimidines. FA binds with a high affinity to the glycosylphosphatidylinositol-linked folate receptor, which is significantly overexpressed on the many cancer cells compared to healthy tissues. Unlike normal cells, which cannot transport folate conjugates of any type, neoplastic cells can easily transport folate conjugates through the cell membrane via folic acid receptor (Low et al., 2004). Moreover, folic acid ligands are nontoxic, nonimmunogenic, easy to attach to the carrier, stable in circulation and inexpensive. All these features led to the establishment of many nanomedicines decorated with the folic acid ligand for specific drug targeting. Nanocarriers functionalized with folic acid usually exhibit enhanced selectivity and cytotoxicity toward tumor cells. For example, folate ligand-modified polyamidoamine (PAMAM) dendrimers conjugated with cytotoxic agent methotrexate demonstrated high accumulation levels in tumor and significant inhibition of tumor growth as compared to non-targeting counterparts when treated mice with folate-positive human KB tumor xenografts (Kukowska-Latallo et al., 2005). Moreover, methotrexate dendrimers displayed lower systemic toxicity and noticeably higher efficacy compared to free methotrexate at an equal cumulative dose. In another study FA-conjugated nanoparticles composed of mixed lipid monolayer shell and biodegradable PLGA core have been fabricated for receptor mediated delivery of anticancer drug docetaxel (Liu et al., 2010). The results demonstrated 50.91% and 93.65% more effectiveness of targeting ability of functionalized particles to human MCF-7 breast tumor cells, in comparison to non-targeted and free docetaxel, respectively. Zhou et al. (2013) designed FA-functionalized PEG-PCL micelles with conjugated via pH-sensitive

TABLE 12.1 Examples of Nanotherapeutics Used for Active Cancer Cells Targeting via Diverse Receptors

Type of receptor	Targeting ligand	Formulation	Drug	Type of cancer cells
Folate receptor	Folic acid	PAMAM dendrimers	Methotrexate	KB
		DLPC/DSPE-PEG2k/DSPE-PEG5k-FOL shell and PLGA core	Docetaxel	MCF-7
		FA-PEG-PCL-hydrazone-DOX	Doxorubicin	Human nasopharyngeal epidermoid carcinoma cell line KB
Epidermal growth factor receptor (EGFR)	EGF	EGF-PEG-PCL	Ellipticine	MDA-MB-468
	EGFR mAb	PLGA-PVA	Rapamycin	MCF-7
	Fab′ fragments of Cetuximab (mAb)	Immunoliposomes-based on Mal-PEG-Chol	Boron anion	$F98_{EGFR}$
HER-2	Trastuzumab	DPPC/TPGS/TPGS-trastuzumab	Docetaxel	SK-BR-3
	Trastuzumab	PLA-TPGS/TPGS-COOH	Docetaxel	SK-BR-3
Transferrin receptor	Transferrin	PEG-PCL polymersomes	Doxorubicin	C6 glioma
	Transferrin	Tf-L-DOX/VER	Doxorubicin, verapamil	K562
	Transferrin	Tf-PEG-PHDCA	Hydroxycamptothecin	KB, K562, S180
Glycoproteins	Galactosamine	N-(2-hydroxypropyl) methacrylamide (HPMA)	Doxorubicin	Liver cancer (phase I clinical study)
	Galactose	PEG-PCL	Doxorubicin	HepG2
	Hyaluronic acid	HA-SS-DOCA	Paclitaxel	MDA-MB-231

linker-hydrozone doxorubicin (FA-PEG-PCL-hydrazone-DOX) (Guo et al., 2013). This nanosystem showed improved cellular uptake and higher cytotoxicity to human nasopharyngeal epidermoid carcinoma (KB) cells, suggesting that FA-functionalized system achieved highly selective anticancer drug delivery.

12.3.1.2 Human Epidermal Receptor

The human epidermal receptor (HER) family of receptor tyrosine kinases can offer two targets for active drug targeting located on cancer cell surfaces. Two receptor tyrosine kinases, epidermal growth factor receptor (EGFR) and human epidermal receptor-2 (HER-2) have been widely studied for anticancer therapy. These receptors mediate a cell signaling pathway for growth and proliferation in response to the binding of the growth factors (Byrne et al., 2008).

The EGFR is a member of tyrosine kinase receptors family. Its activation promotes key processes involved in tumor growth and progression, such as proliferation, angiogenesis invasion and metastasis (Lurje et al., 2009). EGFR is overexpressed in many solid tumors such as non-small cell lung cancer, breast, squamous cell carcinoma of the head and neck, ovarian, kidney, prostate, and pancreatic cancer. Usually EGFR expression characterizes a more advanced disease stage and its presence corresponds directly to the metastatic capabilities in various types of cancer (Byrne et al., 2008). Expression of EGFR on tumor cells is much higher than normal cells making EGFR appropriate candidate for selective drug delivery.

EGFR has six known ligands: EGF, transforming growth factor-α (TGF-α), heparin-binding EGF (HB-EGF), epiregulin, amphiregulin, and betacellulin (Laskin et al., 2004). Any of these ligands can be used as targeting ligand for EGFR specific drug delivery. For example, EGF-conjugated PEG-poly(ε-caprolactone) (EGF-PEG-PCL) micelles loaded with chemotherapeutic agent ellipticine exhibited potent antiproliferative effect against EGFR-overexpressing MDA-MB-468 cells in comparison to cells expressing a low level of EGFR (Lee et al., 2007). EGFR also offers an attractive target for monoclonal antibody-based therapeutic strategies. For example, EGFR mAb-functionalized rapamycin-loaded PLGA-poly(vinyl alcohol) nanosystem (EGFR mAb-PLGA-PVA) displayed 17 times higher uptake by breast cancer cells (MCF-7) than the non-targeting counterparts (Acharya et al., 2009). In addition, targeted nanoparticles showed significant antiproliferative activity toward MCF-7 cells in comparison to free drug rapamycin and non-functionalized nanoparticles. A well-known antiEGFR monoclonal antibody Cetuximab also was employed as targeting moiety. Immunoliposomes decorated with Fab′ fragments of Cetuximab exhibited higher delivery efficacy in EGFR positive cells ($F98_{EGFR}$) than non-targeted IgG immunoliposomes (Pan et al., 2007).

HER-2 is also overexpressed in variety of solid tumors, including breast, lung, gastric and ovarian cancer. HER-2 plays a significant role in the pathogenetic processes of many types of cancers. Utilizing of monoclonal antibodies as antiHER-2 targeting moieties shows high selective capabilities of functionalized nanosystems in cancer therapy. Trastuzumab (Herceptin®) an FDA approved monoclonal antibody have been developed for cancer treatment as well as active anticancer drug delivery (Harries et al., 2002; Li et al., 2012). For example, trastuzumab-conjugated vitamin E TPGS-coated liposomes bearing anticancer drug docetaxel demonstrated longer half-life and greater apoptotic activity as compared to non-targeting nanoparticles and free docetaxel (IC_{50}= 0.08 ± 0.4, 3.74 ± 0.98, 20.23 ± 1.95 μg/ml respectively) in Her2 overexpressing SK-BR-3 cells (Raju et al., 2013). In a similar way, PLA-D-α-tocopheryl poly(ethylene glycol) succinate (PLA-TPGS) was decorated with herceptin as a targeting ligand (Mi et al., 2012). This system showed an obvious targeting delivery of docetaxel in Her2 receptor overexpressed breast cancer cells.

12.3.1.3 Transferrin Receptor

Transferrin receptor (Tf) is a serum nonheme iron-binding glycoprotein that transports iron into proliferating cells via receptor-mediated endocytosis. Tf is involved in iron homeostasis, the regulation of cell growth, and is overexpressed in neoplastic cells much higher than that in normal cells due to increased requirement of iron, that makes this receptor an attractive target for tumor therapy (Daniels et al., 2006; Gatter et al., 1983). Tf has been used as target for drug delivery for various formulations. For example, it was reported that doxorubicin-loaded Tf-conjugated biodegradable PEG-PCL polymersomes exhibited enhanced intracellular delivery and the highest cytotoxicity in C6 cells than those of free drug and non-functionalized polymersomes (Pang et al., 2011). In another study Tf-conjugated liposomes coencapsulated with doxorubicin and verapamil (Tf-L-DOX/VER) showed improved cellular uptake of Tf-L-DOX/VER and greater cytotoxicity in comparison to non-targeted liposomes with doxorubicin and verapamil and Tf-conjugated liposomes loaded with doxorubicin alone in a chronic myelogenous leukemia cell line (K562 cells) (Wu et al., 2007). Pei et al. fabricated Tf-targeted niosomes by conjugating Tf to poly(ethylene glycol) cyanoacrylate-cohexadecyl cyanoacrylate (PEG-PHDCA) for delivery of hydroxycamptothecin (Hong et al., 2009). The results of cell lines experiments revealed that unlike non-targeted noisomes, non-stealth niosomes and free drug, Tf-functionalized niosomes had the greatest intracellular uptake and the highest cytotoxicity in all three carcinoma cell lines (KB, K562, and S180 cells). Further in vivo experiments in S180 tumor-bearing mice displayed significant tumor inhibition as compared to the free drug and non-targeted counterparts (Hong et al., 2010).

12.3.1.4 Glycoproteins

During neoplastic process cancer cells often express different glycoproteins compared to normal cells. Saccharides are able to recognize and specifically bind to carbohydrate moieties attached to glycoproteins (Rutishauser et al., 1975). Thus, saccharides can be attached onto the nanocarriers surface as targeting moieties that are directed to cell-surface carbohydrates. For example, saccharides, such as galactose and galactosamine can be employed as targeting ligands against hepatocellular carcinoma cells that overexpress asialoglycoprotein receptors (ASGP-R) (Li et al., 2008). A number of various drug-loaded nanomedicines have been fabricated for liver cancer therapy. It was reported that doxorubicin conjugated with *N*-(2-hydroxypropyl)methacrylamide polymer modified with galactosamine as a targeting ligand, has been tested for clinical phase I study (Seymour et al., 2002). The results exhibited that targeting the anticancer drug doxorubicin using polymers functionalized with galactosamine can effectively target the liver and can achieve higher drug levels in tumors than using non-galactosylated polymers or free doxorubicin. In another study biodegradable micelles-based on poly(ethylene glycol)-poly(ε-caprolactone) (PEG-PCL) copolymer decorated with galactose have been used for active targeting delivery of doxorubicin in (ASGP-R)-overexpressing HepG2 cells (Zhong et al., 2013). Targeted micelles revealed much lower half maximal inhibitory concentration of doxorubicin, in comparison to free drug and non-targeting counterparts. Flow cytometry showed increased doxorubicin level in HerG2 cells than that of non-targeting control. The data suggests that galactose functionalized micelles have an apparent targetability and significantly enhanced antitumor efficacy toward tumor cells that overexpress an asiaglycoprotein receptor.

Besides saccharides, polysaccharides, in particular, hyaluronic acid (HA), a natural polymer with excellent biodegradability and biocompatibility, also has received great interest as a targeting moiety for specific drug delivery (Mizrahy et al., 2012). Since HA can recognize and bind receptors, such as CD44 and RHAMM, which are overexpressed on the cell surface of several metastatic tumors, many different HA-functionalized formulations have been fabricated (Oh et al., 2010). Thus, HA-modified liposome, micelles, polymersomes have been used as drug carriers for various cytotoxic drugs against different tumor types, including breast cancer, colon cancer, squamous cell carcinoma and glioma. It was demonstrated that high inhibition activity of modified nanomedicines against tumor cells has been achieved by selective targeting via high binding affinity between overexpressed receptors on cancer cells surface and targeting ligand HA. For example, paclitaxel loaded micelles-based on HA-deoxycholic acid (HA-SS-DOCA) significantly enhanced intracellular accumulation and antitumor activity of paclitaxel in human breast MDA-MB-231 (Li et al., 2012).

12.3.2 ACTIVE TUMOR MICROENVIRONMENT TARGETING

Active targeting is essential to improve therapeutic efficacy of drugs and reduce their side effects. Tumor microenvironment-selective compounds are focused on molecular and physiological differences existing between cancer and normal tissues. The main purpose of drug targeting is to suppress the ability of tumor microenvironment to support the neoplastic progression. Due to a large number of various factors, many strategies of active targeting have been developed. One of the approaches includes exploiting molecular drugs, such as peptides, antibodies, and proteins, for specific inhibition of diverse components and pathways, related to cancer development (Allen, 2002). Other approach involves applying nanotherapeutics, such as liposomes, nanoparticles, or micelles, with surface modified by various ligands for targeted delivery of anticancer drugs (Brannon-Peppas et al., 2012; Byrne et al., 2008). It is worth noting that molecular drugs can be used as targeting moiety for nanocarriers. Generally speaking, targeting molecules can play the role of both targeting ligand and drug. Herein, the main constituents of tumor microenvironment and their targeting approaches are presented (Table 12.2).

12.3.2.1 Targeting Cancer-Associated Fibroblasts

Cancer-associated fibroblasts (CAFs) are heterogeneous subpopulation of fibroblasts, which derived from various cells, such as normal fibroblasts, bone marrow-derived stromal cells, endothelial cells, and malignant or normal epithelial cells either via endothelial-mesenchymal transition (EndMT) or epithelial–mesenchymal transition (EMT) (Allen et al., 2011; Marsh et al., 2013; Quante et al., 2011; Räsänen et al., 2010). CAFs are one of the most abundant cell types in tumor microenvironment and play a key role in tumor progression. There is abundant evidence, that CAFs can support tumor growth, by secreting multiple factors. Among them are classical growth factors, such as epidermal growth factor (EGF), hepatocyte growth factor (HGF), and transforming growth factor beta (TGF-β) (Kalluri et al., 2006). Moreover, cancer-associated fibroblasts produce a number of additional factors, which can improve proliferative, metastatic and invasive properties of cancer cells. Among these factors are chemokines and cell surface molecules like integrin α11 or syndecan-1; proteases, like matrix metalloproteinase-2 (MMP2); mitogenic factors, such as secreted frizzled-related protein 1 (SFRP1); insulin-like growth factor 1 (IGF1) and insulin-like growth factor 2 (IGF2) (Östman et al., 2009). In order to facilitate angiogenesis, CAFs release proangiogenic factors,

TABLE 12.2 Examples of Therapeutics Used for Tumor Microenvironment Targeting

Category	Target	Targeting agents/ nanocarriers	Drug's classification
Cancer-associated fibroblasts (CAFs)	Fibroblast activation protein (FAP)	Sibrotuzumab	Humanized monoclonal antibody
		FAP5-DM1-maytansinoid	Monoclonal antibody-cytotoxic agent
		Val-boro Pro; PT630	Chemical inhibitors
		Z-Gly Pro-dox	Prodrug (FAP-specific dipeptide-dox)
	PDGF-PDGFR signaling pathway	Imatinib	Tyrosine kinase inhibitor
	TGFβ-TGFβR signaling pathway	1D11	Anti-TGFβ neutralizing antibody
Angiogenesis	VEGF	Bevacizumab	Humanized anti-VEGF monoclonal antibody
	VEGFR-2	Anti–Flk-1 MAb-NP-90Y	Nanotherapeutics (Monoclonal antibody-radiolabeled NPs)
	VEGFR/PDGFR	Sunitinib	Tyrosine kinase inhibitors
		Sorafenib	
		Pazopanib	
	VEGFR/FGFR	Brivanib	
	VCAM-1	Liposomes modified with anti-VCAM-1 antibody	Nanotherapeutics
Hypoxia	HIF-1α (mRNA)	EZN-2968	Antisense oligodeoxynucleotide
	HIF-1α (FRAP/ mTOR)	Rapamycin	FRAP/mTOR inhibitor
	Hypoxic region	Etoposide loaded chitosan NPs	Nanotherapeutics
	Hypoxic region	Tirapazamine	Bioreductive prodrug
Inflammation	NF-κB activity (IKKβ)	PS-1145	Small-molecule inhibitor of IKKβ
	NF-κB activity (26 s proteasomes)	Bortezomib	Proteasome inhibitor

TABLE 12.2 (Continued)

Category	Target	Targeting agents/ nanocarriers	Drug's classification
	NF-κB activity (NF-κB)	Curcumin	Suppressor of NF-κB–dependent gene expression
	IL-6/JAK/STAT3 signaling pathway (IL-6)	Siltuximab	Monoclonal neutralizing antibody
	IL-6/JAK/STAT3 signaling pathway (JAK)	Ruxolitinib	Small-molecule inhibitor of JAK
	TNF-α signaling pathway	Infliximab	Monoclonal antibody
		Etanercept	Fusion protein, inhibitor of TNF-α receptors
Extracellular matrix (ECM)	$\alpha_v\beta_3$, $\alpha_v\beta_5$ integrins	Nanocarriers modified with RGD peptide	Nanotherapeutics
	LOX enzyme	PLGA NPs modified with anti-LOX antibody	Nanotherapeutics
	MMPs (MT1-MMP)	Liposomes coated with GPLPLR peptide	Nanotherapeutics
		Liposomes coated with anti-MT1-MMP antibody	Nanotherapeutics

such as vascular endothelial growth factor (VEGF), fibroblast growth factors (FGF), and platelet-derived growth factor (PDGF). In addition, CAFs can recruit bone marrow-derived cells or immune cells into the tumor progression (Cirri et al., 2012). Thus, it can be concluded, that the role of CAFs in tumor progression is significant and hence, targeting to various CAFs factors is a promising strategy in cancer therapy.

There are many agents, which can be used for targeting the CAFs' activity. For example, sibrotuzumab, a humanized monoclonal antibody, have been used to inhibit the activity of CAFs-expressed-fibroblast-activated-protein

(FAP), that is not detected in normal fibroblasts, and which contributes in promoting the tumorigenesis. FAP inhibition leads to a retardation of tumor growth through improving tumoral immune response (Togo et al., 2013). Other agents have been demonstrated to have FAP-inhibitive properties, including monoclonal antibody FAP5-DM1 conjugated with cytotoxic agent maytansinoid (Ostermann et al., 2008); FAP chemical inhibitors [Val-boro Pro (Narra et al., 2007) and PT630 (Santos et al., 2009)]; prodrugs, such as doxorubicin conjugated with a FAP-specific dipeptide (Z-Gly Pro) (Huang et al., 2011). Moreover, targeting diverse signaling pathways to reduce the impact of CAFs on tumor progression plays an important role in cancer therapy. Thus, disturbing PDGF-PDGFR signaling pathway by imatinib, a tyrosine-kinase inhibitor, or TGFβ-TGFβR signaling pathway by anti-TGFβ neutralizing antibody (1D11) leads to a reduction in tumor growth and assisted intratumoral drug delivery (Pietras et al., 2008). In addition, targeting SDF-1-CXCR4 and HGF-Met signaling pathways by neutralizing antibodies exhibited decreased tumor growth, angiogenesis and drug resistance (Orimo et al., 2005; Wang et al., 2009; Wilson et al., 2012).

12.3.2.2 Targeting Hypoxia and Angiogenesis

Angiogenesis is a process of new blood vessel formation. Insufficient blood supply during the tumor growth causes the deficiency of nutrients and oxygen to the cancer cells. Moreover, high metabolic activity and oxygen consumption by rapidly proliferating neoplastic cells are complimentary factors of hypoxia in tumor tissues. Hypoxia promotes genomic instability and the expression of proto-oncogenes in cancer cells. In order to satisfy the needs of tumor tissues, the compensatory formation of new blood vessels needs to be developed. Hence, hypoxia induces vascularization by stimulation of hypoxia-inducible transcription factors (HIF), which respond to hypoxic condition in the cellular microenvironment. Among this class of proteins, HIF-1α is the most ubiquitously expressed and its main role in angiogenesis is to up-regulate the proangiogenic factors, such as VEGF, FGF, angiopoietin 2 (Ang-2), and PDGF, and down-regulate antiangiogenic factors, such as thrombospondin (Ruan et al., 2009; Liao et al., 2007).

VEGF is a key proangiogenic factor, and its activation and binding to receptor (VEGFR), is specifically expressed in endothelial cells, and this causes proliferation and migration of endothelial cells. After endothelial cells migrate to surrounding tissues, they begin to divide and form hollow tubes that soon organize a network of blood vessels (Nishida et al., 2006). Non-hypoxia mediated mechanisms, such as up-regulation of VEGF by oncogene activation, are also involved in angiogenesis process. Moreover, VEGF induce antiapoptotic molecules in order to protect the tumor vasculature from apoptosis, and thus cause degradation of the extracellular matrix through promotion of certain enzymes to facilitate tumor angiogenesis (Carmeliet, 2005). In addition, vascular cell adhesion molecule-1 (VCAM-1), an immunoglobulin-like transmembrane glycoprotein that is expressed on the surface of endothelial cancer cells, mediates cell-cell adhesion, and thereby, promotes angiogenesis (Dienst et al., 2005).

Considering the crucial role of aforementioned signaling pathways in tumor angiogenesis, the strategy of disturbing these pathways makes it reasonable for anticancer therapy. One of the first agents used in clinical trials for targeting angiogenesis was recombinant humanized anti-VEGF monoclonal antibody bevacizumab (Shih et al., 2006). It has been approved by U.S. Food and Drug Administration (FDA) in combination with cytotoxic drugs for the treatment of metastatic colon cancer and several other metastatic cancers. Other FDA approved inhibitors of VEGF signaling include small molecule receptor tyrosine kinase inhibitors (e.g., sunitinib, sorafenib, and pazopanib), which offer a clinical benefit in renal cell carcinoma therapy (Fang et al., 2013). However, anti-VEGF therapy demonstrated rather modest therapeutic results. This phenomenon is associated with microenvironmental defense mechanisms, in particular, resistance to anti-VEGF therapy through upregulation of other proangiogenic factors, such as fibroblast growth factor (FGF) (Sounni et al., 2013). Thus, emerging challenges in targeting VEGF have necessitated the expansion of an arsenal of antiangiogenic drugs against other signaling pathways. For example, brivanib, a dual tyrosine kinase inhibitor, demonstrates simultaneous inhibition of VEGF and FGF receptors, which in turn, leads to inhibition of tumor growth in hepatocellular carcinoma (Huynh et al.,

2008). Nanoparticle-based approach also has been applied to improve the therapeutic efficacy of antiangiogenesis therapy. Li et al. (2004) demonstrated high treatment efficiency of nanoparticle-based targeting agents radiolabeled with ^{90}Y. The small molecule integrin antagonist, which binds to the integrin $\alpha_v\beta_3$, and a monoclonal antibody against vascular endothelial growth factor receptor 2 (VEGF-2), have been used to target nanoparticles.

PDGF is also of interest as a target in antiangiogenic treatment. In solid tumors PDGF regulates angiogenesis by promotion pericyte recruitment to blood vessels. Targeting the PDGF by low molecular inhibitors of the receptor kinases, such as imatinib, sunitinib, sorafenib, and pazopanib, makes PDGF a promising in antitumor therapy. However, all of these inhibitors have the lack of specificity, which accordingly leads to side effects. Alternatively, other PDGF signaling antagonists, such as antibodies or DNA aptamers, can bind receptors with high specificity. But the use of these types of antagonists is associated with a number of issues, including their high cost and difficulty for targeted administration (Heldin, 2013).

VCAM-1 was targeted by immunoliposomes, which were decorated with anti-VCAM-1 antibody. Liposomes display specific targeting ability to tumor vasculature in vivo, which make them a promising candidate for targeted drug delivery against tumor vasculature (Gosk et al., 2008).

Hypoxia is another hallmark of malignant progression. It is well established, that hypoxic cells are more resistant to anticancer therapy (Kizaka-Kondoh et al., 2003). Under low oxygen tension conditions, the cytotoxicity of drugs whose activity is mediated by free radicals is decreased (Trédan et al., 2007). Moreover, low oxygen concentration decreases cell proliferation and as a consequence, the activity of these drugs that selectively target highly proliferating cells is also reduced. Therefore, hypoxia-targeted therapy is-based on two main approaches. The first one is to enhance oxygen concentration in tumor tissues by improving oxygenation and combining it with chemotherapy (Kizaka-Kondoh et al., 2003). The second approach involves targeting the hypoxic tumor microenvironment. Hypoxia-targeted strategy is-based on exploitation of bio-reductive prodrugs or HIF-1α-targeted compounds (Guise et al., 2014; Chen et al., 2003). Bioreductive prodrugs are agents, which under hypoxic conditions, turns into active cytotoxic metabolites by enzymatic reduction. For example, tirapazamine, a bireductive prodrug, has been activated under hypoxic conditions and reduced form of this prodrug causes DNA damage (Reddy et al., 2009). Other effective strategy is targeting the HIF-1α signaling pathway components. For instance, by targeting HIF antisense mRNA with EZN-2968 or FRAP/mTOR by rapamycin, can block HIF-1α function and can attenuate tumor growth (Kizaka-Kondoh et al., 2003; Wilson et al., 2011).

Nano-therapeutics also had been used for targeting hypoxia in tumor microenvironment. Lin et al. fabricated hypoxia-activated photoresponsive chitosan nanoparticles for drug delivery (Lin et al., 2013). This system consists of hypoxic sensor or a phototrigger lock, a coumarin phototrigger, and a caged drug (etoposide). The results showed that the anticancer drug release specifically in hypoxic tumor cells, but not in aerobic healthy cells. In another work, Poon et al. demonstrated an ability of layer-by-layer assembled polyelectrolyte nanoparticles to target low pH in the hypoxic tumor microenvironment (Poon et al., 2011).

12.3.2.3 Targeting Cancer-Associated Inflammation

An inflammatory component is present in the microenvironment of many tumors. It is now well established that inflammatory cells, such as cytokines, chemokines, and enzymes, can also promote tumor initiation, progression, and metastasis (Coussens et al., 2002). One of the key players of cancer-associated inflammation is the transcription factor – nuclear factor-kappa B (NF-κB), which induces expression of wide range of agents, including cytokines, adhesion factors, enzymes, and angiogenic factors (Karin, 2006). In addition, NF-κB can activate the expression of antiapoptotic and growth factor genes (Lu et al., 2006). Thereby, NF-κB is also able to mediate inflammatory processes, stimulate cell proliferation, promote tumor invasion, metastasis, and survival of transformed cells. Along with NF-κB, another transcription factor, known as signal transducer activator of transcription3 (STAT3) protein, plays a pivotal role during the inflammation process (Yu et al., 2007). Prior literature suggests that STAT3 contributes to cancer progression by promoting cell proliferation and apoptosis resistance (Becker et al., 2004; Hodge et al., 2004; Wang et al., 2004). Another proinflammatory cytokine TNF-α has protumoral properties.

TNF-α at physiologic doses regulates the production of cytokines, chemokines (e.g., IL-6), proteases, and proangiogenic factors, such as VEGF, FGF, etc. (Balkwill, 2009; Kulbe et al., 2007; Balkwill, 2006). Conversely, high doses of TNF-α may cause damage of tumor vasculature and have an antitumor activity (Balkwill, 2002). During the development of cancer-associated inflammation, it is difficult to underestimate the role of tumor-associated macrophages (TAMs), which are derived from monocytes that are recruited by chemokines (Colotta et al., 2009). TAMs are an essential component in most tumor tissues, that release many tumor-promoting factors, including angiogenic factors (e.g., VEGF) to facilitate blood vessels formation, epidermal growth factors to induce tumor cell proliferation and migration, matrix metalloproteinases (e.g., MMP2, MMP9) to promote tumor cell invasion and metastasis, IL-10 and prostaglandin E_2 to down regulate antitumor response (Allen et al., 2011; Colotta et al., 2009).

It is clear that inflammation associated with neoplastic progression is a complex phenomenon. The microenvironment, which contains a large number of cells and mediators, also are implicated in these inflammatory processes since diverse pathways influence tumor development. Targeting the cancer-associated inflammation certainly will contribute to anticancer therapy generally. Since inflammation is a complex process, agents can be targeted in different pathways. For example, NF-κB activity can be reduced by various compounds, which exert their activity through diverse mechanisms, such as inhibition of IKKβ (NF-κB kinase subunit β), inhibition of proteasomes to disrupt the degradation of IκB or direct targeting of NF-κB-dependent gene expression (Fang et al., 2013). However, sustained NF-κB inhibition can lead to severe side effects caused by immune deficiency (Greten et al., 2007). The IL-6/JAK/STAT3 signaling pathway plays an important role in cancer-associated inflammation (Yu et al., 2009) and can be disrupted at different levels. Through the demonstrated use of siltuximab, as IL-6 ligand-blocking antibody (Sansone et al., 2012) which can neutralize IL-6, and that ruxolitinib (Verstovsek, 2009) that blocks Janus activated kinase (JAK) phosphorylation of STAT3, tumor-associated inflammation can be significantly reduced. In order to reduce inflammation and its protumoral influence, TNF-α antagonists, such as, monoclonal antibody- infliximab, or fusion protein-etanercept, also have been applied. Moreover, inhibition of tumor-associated macrophages (TAMs) through targeting recruitment factors, including CSF-1, CCL2, or MCP-1, makes this approach potentially useful for the treatment of cancer-associated inflammation (Fang et al., 2013).

12.3.2.4 Targeting Extracellular Matrix

The extracellular matrix (ECM) is another major component of the tumor microenvironment, which surrounds cells and consists of various macromolecules, including proteins, glycoproteins, proteoglycans, and polysaccharides (Lu et al., 2012). Under normal conditions, ECM controls tissue development and homeostasis. During tumor progression ECM undergoes significant remodeling largely mediated by matrix-degrading enzymes, such as matrix metalloproteases (MMPs), and proteases of cysteine and serine. ECM degradation by MMPs facilitates angiogenesis and metastasis of tumor cells (Joyce, 2005; Pickup et al., 2014). However, some MMPs exhibit tumor suppressor properties by releasing antiangiogenic proteins such as endostatin, angiostatin, and tumstatin (Joyce, 2005). This evidence may explain the failure of broad-spectrum MMP inhibitors during clinical trials (Coussens et al., 2002). Remodeled ECM contributes to neoplastic progression by interaction between various ECM's proteins (e.g., fibrinogen, vibronectin) and cell surface receptors, such as $\alpha v\beta 3$ and $\alpha v\beta 5$, which are expressed on endothelial and tumor cells (Desgrosellier et al., 2010). Along with composition, ECM's biomechanical properties also impact on tumor progression. For example, overexpression of lysyl oxidase (LOX) by fibroblasts leads to cross-linking collagen fibers and elastin, which, in turn, increases ECM's stiffness and promotes cancer cell invasion and progression (Lu et al., 2012; Xiao et al., 2012). In order to reduce density and rigidity and thereby, improve transport of therapeutic agents into tumor tissues, co-delivery enzymes such as collagenase and gelatinase with drugs loaded nanoparticles has been used (Goodman et al., 2007; Liu et al., 2012).

Other strategies of targeting extracellular matrix are based on inactivation of its components. For example, cyclic or linear derivatives of RGD (Arg–Gly–Asp) peptide have been developed for binding $\alpha_v\beta_3$ and $\alpha_v\beta_5$ integrins. RGD moiety has been widely applied as a targeted ligand of nanocarriers, including liposomes, micelles and nanoparticles, for efficient delivery of anticancer drugs (Danhier et al., 2009; Xiong et al.,

2005). It was also revealed that suppression of LOX enzyme activity results in reduction of tumor growth by decreasing cross-linking of collagen fibers and restoration of normal-like tissues. For this purpose PLGA-based nanoparticles have been used as carriers for anti-LOX antibodies, which exhibited tumor suppressive properties in in vivo experiments (Kanapathipillai et al., 2012). MMPs are considered as a potential aim for anticancer therapy. Nanoparticles-based approaches have been widely used for targeting MMPs. For example, targeting membrane type 1 matrix metalloproteinase (MT1-MMP), which is expressed on endothelial tumor cells, by liposomes coated with GPLPLR peptide (Kondo et al., 2004), or by liposomes modified with anti-MT1-MMP antibody for doxorubicin delivery (Hatakeyama et al., 2007), demonstrated comparatively high antitumor activity.

12.4 CONCLUSION

In the field of medicine, specific drug targeting became an important tool for efficient therapy. Selective exposure on certain organs, tissues and cells can significantly reduce impact on healthy tissues, resulting in decreasing of undesirable side effects. Specific effects on pathological area may reduce administered drug's dose, which in turn leads to decrease in the drug's concentration throughout the body without reduction of therapeutic index. Precise drug targeting offers the possibility to enhance the therapeutic efficacy and minimize damage of healthy tissues.

In this chapter, we focused on treatment of cancer using targeted molecular drugs and nanotherapeutics. Drug targeting approach is based on distinctive features of tumors which can differentiate it from normal tissues. Passive and active drug targeting approaches make them very effective strategies in cancer treatment. Passive targeting is-based on enhanced vascular permeability, which is a distinctive property of angiogenesis in neoplastic progression. In order to achieve efficient extravasation through the leaky vasculature and homogeneous penetration into the tumor site, agents need to have appropriate physicochemical properties, such as size, shape, surface charge. Active targeting is-based on use of agents with high affinity for certain distinctive factors of neoplastic progression. Some of these agents can be used either as an independent drug or targeting moiety of nanocarrier. The purpose of active targeting is to incapacitate the mechanisms, which play a significant role in tumor development. Drugs can be specifically targeted either intracellular or extracellular constituents of tumors, and subsequently may cause cellular dysfunction followed with apoptosis. Usually it is difficult to address passive and active targeting methods independently, since for efficient active targeting, agents need to extravasate and diffuse deep into the tumor tissue.

The utilization of nanocarriers for medical purposes opens new perspectives in the treatment of such diseases as cancer. The nanosized carriers can be designed according to the terms of use. In particular, they can offer such properties as co-loading of various drugs and imaging agents, specific targeting, and controlled drug release. Thus, nanomedicines can provide effective and less toxic treatment. In order to achieve efficient comprehensive treatment, nanomedicines can be combined with radiotherapy or conventional chemotherapy. Moreover, nanomedicine is expected to have a significant progress in cancer immunotherapy.

Despite evident benefits, drug-targeting strategies still have a number of challenges. Especially in the case of cancer, highly accurate targeting is difficult to achieve due to presence of targeted components that exist in both cancer and normal tissues. Thereby, simultaneous specificity to more than one tumor constituents becomes a reasonable and appropriate strategy for drug targeting and delivery. In vivo application of nanotherapeutics is also rather limited. Biocompatibility of many nanocarriers are still remains an issue. For successful further use of nanocarriers it becomes important to study the impact of nanocarriers on biological components. To present day only some of nanotherapeutics received approvals demonstrating their biocompatibility. Another challenge of nanocarriers is that they do easily biodegrade within the tissue. Many nanocarriers can be fabricated through the use of biodegradable polymers, and through this method, challenges to drug targeting can be overcome. For non-biodegradable systems (e.g., inorganic nanoparticles), the following solution can be offered: fabrication of comparatively big carriers from a number of small particles using biodegradable linkers. Such a system can be passively targeted through the facilitation of the EPR effect. Then these carriers can penetrate inside the tumor tissue, while the biodegradable linkers will be degraded by tumor specific enzymes, thus causing the disassembly of these small particles, which can be easily eliminated from the body via renal filtration.

Looking into the past and noting impressive progress made in nanomedicine, in the future of is indeed optimistic, and we can expect the emergence of new types of efficient treatment of different diseases, based on personalized nanomedicine technologies.

KEYWORDS

- **active targeting**
- **cancer therapy**
- **cells receptors**
- **EPR effect**
- **passive targeting**
- **targeting ligands**
- **tumor microenvironment**

REFERENCES

Acharya, S.; Dilnawaz, F.; Sahoo, S. K. Targeted epidermal growth factor receptor nanoparticle bioconjugates for breast cancer therapy. *Biomaterials.* **2009**, *30*(29), 5737–5750.

Allen, M.; Louise Jones, J. Jekyll and Hyde: the role of the microenvironment on the progression of cancer. *The Journal of Pathology.* **2011**, *223*(2), 163–177.

Allen, T. M. Ligand-targeted therapeutics in anticancer therapy. *Nat. Rev. Cancer.* **2002**, *2*(10), 750–763.

Balkwill, F. TNF-α in promotion and progression of cancer. *Cancer Metastasis Rev.* **2006**, *25*(3), 409–416.

Balkwill, F. Tumor necrosis factor or tumor promoting factor? *Cytokine Growth Factor Rev.* **2002**, *13*(2), 135–141.

Balkwill, F. Tumour necrosis factor and cancer. *Nature Reviews Cancer.* **2009**, *9*(5), 361–371.

Bates, D.; Hillman, N.; Williams, B.; Neal, C.; Pocock, T. Regulation of microvascular permeability by vascular endothelial growth factors. *J. Anat.* **2002**, *200*(6), 581–597.

Becker, C.; Fantini, M. C.; Schramm, C.; Lehr, H. A.; Wirtz, S.; Nikolaev, A. et al. TGF-β suppresses tumor progression in colon cancer by inhibition of IL-6 transsignaling. *Immunity.* **2004**, *21*(4), 491–501.

Bergers, G.; Benjamin, L. E. Tumorigenesis and the angiogenic switch. *Nature Reviews Cancer.* **2003**, *3*(6), 401–410.

Bertrand, N.; Wu, J.; Xu, X.; Kamaly, N.; Farokhzad, O. C. Cancer nanotechnology: the impact of passive and active targeting in the era of modern cancer biology. *Adv. Drug Del. Rev.* **2014**, *66*, 2–25.

Blanco, M.; Teijon, C.; Teijón, J.; Olmo, R.; *Targeted Nanoparticles for Cancer Therapy.* INTECH Open Access Publisher, 2012.

Brannon-Peppas, L.; Blanchette, J. O. Nanoparticle and targeted systems for cancer therapy. *Adv. Drug Del. Rev.* **2012**, *64, Supplement*, 206–212.

Byrne, J. D.; Betancourt, T.; Brannon-Peppas, L. Active targeting schemes for nanoparticle systems in cancer therapeutics. *Adv. Drug Del. Rev.* **2008**, *60*(15), 1615–1626.

Carmeliet, P. VEGF as a Key Mediator of Angiogenesis in Cancer. *Oncology.* **2005**, *69* (Suppl. 3), 4–10.

Chauhan, V. P.; Popović, Z.; Chen, O.; Cui, J.; Fukumura, D.; Bawendi, M. G. et al. Fluorescent Nanorods and Nanospheres for Real-Time In Vivo Probing of Nanoparticle Shape-Dependent Tumor Penetration. *Angew. Chem.* **2011**, *123*(48), 11619–11622.

Chen, J.; Zhao, S.; Nakada, K.; Kuge, Y.; Tamaki, N.; Okada, F. et al. Dominant-negative hypoxia-inducible factor-1α reduces tumorigenicity of pancreatic cancer cells through the suppression of glucose metabolism. *The American Journal of Pathology.* **2003**, *162*(4), 1283–1291.

Cirri, P.; Chiarugi, P. Cancer-associated-fibroblasts and tumor cells: a diabolic liaison driving cancer progression. *Cancer Metastasis Rev.* **2012**, *31*(1–2), 195–208.

Colotta, F.; Allavena, P.; Sica, A.; Garlanda, C.; Mantovani, A. Cancer-related inflammation, the seventh hallmark of cancer: links to genetic instability. *Carcinogenesis.* **2009**, *30*(7), 1073–1081.

Coussens, L. M.; Fingleton, B.; Matrisian, L. M. Matrix metalloproteinase inhibitors and cancer—trials and tribulations. *Science.* **2002**, *295*(5564), 2387–2392.

Coussens, L. M.; Werb, Z. Inflammation and cancer. *Nature.* **2002**, *420*(6917), 860–867.

Danhier, F.; Feron, O.; Préat, V. To exploit the tumor microenvironment: Passive and active tumor targeting of nanocarriers for anticancer drug delivery. *J. Control. Release.* **2010**, *148*(2), 135–146.

Danhier, F.; Vroman, B.; Lecouturier, N.; Crokart, N.; Pourcelle, V.; Freichels, H. et al. Targeting of tumor endothelium by RGD-grafted PLGA-nanoparticles loaded with paclitaxel. *J. Control. Release.* **2009**, *140*(2), 166–173.

Daniels, T. R.; Delgado, T.; Helguera, G.; Penichet, M. L. The transferrin receptor, Part II: targeted delivery of therapeutic agents into cancer cells. *Clin. Immunol.* **2006**, *121*(2), 159–176.

Daniels, T. R.; Delgado, T.; Rodriguez, J. A.; Helguera, G.; Penichet, M. L. The transferrin receptor part I: Biology and targeting with cytotoxic antibodies for the treatment of cancer. *Clin. Immunol.* **2006**, *121*(2), 144–158.

Decuzzi, P.; Pasqualini, R.; Arap, W.; Ferrari, M. Intravascular delivery of particulate systems: does geometry really matter? *Pharm. Res.* **2009**, *26*(1), 235–243.

Desgrosellier, J. S.; Cheresh, D. A. Integrins in cancer: biological implications and therapeutic opportunities. *Nature Reviews Cancer.* **2010**, *10*(1), 9–22.

Dienst, A.; Grunow, A.; Unruh, M.; Rabausch, B.; Nör, J. E.; Fries, J. W. et al. Specific occlusion of murine and human tumor vasculature by VCAM-1–targeted recombinant fusion proteins. *J. Natl. Cancer Inst.* **2005**, *97*(10), 733–747.

Dreher, M. R.; Liu, W.; Michelich, C. R.; Dewhirst, M. W.; Yuan, F.; Chilkoti, A. Tumor vascular permeability, accumulation, and penetration of macromolecular drug carriers. *J. Natl. Cancer Inst.* **2006**, *98*(5), 335–344.

Fang, H.; DeClerck, Y. A. Targeting the Tumor Microenvironment: From Understanding Pathways to Effective Clinical Trials. *Cancer Res.* **2013**, *73*(16), 4965–4977.

Gatter, K. C.; Brown, G.; Trowbridge, I.; Woolston, R.; Mason, D. Transferrin receptors in human tissues: their distribution and possible clinical relevance. *J. Clin. Pathol.* **1983**, *36*(5), 539–545.

Goodman, T. T.; Olive, P. L.; Pun, S. H. Increased nanoparticle penetration in collagenase-treated multicellular spheroids. *International Journal of nanomediCine.* **2007**, *2*(2), 265.

Gosk, S.; Moos, T.; Gottstein, C.; Bendas, G. VCAM-1 directed immunoliposomes selectively target tumor vasculature in vivo. *Biochimica et Biophysica Acta (BBA)-Biomembranes.* **2008**, *1778*(4), 854–863.

Gref, R.; Lück, M.; Quellec, P.; Marchand, M.; Dellacherie, E.; Harnisch, S. et al. 'Stealth' corona-core nanoparticles surface modified by polyethylene glycol (PEG): influences of the corona (PEG chain length and surface density) and of the core composition on phagocytic uptake and plasma protein adsorption. *Colloids Surf. B. Biointerfaces.* **2000**, *18*(3), 301–313.

Greten, F. R.; Arkan, M. C.; Bollrath, J.; Hsu, L.-C.; Goode, J.; Miething, C. et al. NF-κB is a negative regulator of IL-1β secretion as revealed by genetic and pharmacological inhibition of IKKβ. *Cell.* **2007**, *130*(5), 918–931.

Gu, F. X.; Karnik, R.; Wang, A. Z.; Alexis, F.; Levy-Nissenbaum, E.; Hong, S. et al. Targeted nanoparticles for cancer therapy. *Nano Today.* **2007**, *2*(3), 14–21.

Guise, C. P.; Mowday, A. M.; Ashoorzadeh, A.; Yuan, R.; Lin, W.-H.; Wu, D.-H. et al. Bioreductive prodrugs as cancer therapeutics: targeting tumor hypoxia. *Chin. J. Cancer.* **2014**, *33*(2), 80.

Guo, X.; Shi, C.; Wang, J.; Di, S.; Zhou, S. pH-triggered intracellular release from actively targeting polymer micelles. *Biomaterials.* **2013**, *34*(18), 4544–4554.

Haley, B.; Frenkel, E. Nanoparticles for drug delivery in cancer treatment. *Urol. Oncol.* **2008**, *26*(1), 57–64.

Harries, M.; Smith, I. The development and clinical use of trastuzumab (Herceptin). *Endocr. Relat. Cancer.* **2002**, *9*(2), 75–85.

Hashizume, H.; Baluk, P.; Morikawa, S.; McLean, J. W.; Thurston, G.; Roberge, S. et al. Openings between defective endothelial cells explain tumor vessel leakiness. *The American Journal of Pathology.* **2000**, *156*(4), 1363–1380.

Hatakeyama, H.; Akita, H.; Ishida, E.; Hashimoto, K.; Kobayashi, H.; Aoki, T. et al. Tumor targeting of doxorubicin by anti-MT1-MMP antibody-modified PEG liposomes. *Int. J. Pharm.* **2007**, *342*(1–2), 194–200.

Heldin, C.-H. Targeting the PDGF signaling pathway in tumor treatment. *Cell Commun Signal.* **2013**, *11*(1), 97.

Heldin, C.-H.; Rubin, K.; Pietras, K.; Ö stman, A. High interstitial fluid pressure—an obstacle in cancer therapy. *Nature Reviews Cancer.* **2004**, *4*(10), 806–813.

Hodge, D. R.; Xiao, W.; Wang, L. H.; Li, D.; Farrar, W. L. Activating mutations in STAT3 and STAT5 differentially affect cellular proliferation and apoptotic resistance in multiple myeloma cells. *Cancer Biol. Ther.* **2004**, *3*(2), 188–194.

Honary, S.; Zahir, F. Effect of zeta potential on the properties of nano-drug delivery systems-a review (Part 1). *Tropical Journal of Pharmaceutical Research.* **2013**, *12*(2), 255–264.

Honary, S.; Zahir, F. Effect of zeta potential on the properties of nano-drug delivery systems-a review (Part 2). *Tropical Journal of Pharmaceutical Research.* **2013**, *12*(2), 265–273.

Hong, M.; Zhu, S.; Jiang, Y.; Tang, G.; Pei, Y. Efficient tumor targeting of hydroxycamptothecin loaded PEGylated niosomes modified with transferrin. *J. Control. Release.* **2009**, *133*(2), 96–102.

Hong, M.; Zhu, S.; Jiang, Y.; Tang, G.; Sun, C.; Fang, C. et al. Novel antitumor strategy: PEG-hydroxycamptothecin conjugate loaded transferrin-PEG-nanoparticles. *J. Control. Release.* **2010**, *141*(1), 22–9.

Huang, S.; Fang, R.; Xu, J.; Qiu, S.; Zhang, H.; Du, J. et al. Evaluation of the tumor targeting of a FAPα-based doxorubicin prodrug. *J. Drug Target.* **2011**, *19*(7), 487–496.

Huynh, H.; Ngo, V. C.; Fargnoli, J.; Ayers, M.; Soo, K. C.; Koong, H. N. et al. Brivanib Alaninate, a Dual Inhibitor of Vascular Endothelial Growth Factor Receptor and Fibroblast Growth Factor Receptor Tyrosine Kinases, Induces Growth Inhibition in Mouse Models of Human Hepatocellular Carcinoma. *Clin. Cancer Res.* **2008**, *14*(19), 6146–6153.

Jain, R. K. Transport of molecules across tumor vasculature. *Cancer Metastasis Rev.* **1987**, *6*(4), 559–593.

Jain, R. K. Transport of molecules in the tumor interstitium: a review. *Cancer Res.* **1987**, *47*(12), 3039–3051.

Jain, R. K.; Stylianopoulos, T. Delivering nanomedicine to solid tumors. *Nature reviEws Clinical Oncology.* **2010**, *7*(11), 653–664.

Jain, R. K.; Stylianopoulos, T. Delivering nanomedicine to solid tumors. *Nature Reviews. Clinical Oncology.* **2010**, *7*(11), 653–664.

Joyce, J. A. Therapeutic targeting of the tumor microenvironment. *Cancer Cell.* **2005**, *7*(6), 513–520.

Kalluri, R.; Zeisberg, M. Fibroblasts in cancer. *Nat. Rev. Cancer.* **2006**, *6*(5), 392–401.

Kanapathipillai, M.; Mammoto, A.; Mammoto, T.; Kang, J. H.; Jiang, E.; Ghosh, K. et al. Inhibition of mammary tumor growth using lysyl oxidase-targeting nanoparticles to modify extracellular matrix. *Nano Lett.* **2012**, *12*(6), 3213–3217.

Karin, M. Nuclear factor-κB in cancer development and progression. *Nature.* **2006**, *441*(7092), 431–436.

Kizaka-Kondoh, S.; Inoue, M.; Harada, H.; Hiraoka, M. Tumor hypoxia: a target for selective cancer therapy. *Cancer Sci.* **2003**, *94*(12), 1021–1028.

Kondo, M.; Asai, T.; Katanasaka, Y.; Sadzuka, Y.; Tsukada, H.; Ogino, K. et al. Anti-neovascular therapy by liposomal drug targeted to membrane type-1 matrix metalloproteinase. *Int. J. Cancer.* **2004**, *108*(2), 301–306.

Kukowska-Latallo, J. F.; Candido, K. A.; Cao, Z.; Nigavekar, S. S.; Majoros, I. J.; Thomas, T. P. et al. Nanoparticle targeting of anticancer drug improves therapeutic response in animal model of human epithelial cancer. *Cancer Res.* **2005**, *65*(12), 5317–5324.

Kulbe, H.; Thompson, R.; Wilson, J. L.; Robinson, S.; Hagemann, T.; Fatah, R. et al. The inflammatory cytokine tumor necrosis factor-α generates an autocrine tumor-promoting network in epithelial ovarian cancer cells. *Cancer Res.* **2007**, *67*(2), 585–592.

Laskin, J. J.; Sandler, A. B. Epidermal growth factor receptor: a promising target in solid tumors. *Cancer Treat. Rev.* **2004**, *30*(1), 1–17.

Lee, H.; Hu, M.; Reilly, R. M.; Allen, C. Apoptotic epidermal growth factor (EGF)-conjugated block copolymer micelles as a nanotechnology platform for targeted combination therapy. *Mol. Pharm.* **2007**, *4*(5), 769–781.

Li, J.; Chen, F.; Cona, M. M.; Feng, Y.; Himmelreich, U.; Oyen, R. et al. A review on various targeted anticancer therapies. *Target. Oncol.* **2012**, *7*(1), 69–85.

Li, J.; Huo, M.; Wang, J.; Zhou, J.; Mohammad, J. M.; Zhang, Y. et al. Redox-sensitive micelles self-assembled from amphiphilic hyaluronic acid-deoxycholic acid conjugates for targeted intracellular delivery of paclitaxel. *Biomaterials.* **2012**, *33*(7), 2310–2320.

Li, L.; Wartchow, C. A.; Danthi, S. N.; Shen, Z.; Dechene, N.; Pease, J. et al. A novel antiangiogenesis therapy using an integrin antagonist or anti–Flk-1 antibody coated 90Y-labeled nanoparticles. *International Journal of Radiation Oncology Biology Physics.* **2004**, *58*(4), 1215–1227.

Li, Y.; Huang, G.; Diakur, J.; Wiebe, L. I. Targeted delivery of macromolecular drugs: asialoglycoprotein receptor (ASGPR) expression by selected hepatoma cell lines used in antiviral drug development. *Curr Drug Deliv.* **2008**, *5*(4), 299–302.

Liao, D.; Johnson, R. Hypoxia: A key regulator of angiogenesis in cancer. *Cancer Metastasis Rev.* **2007**, *26*(2), 281–290.

Lieleg, O.; Baumgärtel, R. M.; Bausch, A. R. Selective filtering of particles by the extracellular matrix: an electrostatic bandpass. *Biophys. J.* **2009**, *97*(6), 1569–1577.

Lin, Q.; Bao, C.; Yang, Y.; Liang, Q.; Zhang, D.; Cheng, S. et al. Highly discriminating photorelease of anticancer drugs-based on hypoxia activatable phototrigger conjugated chitosan nanoparticles. *Adv. Mater.* **2013**, *25*(14), 1981–1986.

Liu, Q.; Li, R.-T.; Qian, H.-Q.; Yang, M.; Zhu, Z.-S.; Wu, W. et al. Gelatinase-stimuli strategy enhances the tumor delivery and therapeutic efficacy of docetaxel-loaded poly (ethylene glycol)-poly (varepsilon-caprolactone) nanoparticles. *Int J Nanomedicine.* **2012**, *7*, 281–295.

Liu, Y.; Li, K.; Pan, J.; Liu, B.; Feng, S.-S. Folic acid conjugated nanoparticles of mixed lipid monolayer shell and biodegradable polymer core for targeted delivery of Docetaxel. *Biomaterials.* **2010**, *31*(2), 330–338.

Low, P. S.; Antony, A. C. Folate receptor-targeted drugs for cancer and inflammatory diseases. *Adv. Drug Del. Rev.* **2004**, *56*(8), 1055–1058.

Lu, H.; Ouyang, W.; Huang, C. Inflammation, a key event in cancer development. *Mol. Cancer Res.* **2006**, *4*(4), 221–233.

Lu, P.; Weaver, V. M.; Werb, Z. The extracellular matrix: a dynamic niche in cancer progression. *The Journal of cell biology.* **2012**, *196*(4), 395–406.

Lurje, G.; Lenz, H.-J. EGFR signaling and drug discovery. *Oncology.* **2009**, *77*(6), 400–410.

Maeda, H.; Bharate, G. Y.; Daruwalla, J. Polymeric drugs for efficient tumor-targeted drug delivery-based on EPR-effect. *Eur. J. Pharm. Biopharm.* **2009**, *71*(3), 409–419.

Maeda, H.; Sawa, T.; Konno, T. Mechanism of tumor-targeted delivery of macromolecular drugs, including the EPR effect in solid tumor and clinical overview of the prototype polymeric drug SMANCS. *J. Control. Release.* **2001**, *74*(1–3), 47–61.

Marsh, T.; Pietras, K.; McAllister, S. S. Fibroblasts as architects of cancer pathogenesis. *Biochimica et Biophysica Acta (BBA) – Molecular Basis of Disease.* **2013**, *1832*(7), 1070–1078.

Matsumura, Y.; Gotoh, M.; Muro, K.; Yamada, Y.; Shirao, K.; Shimada, Y. et al. Phase I and pharmacokinetic study of MCC-465, a doxorubicin (DXR) encapsulated in PEG immunoliposome, in patients with metastatic stomach cancer. *Ann. Oncol.* **2004**, *15*(3), 517–525.

Matsumura, Y.; Maeda, H. A new concept for macromolecular therapeutics in cancer chemotherapy: mechanism of tumoritropic accumulation of proteins and the antitumor agent smancs. *Cancer Res.* **1986**, *46*(12 Pt 1), 6387–6392.

Mi, Y.; Liu, X.; Zhao, J.; Ding, J.; Feng, S.-S. Multimodality treatment of cancer with herceptin conjugated, thermomagnetic iron oxides and docetaxel loaded nanoparticles of biodegradable polymers. *Biomaterials.* **2012**, *33*(30), 7519–7529.

Mizrahy, S.; Peer, D. Polysaccharides as building blocks for nanotherapeutics. *Chem. Soc. Rev.* **2012**, *41*(7), 2623–2640.

Morachis, J. M.; Mahmoud, E. A.; Almutairi, A. Physical and Chemical Strategies for Therapeutic Delivery by Using Polymeric Nanoparticles. *Pharmacol. Rev.* **2012**.

Narra, K.; Mullins, S. R.; Lee, H.-O.; Strzemkowski-Brun, B.; Magalong, K.; Christiansen, V. J. et al. Phase II trial of single agent Val-boroPro (talabostat) inhibiting fibroblast activation protein in patients with metastatic colorectal cancer. *Cancer Biol. Ther.* **2007**, *6*(11), 1691–1699.

Nishida, N.; Yano, H.; Nishida, T.; Kamura, T.; Kojiro, M. Angiogenesis in Cancer. *Vascular Health and Risk Management.* **2006**, *2*(3), 213–219.

Noguchi, Y.; Wu, J.; Duncan, R.; Strohalm, J.; Ulbrich, K.; Akaike, T. et al. Early phase tumor accumulation of macromolecules: a great difference in clearance rate between tumor and normal tissues. *Jpn. J. Cancer Res.* **1998**, *89*(3), 307–314.

Nori, A.; Jensen, K. D.; Tijerina, M.; Kopečková, P.; Kopeček, J. Tat-Conjugated Synthetic Macromolecules Facilitate Cytoplasmic Drug Delivery To Human Ovarian Carcinoma Cells. *Bioconjug. Chem.* **2003**, *14*(1), 44–50.

Oh, E. J.; Park, K.; Kim, K. S.; Kim, J.; Yang, J.-A.; Kong, J.-H. et al. Target specific and long-acting delivery of protein, peptide, and nucleotide therapeutics using hyaluronic acid derivatives. *J. Control. Release.* **2010**, *141*(1), 2–12.

Orimo, A.; Gupta, P. B.; Sgroi, D. C.; Arenzana-Seisdedos, F.; Delaunay, T.; Naeem, R. et al. Stromal Fibroblasts Present in Invasive Human Breast Carcinomas Promote Tumor Growth and Angiogenesis through Elevated SDF-1/CXCL12 Secretion. *Cell.* **2005**, *121*(3), 335–348.

Ostermann, E.; Garin-Chesa, P.; Heider, K. H.; Kalat, M.; Lamche, H.; Puri, C. et al. Effective Immunoconjugate Therapy in Cancer Models Targeting a Serine Protease of Tumor Fibroblasts. *Clin. Cancer Res.* **2008**, *14*(14), 4584–4592.

Östman, A.; Augsten, M. Cancer-associated fibroblasts and tumor growth – bystanders turning into key players. *Curr. Opin. Genet. Dev.* **2009**, *19*(1), 67–73.

Padera, T. P.; Stoll, B. R.; Tooredman, J. B.; Capen, D.; di Tomaso, E.; Jain, R. K. Pathology: cancer cells compress intratumor vessels. *Nature.* **2004**, *427*(6976), 695–695.

Pan, X.; Wu, G.; Yang, W.; Barth, R. F.; Tjarks, W.; Lee, R. J. Synthesis of cetuximab-immunoliposomes via a cholesterol-based membrane anchor for targeting of EGFR. *Bioconjug. Chem.* **2007**, *18*(1), 101–108.

Pang, Z.; Gao, H.; Yu, Y.; Guo, L.; Chen, J.; Pan, S. et al. Enhanced intracellular delivery and chemotherapy for glioma rats by transferrin-conjugated biodegradable polymersomes loaded with doxorubicin. *Bioconjug. Chem.* **2011**, *22*(6), 1171–1180.

Pickup, M. W.; Mouw, J. K.; Weaver, V. M. The extracellular matrix modulates the hallmarks of cancer. *EMBO reports.* **2014**, e201439246.

Pietras, K.; Pahler, J.; Bergers, G.; Hanahan, D. Functions of Paracrine PDGF Signaling in the Proangiogenic Tumor Stroma Revealed by Pharmacological Targeting. *PLoS Med.* **2008**, *5*(1), e19.

Poon, Z.; Chang, D.; Zhao, X.; Hammond, P.; Layer-by-Layer Nanoparticles with a pH-Sheddable Layer for Vivo: 2011.

Quante, M.; Tu, S. P.; Tomita, H.; Gonda, T.; Wang, S. S. W.; Takashi, S. et al. Bone Marrow-Derived Myofibroblasts Contribute to the Mesenchymal Stem Cell Niche and Promote Tumor Growth. *Cancer Cell.* **2011**, *19*(2), 257–272.

Raju, A.; Muthu, M. S.; Feng, S.-S. Trastuzumab-conjugated vitamin E TPGS liposomes for sustained and targeted delivery of docetaxel. *Expert Opinion on Drug Delivery.* **2013**, *10*(6), 747–760.

Räsänen, K.; Vaheri, A. Activation of fibroblasts in cancer stroma. *Exp. Cell Res.* **2010**, *316*(17), 2713–2722.

Reddy, S. B.; Williamson, S. K. Tirapazamine: a novel agent targeting hypoxic tumor cells. **2009**.

Ruan, K.; Song, G.; Ouyang, G. Role of hypoxia in the hallmarks of human cancer. *J. Cell. Biochem.* **2009**, *107* (6), 1053–1062.

Rutishauser, U.; Sachs, L. Cell-to-cell binding induced by different lectins. *The Journal of Cell Biology.* **1975**, *65*(2), 247–257.

Sansone, P.; Bromberg, J. Targeting the interleukin-6/Jak/stat pathway in human malignancies. *J. Clin. Oncol.* **2012**, *30*(9), 1005–1014.

Santos, A. M.; Jung, J.; Aziz, N.; Kissil, J. L.; Puré, E. Targeting fibroblast activation protein inhibits tumor stromagenesis and growth in mice. *The Journal of Clinical Investigation.* **2009**, *119*(12), 3613–3625.

Seymour, L. W.; Ferry, D. R.; Anderson, D.; Hesslewood, S.; Julyan, P. J.; Poyner, R. et al. Hepatic drug targeting: phase I evaluation of polymer-bound doxorubicin. *J. Clin. Oncol.* **2002**, *20*(6), 1668–1676.

Shih, T.; Lindley, C. Bevacizumab: An angiogenesis inhibitor for the treatment of solid malignancies. *Clin. Ther.* **2006**, *28*(11), 1779–1802.

Shukla, S.; Ablack, A. L.; Wen, A. M.; Lee, K. L.; Lewis, J. D.; Steinmetz, N. F. Increased tumor homing and tissue penetration of the filamentous plant viral nanoparticle Potato virus X. *Mol. Pharm.* **2012**, *10*(1), 33–42.

Sounni, N. E.; Noel, A. Targeting the Tumor Microenvironment for Cancer Therapy. *Clin. Chem.* **2013**, *59*(1), 85–93.

Sun, T.; Zhang, Y. S.; Pang, B.; Hyun, D. C.; Yang, M.; Xia, Y. Engineered Nanoparticles for Drug Delivery in Cancer Therapy. *Angew. Chem. Int. Ed.* **2014**, *53*(46), 12320–12364.

Suzuki, R.; Takizawa, T.; Kuwata, Y.; Mutoh, M.; Ishiguro, N.; Utoguchi, N. et al. Effective antitumor activity of oxaliplatin encapsulated in transferrin–PEG-liposome. *Int. J. Pharm.* **2008**, *346*(1–2), 143–150.

Swartz, M. A.; Fleury, M. E. Interstitial flow and its effects in soft tissues. *Annu. Rev. Biomed. Eng.* **2007**, *9*, 229–256.

Togo, S.; Polanska, U. M.; Horimoto, Y.; Orimo, A. Carcinoma-associated fibroblasts are a promising therapeutic target. *Cancers (Basel).* **2013**, *5*(1), 149–169.

Torchilin, V. P. Drug targeting. *Eur. J. Pharm. Sci.* **2000**, *11, Supplement 2*, S81–S91.

Trédan, O.; Galmarini, C. M.; Patel, K.; Tannock, I. F. Drug resistance and the solid tumor microenvironment. *J. Natl. Cancer Inst.* **2007**, *99*(19), 1441–1454.

Verstovsek, S. Therapeutic potential of JAK2 inhibitors. *ASH Education Program Book.* **2009**, *1*, 636–642.

Wang, T.; Niu, G.; Kortylewski, M.; Burdelya, L.; Shain, K.; Zhang, S. et al. Regulation of the innate and adaptive immune responses by Stat-3 signaling in tumor cells. *Nat. Med.* **2004**, *10*(1), 48–54.

Wang, W.; Li, Q.; Yamada, T.; Matsumoto, K.; Matsumoto, I.; Oda, M. et al. Crosstalk to Stromal Fibroblasts Induces Resistance of Lung Cancer to Epidermal Growth Factor Receptor Tyrosine Kinase Inhibitors. *Clin. Cancer Res.* **2009**, *15*(21), 6630–6638.

Wilson, T. R.; Fridlyand, J.; Yan, Y.; Penuel, E.; Burton, L.; Chan, E. et al. Widespread potential for growth-factor-driven resistance to anticancer kinase inhibitors. *Nature.* **2012**, *487*(7408), 505–509.

Wilson, W. R.; Hay, M. P. Targeting hypoxia in cancer therapy. *Nature Reviews Cancer.* **2011**, *11*(6), 393–410.

Wu, J.; Lu, Y.; Lee, A.; Pan, X.; Yang, .X.; Zhao, X. et al. Reversal of multidrug resistance by transferrin-conjugated liposomes coencapsulating doxorubicin and verapamil. *J. Pharm. Pharm. Sci.* **2007**, *10*(3), 350–357.

Xiao, Q.; Ge, G. Lysyl oxidase, extracellular matrix remodeling and cancer metastasis. *Cancer Microenviron.* **2012**, *5*(3), 261–273.

Xiong, X. B.; Huang, Y.; LU, W. L.; Zhang, X.; Zhang, H.; Nagai, T. et al. Intracellular delivery of doxorubicin with RGD-modified sterically stabilized liposomes for an improved antitumor efficacy: In vitro and in vivo. *J. Pharm. Sci.* **2005**, *94*(8), 1782–1793.

Yu, H.; Kortylewski, M.; Pardoll, D. Crosstalk between cancer and immune cells: role of STAT3 in the tumor microenvironment. *Nature Reviews Immunology.* **2007**, *7*(1), 41–51.

Yu, H.; Pardoll, D.; Jove, R. STATs in cancer inflammation and immunity: a leading role for STAT3. *Nature Reviews Cancer.* **2009**, *9*(11), 798–809.

Zhong, Y.; Meng, F.; Deng, C.; Zhong, Z. Ligand-Directed Active Tumor-Targeting Polymeric Nanoparticles for Cancer Chemotherapy. *Biomacromolecules.* **2014**, *15*(6), 1955–1969.

Zhong, Y.; Yang, W.; Sun, H.; Cheng, R.; Meng, F.; Deng, C. et al. Ligand-directed reduction-sensitive shell-sheddable biodegradable micelles actively deliver doxorubicin into the nuclei of target cancer cells. *Biomacromolecules.* **2013**, *14*(10), 3723–3730.

INDEX

B

D

E

F

G

H

I

J

K

L

M

N

O

P

U

V

W

X

Y

Z